The Impact of Food Bio-Actives on Gut Health

Kitty Verhoeckx

Editor-in-Chief

Paul Cotter • Iván López-Expósito
Charlotte Kleiveland • Tor Lea
Alan Mackie • Teresa Requena
Dominika Swiatecka • Harry Wichers

Editors

The Impact of Food Bio-Actives on Gut Health

In Vitro and Ex Vivo Models

 Springer Open

EUROPEAN COOPERATION
IN SCIENCE AND TECHNOLOGY

Editor-in-Chief
Kitty Verhoeckx
TNO, Zeist
The Netherlands

Editors
Paul Cotter
Cork, Ireland

Charlotte Kleiveland
Norwegian University of Life Sciences
As, Norway

Alan Mackie
Institute of Food Research Norwich
Research Park
Norwich, Norfolk, UK

Dominika Swiatecka
Polish Academy of Sciences
Olsztyn, Poland

Iván López-Expósito
Instituto de Investigación en Ciencias de la
Alimentación (CIAL, CSIC-UAM)
Madrid, Spain

Tor Lea
Norwegian University of Life Sciences
As, Norway

Teresa Requena
Universidad Autonoma de Madrid
Madrid, Spain

Harry Wichers
Wageningen University
Wageningen, The Netherlands

ISBN 978-3-319-15791-7   ISBN 978-3-319-16104-4   (eBook)
DOI 10.1007/978-3-319-16104-4

Library of Congress Control Number: 2015937183

Springer Cham Heidelberg New York Dordrecht London

Preface

This book is the final product of the InfoGest FA1005 COST Action. InfoGest is an international network that aims at "Improving Health Properties of Food by Sharing our Knowledge on the Digestive Process". The specific objectives of the network are to:

- Compare the existing digestion models, harmonise the methodologies and propose guidelines for performing new experiments
- Validate in vitro models towards in vivo data (animal and/or human)
- Identify the beneficial/deleterious components that are released in the gut during food digestion
- Demonstrate the effect of these compounds on human health
- Determine the effect of the matrix structure on the bioavailability of food nutrients and bioactive molecules

InfoGest is supported for 4 years (June 2011–May 2015) by European COST funds and gathers more than 320 scientists from 34 countries (primarily within Europe but also Canada, Australia, Argentina and New Zealand). Connections between academic partners and industry are also strengthened through the participation of more than 40 food companies (large groups as well as SMEs). InfoGest has released several reviews and opinion papers on the topic of food digestion and related topics (e.g. health effects, bioavailability) and has proposed a consensus in vitro digestion model to the scientific community (Minekus et al. 2014). A standardised static in vitro digestion method suitable for food—an international consensus. Food & Function 5:1113–1124). It has also created the International

Conference on Food Digestion that every year gathers around 200 scientists from all over the world.

In this book, which was coordinated by Kitty Verhoeckx, we describe the in vitro and ex vivo models that can be used to investigate beneficial or detrimental effects of digested food products and highlight the advantages and limitations of each one of them. It is hoped that the details provided, and the citations included, will allow you to identify the model(s) that best suit your needs.

We hope that you will enjoy reading this book and will learn a bit more about the complexity of the digestive process.

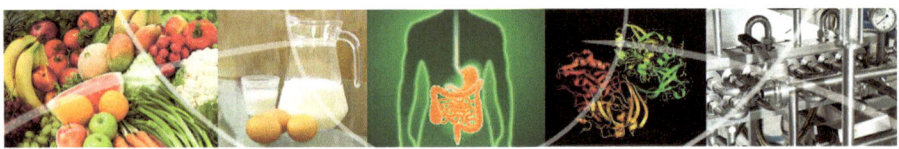

Rennes, France   Didier Dupont
Zeist, The Netherlands   Kitty Verhoeckx

General Introduction

This book is a product of the InfoGest COST Action FA 1005. This Action was granted by COST in the domain of Food and Agriculture and coordinated by the French National Institute for Agricultural Research (INRA).

Every day our gastrointestinal tract (GIT) is exposed to many different food components. These food components can be processed, digested and eventually transported across the intestinal tract and can have a variety of direct and indirect effects, both positive and negative, on our health. In vivo assessments of the impacts of food components (or food bioactives) on health are not trivial and are not practical when examining more than just a few components. Thus, in many instances it is necessary to employ in vitro and ex vivo models as part of a funnelled approach to identify the components that merit particular attention.

In this book, we describe the in vitro and ex vivo models which can be used to investigate beneficial or detrimental effects of digested food products. The models that we describe in this book include those used to study digestion and fermentation in the small and large intestine (Parts I and VI), models used to investigate absorption (e.g. Ussing chamber, epithelial cell systems) in Part II and Part V and the immune and enteroendocrine responses (e.g. macrophages, dendritic cells, co-cultures) in Parts III and IV. For each model, we provide you with background information, a general protocol with tips and tricks concerning their proper use, readouts provided by the systems, the applicability of the model with respect to food research and pros and cons of the model. Indeed, this book has been prepared for the particular benefit of students/researchers who are not experienced in the use of these models but are considering their use. It is hoped that the details provided, and the citations included, will allow you to identify the model(s) that best suit(s) your needs.

Digestion and Absorption

The GIT represents the largest interface between our body and the environment and, when functioning correctly, absorbs nutrients while providing protection from harmful components. By the broadest definition, the GIT extends from the mouth to the anus and can be divided into the upper and the lower tracts. The upper tract consists of the oral cavity, oesophagus, stomach, duodenum, jejunum and the ileum. The latter three together represent the small intestine. The lower tract comprises the large intestine consisting of the cecum, colon, rectum and anal canal (see Fig. 1).

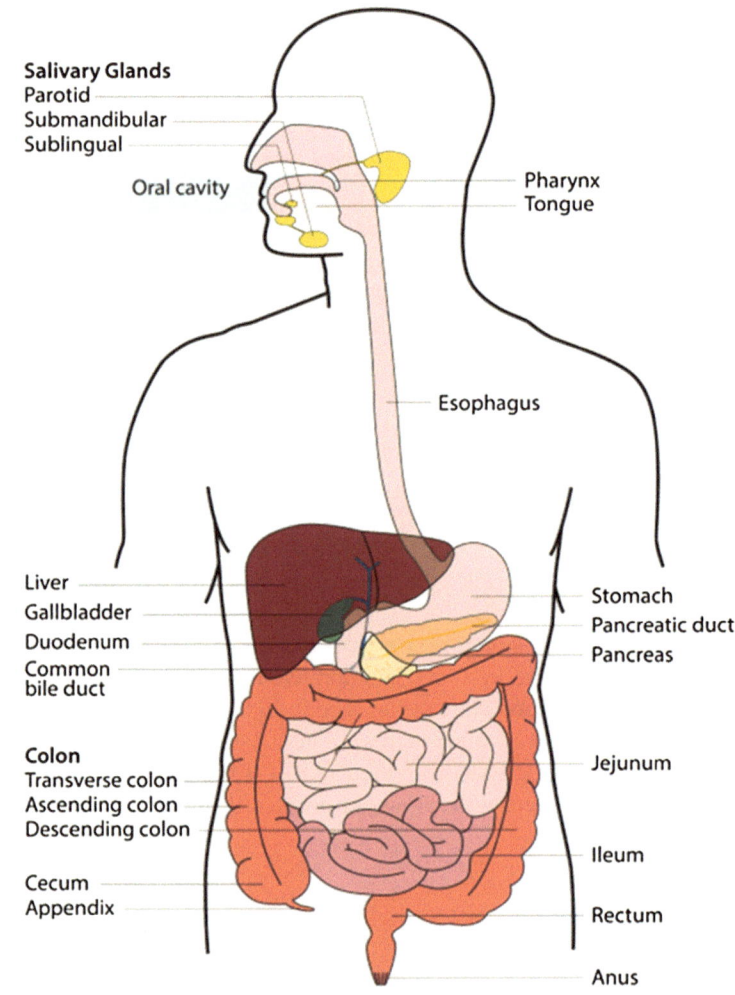

Fig. 1 Gastrointestinal tract

Every part of the GIT has its own function in food processing. The mouth is responsible for mastication and mixing of the food with saliva, which contains a complex array of components including amylase, an enzyme that catalyses the hydrolysis of starch into sugars. After the formation of a food bolus, the food is transported through the oesophagus to the stomach where it is further processed. The food bolus is mixed with enzymes like protease and lipase, which break down proteins and lipids, respectively. Acid is also secreted into the stomach and this will gradually lower the pH of the content and aid in hydrolysis. The food bolus is broken down into chyme, which is gradually transported to the small intestine. In the duodenum, the low pH of the stomach is neutralised by bicarbonate and digestive juices from the pancreas (digestive enzymes like trypsin and chymotrypsin) and the gall bladder (bile acids) are introduced. The digestive enzymes further break down the proteins, lipids and starch, while the bile acids help emulsify the products of lipid hydrolysis into micelles (Withney 2008; Wickham et al. 2009). More information on digestion and digestion models can be found in Part I. The final stage in the digestion of dietary carbohydrates and proteins occurs right on the surface of small intestinal enterocytes by brush boarder enzymes (Shimizu 2004). These enzymes, including maltase, sucrose-isomaltase, lactase and peptidases, are integral membrane proteins that are present in enterocytes. The nutrients produced are mainly absorbed by the enterocytes of the jejunum and to a lesser extent in the ileum (Withney 2008). One of the main functions of the large intestine is the absorption of water. In addition, while it is known that the GIT contains a large microbial population, the concentration of this population is greatest in the large intestine. Many of these microorganisms contribute to the digestion of food components, including prebiotics such as complex polysaccharides which cannot be digested by human enzymes (Flint 2012).

After absorption of nutrients by enterocytes, the compounds especially peptides can be further degraded by intracellular proteases before they enter the bloodstream or the lymphatic system. Water-soluble nutrients are mostly released into the bloodstream and end up in the liver via the hepatic portal vein. Fat-soluble nutrients are transported into the lymph after assembly into chylomicrons. After reprocessing, these compounds also end up in the blood (Withney 2008). Further details on transport mechanisms can be found in Chap. 24.

Cells Present in the Intestine

As already mentioned, the GIT is the largest interface between the body and the environment, and, for this reason, it also serves as a point of communication between the environment and the host immune system (Brandtzaeg 2011; Faria et al. 2013). This interface consists of a single epithelial layer folded into crypts and villi to increase the surface area of the gut (Cummins and Thompson 2002; Ismail and Hooper 2005). The colon does not contain villi. The intestinal epithelial layer is composed of several distinct cell types, originating from multipotent stem cells present in the crypts (see Fig. 2). The most abundant are the enterocytes that have an absorptive function. Interlaced between the enterocytes are mucin-secreting goblet cells and peptide hormone exporting enteroendocrine cells (see also Part IV) (Ismail and Hooper 2005; Snoeck et al. 2005). During their migration to the top of the villi, enterocytes, goblet cells and enteroendocrine cells differentiate and eventually die (apoptosis) when they reach the top of the villi. A fourth cell type, the Paneth cells, migrates downwards to the crypt base. The Paneth cells secrete digestive enzymes, growth factors and antimicrobial peptides (AMPs) such as cryptdins or defensins (Snoeck et al. 2005). For more information on epithelial cells and ex vivo cell systems, see Part II and Part IV.

Given its large surface area and the number of antigens and microorganisms to which the GIT is exposed, it is not surprising that it contains the highest number of lymphoid (immune) cells in the entire body (Faria et al. 2013). The gut immune system consists of inductive sites where antigen recognition and primary adaptive immune responses take place and effector sites that harbour, amongst other cells, activated T- and B-cells and memory cells. The main inductive sites are gut-associated lymphoid tissues (GALT) such as Peyer's patches (PP), isolated lymphoid follicles (ILF) and the mesenteric lymph nodes (mLNs). The lamina propria (LP) and the epithelium constitute the main effector sites (Pabst and Mowat 2012). In the GALT and at the effector sites, a wide range of immune cells are present (Faria et al. 2013). Dendritic cells (DCs), which are recognised as an important link between innate and adaptive immunity, take up, process and present antigens to T-cells (Willart et al. 2013). DCs are found in all organised intestinal lymphoid tissues. In the sub-epithelial dome (SED) of Peyer's patches, they capture antigens that are transported into the SED by specialised epithelial cells called microfold cells (M-cells) (Shreedhar et al. 2003). Next to DCs a mixture of T and B lymphocytes, plasma cells and macrophages, a second type of antigen-presenting cells, are present in the SED (Pabst 1987). In the lamina propria (LP), a range of different immune cells can be found, typically DCs, macrophages, plasma cells, memory B- and T-cells, mast cells, eosinophils and cytotoxic natural killer cells (NK cells) (Macdonald and Monteleone 2005; Peterson and Artis 2014). Also, LP contains additional innate immune cell populations not found in peripheral blood. These cells, called innate lymphoid cells (ILC), are potent cytokine producers, much like the classical T helper cell subsets. Recent evidence suggests important functions of ILCs in the maintenance of barrier integrity and mucosal homeostasis. The only immune cells that are virtually absent in the healthy intestine are neutrophils and basophils. Both cell types, however, will infiltrate intestinal tissues in case of inflammation (Stone et al. 2010; Ismail and Hooper 2005). Generally, the small intestine contains more immune cells than the colon. More details on the different immune cell types present in the intestine can be found in Part III.

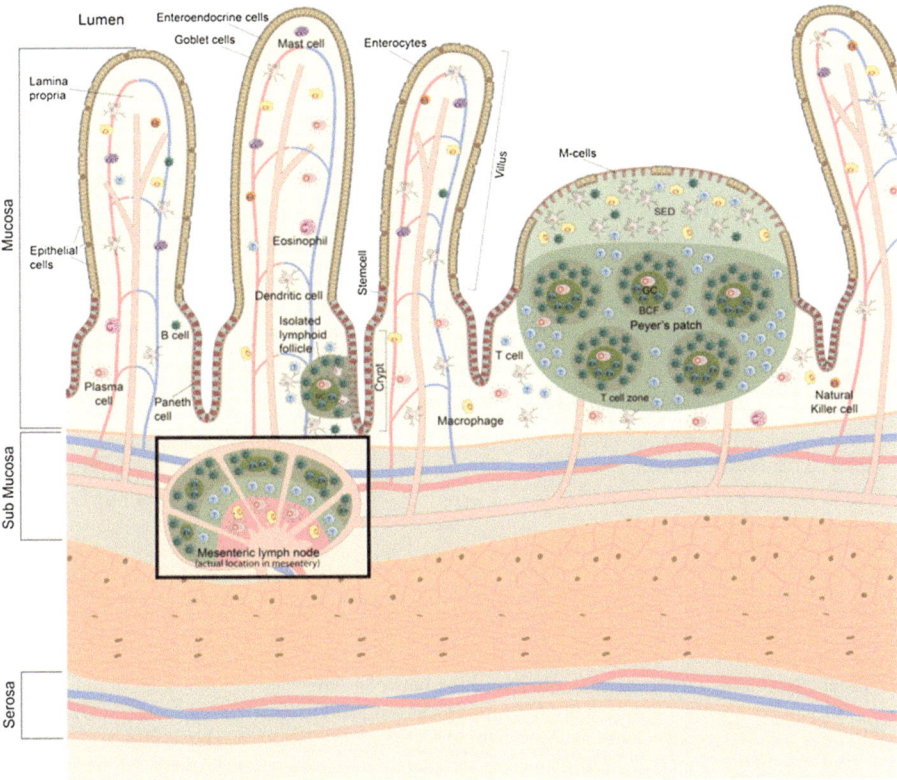

Fig. 2 Intestinal cells and structures in the small intestine. *Note*: Normally you will not find the mesenteric lymph node in the submucosa, but in humans they are situated close to the serosa on the mesenteric side of the gut. But for simplicity and to be complete, we added the node in the picture

Role of Microbiota

As noted above, the investigation of the impact of foods and food components on host health cannot be said to be complete without considering the role of the host microbiota. Our understanding of the interplay between the human GIT microbiota and digestion as well as health has increased dramatically in recent years as a consequence of the development of culture-independent, DNA-sequencing-based approaches to study these populations. It is now clear that diet has a considerable influence on the composition and function of the gut microbiota (Yatsunenko et al. 2012; Claesson et al. 2012). These microbial populations in turn impact on the extraction of energy from food, including the fermentation of complex carbohydrates and proteins to produce short-chain fatty acids and other metabolites (Russell et al. 2013). The relationship between gut microbes and diet can also have a significant impact on mucosal and systemic immune responses (Hooper et al. 2002). It is thus notable that, among the model gut systems that are described in this book (see Part VI), several can be populated with human GIT microbial populations to facilitate investigations into the impact gut microbes in food digestion.

In conclusion, while, in many instances, in vivo human or animal data represent the "gold standard" with respect to carrying out studies on food and health, frequently, due to financial, ethical or practical (e.g. high-throughput screens) reasons, in vitro and/or ex vivo models are more appropriate. For this reason, in this book, we focus on in vitro and ex vivo models, with the requirement that the models should be closely related to human physiology and should be readily available to the scientific community. We hope that this book will be your first port of call with respect to determining which in vitro and ex vivo assays best serve your needs when studying the health effects of food bioactives in the gut.

Zeist, The Netherlands   Kitty Verhoeckx
Cork, Ireland   Paul D. Cotter

References

Brandtzaeg P (2011) The gut as communicator between environment and host: immunological consequences. Eur J Pharmacol 668(Suppl 1):S16–S32. doi:10.1016/j.ejphar.2011.07.006

Claesson MJ, Jeffery IB, Conde S, Power SE, O'Connor EM, Cusack S, Harris HMB, Coakley M, Lakshminarayanan B, O'Sullivan O, Fitzgerald GF, Deane J, O'Connor M, Harnedy N, O'Connor K, O'Mahony D, van Sinderen D, Wallace M, Brennan L, Stanton C, Marchesi JR, Fitzgerald AP, Shanahan F, Hill C, Ross RP, O'Toole PW (2012) Gut microbiota composition correlates with diet and health in the elderly. Nature 488(7410):178–184

Cummins AG, Thompson FM (2002) Effect of breast milk and weaning on epithelial growth of the small intestine in humans. Gut 51(5):748–754

Faria AM, Gomes-Santos AC, Goncalves JL, Moreira TG, Medeiros SR, Dourado LP, Cara DC (2013) Food components and the immune system: from tonic agents to allergens. Front Immunol 4:102. doi:10.3389/fimmu.2013.00102

Flint HJ (2012) The impact of nutrition on the human microbiome. Nutr Rev 70:S10–S13. doi:10.1111/j.1753-4887.2012.00499.x

Hooper LV, Midtvedt T, Gordon JI (2002) How host-microbial interactions shape the nutrient environment of the mammalian intestine. Annu Rev Nutr 22(1):283–307. doi:10.1146/annurev.nutr.22.011602.092259

Ismail AS, Hooper LV (2005) Epithelial cells and their neighbors. IV. Bacterial contributions to intestinal epithelial barrier integrity. Am J Physiol Gastrointest Liver Physiol 289(5):G779–G784. doi:10.1152/ajpgi.00203.2005

Macdonald TT, Monteleone G (2005) Immunity, inflammation, and allergy in the gut. Science 307(5717):1920–1925. doi:10.1126/science.1106442

Pabst R (1987) The anatomical basis for the immune function of the gut. Anat Embryol (Berl) 176(2):135–144

Pabst O, Mowat AM (2012) Oral tolerance to food protein. Mucosal Immunol 5(3):232–239. doi:10.1038/mi.2012.4

Peterson LW, Artis D (2014) Intestinal epithelial cells: regulators of barrier function and immune homeostasis. Nat Rev Immunol 14(3):141–153. doi:10.1038/nri3608

Russell WR, Hoyles L, Flint HJ, Dumas M-E (2013) Colonic bacterial metabolites and human health. Curr Opin Microbiol 16(3):246–254

Shimizu M (2004) Food-derived peptides and intestinal functions. Biofactors 21(1–4):43–47

Shreedhar VK, Kelsall BL, Neutra MR (2003) Cholera toxin induces migration of dendritic cells from the subepithelial dome region to T- and B-cell areas of Peyer's patches. Infect Immun 71(1):504–509

Snoeck V, Goddeeris B, Cox E (2005) The role of enterocytes in the intestinal barrier function and antigen uptake. Microbes Infect 7(7–8):997–1004. doi:10.1016/j.micinf.2005.04.003

Stone KD, Prussin C, Metcalfe DD (2010) IgE, mast cells, basophils, and eosinophils. J Allergy Clin Immunol 125(2 Suppl 2):S73–S80. doi:10.1016/j.jaci.2009.11.017

Wickham M, Faulks R, Mills C (2009) In vitro digestion methods for assessing the effect of food structure on allergen breakdown. Mol Nutr Food Res 53(8):952–958. doi:10.1002/mnfr.200800193

Willart MA, Poulliot P, Lambrecht BN, Kool M (2013) PAMPs and DAMPs in allergy exacerbation models. Methods Mol Biol 1032:185–204. doi:10.1007/978-1-62703-496-8_15

Withney E, Rolfes SR (ed) (2008) Understanding nutrition, 11th edn. Thomas Wadsworth, Belmont

Yatsunenko T, Rey FE, Manary MJ, Trehan I, Dominguez-Bello MG, Contreras M, Magris M, Hidalgo G, Baldassano RN, Anokhin AP, Heath AC, Warner B, Reeder J, Kuczynski J, Caporaso JG, Lozupone CA, Lauber C, Clemente JC, Knights D, Knight R, Gordon JI (2012) Human gut microbiome viewed across age and geography. Nature 486(7402):222–227

Contents

Part VI   In Vitro Fermentation Models: General Introduction

Part I
Gastrointestinal Digestion Models, General Introduction

General Introduction

This book is a product of the Infogest COST Action chaired by Didier Dupont, which had as one of its main aims to harmonise in vitro approaches to digestion. Thus, as vice-chair of the action I have become aware of the wide range of models and applications that are being used by the scientific community. These range from simple enzyme reactions such as hydrolysis of proteins by pepsin under rather non-physiological conditions in a beaker to the complex sophistication of the TIM-1 and other similar dynamic models. Their application has been equally wide ranging. The pharmaceutical industry typically use these models as dissolution tests for new formulations; resistance to proteolysis has been used by food allergy researchers as a risk factor for a protein being an allergen (Astwood et al. 1996); others have used in vitro digestion to assess the bioaccessibility of soil contaminants (Oomen et al. 2002) and there are many other examples. Despite this diversity in both methods and applications, one can draw a few general conclusions about the design of gastrointestinal (GI) models. Firstly, the model should be as simple as possible but not so simple that the results do not provide information relevant to the "real life" situation. Secondly, what has been done previously is not always the best or indeed most relevant approach. Finally, digestion is not a goal in itself and the way that samples of digesta are collected is very dependent on the type of measurement to be made.

The digestive tract in humans and indeed other mammals is highly complex because of the need to efficiently extract the optimum amount of nutrients and bioactives from the food consumed, whilst at the same time keeping out pathogens and toxic compounds. This requirement has led to the evolution of a complex multilayered system of control involving a number of distinct compartments. These are the oral compartment (mouth) where the initial sensory input from the food is acquired; the gastric compartment (stomach) where food is stored, partially digested and partially sterilised; the duodenum, jejunum and ileum (small intestine) which is the primary site of digestion and absorption and finally the cecum and colon (large intestine) where fermentation breaks down some dietary fibre into an absorbable form and other bioactive compounds are metabolised by the gut flora and absorbed.

There are clearly a number of aspects of digestion that are not readily reproduced by any of the currently available digestion models. For example, at the macroscopic scale some foods may separate out in the stomach leading to different retention times caused by various parameters such as the viscosity of the gastric contents the size of the particles remaining in the stomach and the sensing on nutrients in the small intestine. These parameters are difficult to measure in vivo and so a number of simplifications are generally used. For example a good approximation that can be used in a dynamic model of the gastric phase is that 1–2 kcal/min are generally emptied from the gastric compartment when under energy-control (van Aken 2010). This is of course difficult to follow in vitro if there is any phase separation of the food. At the other end of the scale, transport from the intestinal lumen to the epithelium through the mucus layer and then absorption are also complex processes that are difficult to mimic in vitro (Mackie et al. 2012).

When conducting experiments aimed at understanding the digestion of food or the bioaccessibility of specific compounds we need to understand what aspects of the GI tract are important, where they are digested and absorbed and what environment they may be exposed to. In the models outlined in the rest of this chapter the reader will be presented with a range of models of differing complexity with the aim of guiding those new to the field. The aim is to provide sufficient information so that the correct decision can be made about the level of complexity needed to answer a specific digestion related problem and what facilities are available in different leading groups around the world.

References

Astwood JD, Leach JN, Fuchs RL (1996) Stability of food allergens to digestion in vitro. Nat Biotechnol 14(10):1269–1273. doi:10.1038/nbt1096-1269

Mackie AR, Round AN, Rigby NM, Macierzanka A (2012) The role of the mucus barrier in digestion. Food Dig 3(1):8–15. doi:10.1007/s13228-012-0021-1

Oomen AG, Hack A, Minekus M, Zeijdner E, Cornelis C, Schoeters G, Verstraete W, Van de Wiele T, Wragg J, Rompelberg CJM, Sips A, Van Wijnen JH (2002) Comparison of five in vitro digestion models to study the bioaccessibility of soil contaminants. Environ Sci Technol 36(15):3326–3334. doi:10.1021/es010204v

van Aken GA (2010) Relating food emulsion structure and composition to the way it is processed in the gastrointestinal tract and physiological responses: what are the opportunities? Food Biophys 5(4):258–283. doi:10.1007/s11483-010-9160-5

Chapter 1
Static Digestion Models: General Introduction

Amparo Alegría, Guadalupe Garcia-Llatas, and Antonio Cilla

Abstract Several in vitro methods have been developed to simulate the physiological conditions of the human gastrointestinal digestion, the simplest being the static methods. The following chapter clarifies the concepts of bioaccessibility and dialyzability, and describes the conditions (pH, enzymes, agitation, etc.) to be applied in oral, gastric and intestinal phases when assessing a food component (nutrient, bioactive or toxin) or a food product, in a single or multi-phase model. The advantages and disadvantages of the static models vs. dynamic and in vivo models are discussed, and a review of specific conditions applied on nutrients (minerals, vitamins, proteins, fatty acids, etc.) and bioactive compounds (carotenoids, plant sterols, etc.) from recent studies is provided. Currently, it must be considered that, although the static digestion conditions must be adapted according to the component or food sample to be studied, a harmonization and standardization of the models are needed in order to establish suitable correlations among in vitro and in vivo assays, as it has been defined for some food components (carotenoids, proteins and minerals).

Keywords Static models • Bioaccessibility • Dialyzability • Gastrointestinal • In vitro digestion

1.1 Definition of Concepts: Bioavailability, Bioaccessibility and Bioactivity

The term bioavailability can be defined as the fraction of ingested component available at the site of action for utilization in normal physiological functions, and is determined through in vivo assays (Guerra et al. 2012). Bioavailability is the result of three main steps: digestibility and solubility of the element in the gastrointestinal tract; absorption of the element by the intestinal cells and transport into the

A. Alegría • G. Garcia-Llatas • A. Cilla (✉)
Nutrition and Food Science Area, Faculty of Pharmacy, University of Valencia,
Av. Vicente Andrés Estellés s/n, 46100 Burjassot, Valencia, Spain
e-mail: antonio.cilla@uv.es

© The Author(s) 2015
K. Verhoeckx et al. (eds.), *The Impact of Food Bio-Actives on Gut Health*,
DOI 10.1007/978-3-319-16104-4_1

circulation; and incorporation from the circulation to the functional entity or target (Wienk et al. 1999; Etcheverry et al. 2012).

Bioavailability furthermore includes two additional terms: bioaccessibility and bioactivity. Bioaccessibility has been defined as the fraction of a compound that is released from its food matrix within the gastrointestinal tract and thus becomes available for intestinal absorption (typically established from in vitro procedures). It includes the sequence of events that take place during food digestion for transformation into potentially bioaccessible material but excludes absorption/assimilation through epithelial tissue and pre-systemic metabolism (both intestinal and hepatic). Bioactivity in turn includes events linked to how the nutrient or bioactive compound is transported and reaches the target tissue, how it interacts with biomolecules, the metabolism or biotransformation it may experience, and the generation of biomarkers and the physiological responses induced. Although bioavailability and bioaccessibility are often used indistinctly, it must be clarified that bioavailability includes bioactivity (Etcheverry et al. 2012).

1.2 Static Methods

1.2.1 Solubility/Dialyzability

In the past two to three decades, several in vitro methods have been developed to simulate the physiological conditions (temperature, agitation, pH, enzyme and chemical composition) and the sequence of events that occur during digestion in the human gastrointestinal tract. Static methods (also called biochemical methods) are the simplest techniques in this respect and include two or three digestion steps (oral, gastric, and intestinal) whose products remain largely immobile in a single static bioreactor. These methods simulate a limited number of parameters of physiological digestion (to be described below), and do not mimic physical processes such as shearing, mixing, hydration, changes in conditions over time, or peristalsis (Fernández-García et al. 2009; Wickham et al. 2009).

In a first step, simulated gastrointestinal digestion is applied to homogenized foods or isolated compounds in a closed system, followed by determination of the amount of soluble compound present in the supernatant obtained by centrifugation or filtration (solubility methods). The amount of solubilized component can be used as a measure of the bioaccessibility of a nutrient or bioactive component. An important alternative methodological approach compared with previous systems is the introduction of a dialysis bag containing sodium bicarbonate, after gastric digestion of the food sample, and dialysis of soluble components across a semi-permeable membrane without removal of the dialyzed compounds. The use of a dialysis bag of a specific pore size also permits discrimination between high and low molecular weight components (Ekmekcioglu 2002; Etcheverry et al. 2012).

1.2.2 Digestion Conditions

It is necessary to take into account in vivo conditions when applying in vitro digestion methods, in order to maximally reproduce them. In this sense, Ekmekcioglu (2002) summarized relevant aspects for bioavailability studies using in vitro models like peptic and pancreatic digestion (chemical and enzymatic composition of saliva, gastric juice, duodenal and bile juice, incubation time, temperature), adjustment of pH, peristaltic frequency (shaking or agitation), osmolality, serosal composition, and permeability characteristics of the enterocyte monolayer, based on the physiological conditions.

The entire process in the mouth lasts from a few seconds to minutes, and since the salivary pH value is close to neutral, significant compound dissolution from food samples is not expected in this stage. This is why most methods only include the gastric and intestinal phases of digestion, and oral processing is perhaps the most difficult to simulate for solid foods. In place of such processing, use is normally made of a homogenization step, though this does not create a bolus. In the case of liquid foods or isolated food components, the homogenization and bolus formation phase is not performed, though salivary amylase may be added (Moreda-Piñeiro et al. 2011). For example, in studies on the hydrolysis of proteins there is no significant enzyme action in the mouth (Wickham et al. 2009), although an oral phase has been applied for other components (carotenoids, plant sterols and minerals). Some examples are shown in Table 1.1.

The gastric phase is performed with HCl or HCl-pepsin under fixed pH and temperature conditions, for a set period of time. Food is homogenized in aqueous solution and typically pepsin is added following adjustment to pH 1–2. The sample is then incubated at 37 °C during 1–3 h, holding the pH constant. In the case of infant food the samples are acidified to pH 4. A recent review has published a compilation of infant digestive conditions of gastric and duodenal phases with the aim of defining them for in vitro methods (Bourlieu et al. 2014).

Regarding the gastric enzymes, a minimum amount of 4,000–5,000 IU of pepsin seems to be necessary for optimal protein digestion (Ekmekcioglu 2002; Etcheverry et al. 2012). Wickham et al. (2009) reported that the pepsin digestion protocols that have been employed involve pepsin activities in the range of 8–12 units per mg of test protein, which may be considered far in excess of values likely to be found in the stomach. The protein dietary intake for an adult (around 75 g in 24 h) would yield a ratio of ~3 mg protein/unit pepsin secreted, compared to ~3 µg protein/unit pepsin during digestion assays. Some authors add mucin in the gastric step in order to better simulate the physiological secretions. Gastric emptying times depend on meal composition, and in this regard meals with high fibre and fat contents can delay gastric emptying. Table 1.1 shows the conditions such as enzymes, pH and gastric emptying times recently used in some studies on different foods.

Intestinal digestion needs subsequent neutralization (usually with NaOH or NaHCO$_3$), and incubation with pancreatic enzymes such as lipase, amylase,

Table 1.1 Conditions used in some studies for in vitro digestion

Analyte	Sample	Oral phase	Gastric phase	Intestinal phase [including micellization (M) and/or dialyzability (D)]	References
Vitamin C, carotenoids and phenolic compounds	Blend of fruit juices (orange, kiwi, pineapple) (200 mL)	–	PS (0.2 g)/pH = 2/2 h	D: cellulose dialysis membrane (12,000 Da), PC (5 mL) (4 g/L), BS (25 g/L) mixture/pH = 7.5/2 h	Rodríguez-Roque et al. (2013)
Lycopene and β-carotene	Broccoli, carrots and tomato soup (5 g)	–	5 mL PS solution (1 g of 2,190 U/mg solid dissolved in 50 mL)/pH = 4 and 2/0.5 h	M: microfiltration (0.22 μm) of supernatants after centrifugation. Intestinal solution (3 mL): (0.4 %) (w/v) PC, (2.5 %) (w/v) BS extract/pH = 6.9/2 h	Alminger et al. (2012)
β-Cryptoxanthin	β-Cryptoxanthin-enriched milk-based fruit drinks (10 g)	Saliva solution (9 mL): α-amylase (145 mg)/pH = 6.5/5 min	Gastric juice (13.5 mL): mucin (1 g), BSA (1 g), and PS (1 g)/pH = 1.1/1 h	Duodenal juice (25 mL): PC (3 g), BS solution (containing bile (0.6 g), pancreatic LP (1 unit), CLP (12.5 μg), CE (5 units), PL-A2 (50 μL), taurocholate salts (19.9 mg))/pH = 6.8/2 h	Granado-Lorencio et al. (2011)
Sialic acid and gangliosides	Infant formulas and human milk (100 mL)	–	PS (0.02 g PS/g sample)/pH = 4/2 h	PC (0.005 g/g of sample), BS (0.03 g/g sample)/pH = 6.5/2 h	Lacomba et al. (2011)
Plant sterols and their oxidation products	Plant sterol-enriched beverages milk and/or fruit based beverages (20 g)	Saliva solution (9 mL): α-amylase (0.19 mg)/pH = 6.5/5 min	Gastric juice (13.5 mL): mucin, BSA, PS/pH = 1.07/1 h	Duodenal juice (25 mL), BS solution (9 mL), pancreatic LP (1 unit), CLP (12.5 μg), CE (5 units), PL-A2 (501.2 units), sodium taurocholate (0.02 mg)/pH = 6.8/2 h	Alemany et al. (2013)
Al, Ba, Cd, Cr, Cu, Mg, Mn and P	Chocolate drink powder (2.25 g)	Salivary fluid (1.5 mL): α-amylase (0.6 g/L), mucin (0.05 g/L)/pH = 7/5 min	Gastric fluid (3 mL): BSA, PS (5 g/L), mucin (6 g/L)/pH = 1/2 h	Duodenal fluid (3 mL): BSA (2 g/L), PC (18 g/L), LP (3 g/L), BS (1.5 mL)/pH = 8/2 h	Peixoto et al. (2013)

Zn	Infant formulas and other infants foods (5–80 g)	—	PS (0.02 g/g sample)/pH=2/2 h	D: dialysis bag (10–12,000 Da). PC (0.03 g/g sample), BS extract (0.03 g/g/pH=5/2 h	Perales et al. (2006)
EPA and DHA	Salmon oil	—	—	Duodenal fluid (8.1 mL): PC (1.2 mg/mL) and BS (11.8 mM)/pH=7/20–110 min. Human duodenal juice (8.1 mL) (1.37 mg protein/mL)/pH=7/20–110 min	Aarak et al. (2013)
Ca, Fe, Zn and Cu	School meals (dishes habitually included in a school menu) (30–40 g)	—	PS (0.5 g/100 g sample)/pH=2/2 h	D: dialysis bags (10–12,000 Da). Pancreatic (0.4 %, w/v)-BS (2.5 %, w/v) mixture (7.5 mL)/pH=7.5/2 h. Pancreatic (0.4 %, w/v)-BS (2.5 %, w/v) mixture (18.8 mL)/pH=5/2 h	Cámara et al. (2005)
Allergenic proteins (bovine β-Lg, β-CN and hen's egg OVA)	Resembling food system	—	PS (182 units-adult- or 22.75 units-infant- of PS/mg of protein)/pH=2.5 (adult) or 3 (infant)/1 h	*Adult*: Sodium taurocholate and glycodeoxycholate (4 mM), 0.4 units/mg of protein C-TRP (activity 40 units/mg of protein), 34.5 units/mg of protein TRP (activity 13,800 units/mg of protein)/pH=6.5/0.5 h. *Infant*: BS concentration reduced by 4-factor, TRP and C-TRP by a 10-factor	Dupont et al. (2010)
Soluble nitrogen (protein digestibility)	Wheat and organic spelt bakery products (sample containing 250 mg protein)	—	PS (3,460 units/mg protein)/pH=1.9/0.5 h	TRP (15,450 units/mg protein), C-TRP (51 units/mg protein), peptidase (102 units/mg solid)/pH=7.98. PC (activity ≥ equivalent to 1×USP specification)/pH=7.5/6 h	Abdel-Aal (2008)

BS bile salt, BSA bovine serum albumin, CE cholesterol esterase, CLP colipase, C-TRP chymotrypsin, DHA docosahexaenoic acid, EPA eicosapentaenoic acid, LP lipase, OVA ovalbumin, PC pancreatin, PL phospholipase, PS pepsin, TRP trypsin, β-CN β-casein, β-Lg β-lactoglobulin ribonuclease and protease with or without bile salts as emulsifiers (see Table 1.1). Since the majority of nutrients are absorbed in the jejunum (pH 6.7–8.8) and ileum (6.8–7.7), most intestinal digestion studies adjust the pH to 6.5–7.5 at 37 °C for 1–5 h (Ekmekcioglu 2002). Lipophilic compounds (carotenoids, plant sterols, etc.) partition into liposomes and micellar phases during intestinal digestion. Consequently, human pancreatic lipase and other specific enzymes (cholesterol esterase, phospholipase A2, co-lipase, etc.) are added to achieve more physiological conditions (Table 1.1). Other components such as phospholipids and calcium are also used in various in vitro models (Hur et al. 2011). Wickham et al. (2009) indicated that the colloidal phases should be included within the design of static digestion models used to assess the digestibility of protein allergens, because the multi-phase nature of the gastric and duodenal environments could play an important role in terms of allergenic protein potential, and thus in the conduction of risk assessments. These authors reviewed the studies on the role of physiological surfactants found in the gastric and the duodenal compartments in relation to potential allergens.

In studies that have used static methods, the choice of enzymes and incubation conditions is conditioned by the study objective. Thus, the application of such methods to a single nutrient has conditioned the use of a single enzyme, e.g., protein-pepsin, starch-amylase or lipid-lipase. Using a single purified enzyme offers the advantage of making standardization of the in vitro model easier; thereby allowing results to be obtained that are more reproducible among different laboratories. However, the digestion of a nutrient is influenced by other food components, and consequently the use of complex mixtures of enzymes affords results that more closely reflect the actual in vivo situation than the utilization of single purified enzymes. As an example, if protein digestion is carried out with three enzymes (trypsin, chymotrypsin and peptidase) in a single-step digestion process, greater protein digestibility (39–66 %) is obtained than in the case of a two-step digestion process with several enzymes (pepsin and pancreatic enzymes) (Abdel-Aal 2008; Hur et al. 2011). The enzymes are collected from human subjects, though a number of studies consider that it is possible to replace human pepsin, pancreatic lipase and co-lipase with porcine enzymes (Hur et al. 2011). Aarak et al. 2013 compared in vitro models using human and porcine intestinal enzymes applied to eicosapentaenoic acid (EPA) and docosahexaenoic acid (DHA) release from salmon, using only a duodenal digestion step. Results show that the human lipolytic enzyme system produces a comparatively higher release of EPA and DHA.

During peptic and pancreatic digestion, food samples are often incubated in a continuously shaking water-bath, although not all studies indicate the conditions used. A recently developed static device (Chen et al. 2011) allows agitation with a spherical probe, applying vertical movement within the vessel to create a flow pattern similar to that of the contraction waves of the stomach wall.

1.3 Applications: Advantages and Disadvantages

Static models are particularly useful where there is limited digestion (e.g., gastric and/or intestinal steps), but are less applicable in total digestion studies, including colonic fermentation. These methods can be used to evaluate the influence of digestion conditions, and to carry out studies on the positive or negative effect of food structure (particle size, addition of emulsifiers, etc.), food composition (food fortification, etc.), dietetic factors (interactions between food components such as fibre, minerals, etc.) and food processing (thermal and non-thermal treatment, fermentation, etc.) upon nutrient and bioactive compound bioaccessibility, in order to establish the nutritional value of foods and improve food formulation/design. In conclusion, static models are predominantly used for digestion studies on simple foods and isolated or purified food components. Such studies not only contribute to improve food properties (nutritional or sensory) but also constitute preliminary trials producing evidence referred to possible nutrition and health claims, since it must be shown that the substance is digested and available to be used by the body (Fernández-García et al. 2009). An overview on different characteristics and conditions of the static models is represented in Fig. 1.1. A recent review assesses the importance of in vitro methods in nutritional, toxicological, pharmaceutical, and microbiological studies (Guerra et al. 2012).

Data from human intervention studies (in vivo assays) constitute the reference methods, whereas bioaccessibility studies (based on in vitro methods) are used as surrogates for predictive purposes. A number of disadvantages, such as limitations in experimental design, difficulties in data interpretation, high cost of equipment and labour, ethical constraints, inter-individual variations, and the lack of certified reference standards to compare data among studies limit the utility of in vivo methods (Fernández-García et al. 2009). In contrast, in vitro models are reproducible, since they allow better control of the experimental variables than animal or human studies, provided they are adequately validated and standardized, with the use of reference material if needed. In general, they are rapid and simple methods, since they only need materials that are routinely available in the laboratory, and are therefore relatively inexpensive and cost-effective. Furthermore, in vitro models allow a reduction of the sample size when this is a limiting factor. Static systems evaluate the aforementioned term "bioaccessibility", and can be used to establish trends in relative bioaccessibility, comparing the solubility of a component in different foods as a screening or categorizing tool. However, it is generally recognized that not all soluble or dialyzable compounds are absorbable.

Nevertheless, despite their potential and broad applicability, none of the static models reproduce the dynamic environment of the intestine. They cannot assess uptake or absorption, or transport kinetics, or measure nutrient or food component competition at the site of absorption as occurs in vivo. They are models lacking the complex mucosal barrier with all its regulatory processes, particularly hormonal and nervous control, feedback mechanisms, mucosal cell activity, complexity of peristaltic movements, gastric emptying or continuous changes in pH and secretion

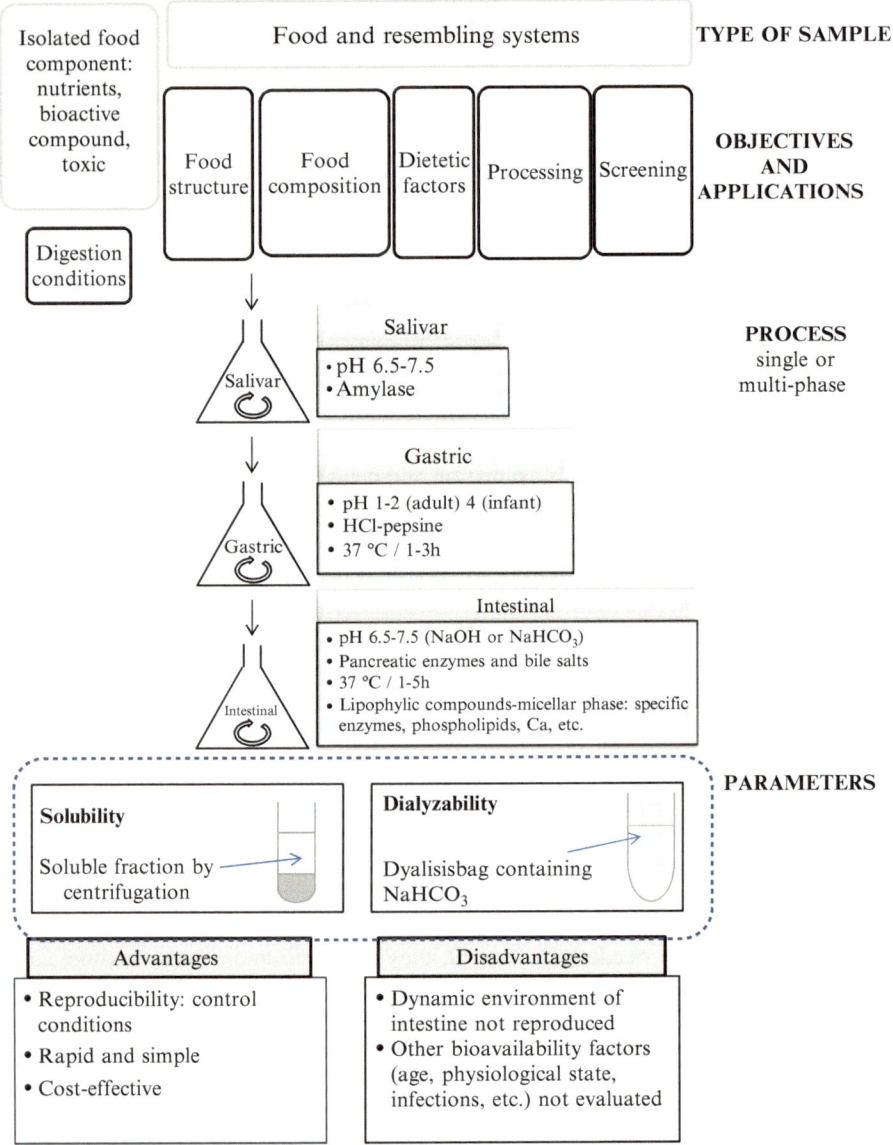

Fig. 1.1 Overview on static model conditions

flow rates, involvement of the local immune system, effects of the intestinal micro-flora and liver metabolism. Furthermore, factors that influence the bioavailability of a nutrient or bioactive compound, such as nutritional status, age, genotype, physio-logical state (e.g., pregnancy, lactation and obesity), or chronic and acute infections cannot be evaluated in static in vitro assays (see Fig. 1.1) (Etcheverry et al. 2012).

1.4 Static Versus In Vivo Digestion: Conclusions

Few studies have evaluated the in vivo–in vitro correlation of results obtained in foods, and it is therefore difficult to properly assess the accuracy of the current in vitro assays. Some reviews, such as that published by Fernández-García et al. (2009), consider that a correlation effectively exists in the case of carotenoids, and that a measure of bioaccessibility might be sufficient as an estimation of how bioavailable a carotenoid is from the food in question (Etcheverry et al. 2012). Likewise, Butts et al. (2012), in reference to amino acid and protein digestibility, affirm that simple in vitro digestion methods have the potential to give useful measures of the in vivo behaviour. Van Campen and Glahn (1999) reviewed static solubility and dialyzability methods for estimating the availability of essential mineral elements, and found these methods to be in reasonable agreement with human absorption data, especially for iron. The authors also indicated that there has been much less development of in vitro methods for other trace minerals (zinc, copper, manganese, selenium) than for iron. A recent review (Etcheverry et al. 2012) compiles in vitro bioaccessibility and bioavailability methods applied to different nutrients, and recommends concrete methods for each nutrient. The need for more validation studies of in vivo–in vitro results is also discussed.

In conclusion, investigators who use static methods must consider how to adapt the static digestion conditions according to the composition of the sample and/or to food components under study—seeking a balance between technical simplification and accuracy, and always retaining the in vivo situation as reference with a view to maximally reproducing the physiological situation through the static model. In addition, it is necessary to know and assess the advantages and disadvantages of static in vitro digestion models for different food samples. Thus, there is urgent need for harmonization and standardization of the in vitro techniques, particularly the static methods. Furthermore, these methods must be validated with proper assessments of gastrointestinal human physiology, in order to afford improved study designs. In this context, although such methods are unable to reproduce all the conditions found in the in vivo setting, their validation at least will allow the comparison of preliminary results among laboratories, prior to the conduction of more advanced studies (dynamic in vitro studies, the use of cell cultures, or in vivo experimentation).

References

Aarak KE, Kirkhus B, Holm H et al (2013) Release of EPA and DHA from salmon oil – a comparison of in vitro digestion with human and porcine gastrointestinal enzymes. Br J Nutr 110: 1402–1410

Abdel-Aal ESM (2008) Effects of baking on protein digestibility of organic spelt products determined by two in vitro digestion methods. LWT 41:1282–1288

Alemany L, Cilla A, Garcia-Llatas G et al (2013) Effect of simulated gastrointestinal digestion on plant sterols and their oxides in enriched beverages. Food Res Int 52:1–7

Alminger M, Svelander C, Wellner A et al (2012) Applicability of in vitro models in predicting the in vivo bioavailability of lycopene and β-carotene from differently processed soups. Food Nutr Sci 3:477–489

Bourlieu C, Ménard O, Bouzerzour K et al (2014) Specificity of infant digestive conditions: some clues for developing relevant in vitro models. Crit Rev Food Sci Nutr 54:1427–1457

Butts CA, Monro JA, Moughan PJ (2012) In vitro determination of dietary protein and amino acid digestibility for humans. Br J Nutr 108:S282–S287

Cámara F, Amaro MA, Barberá R et al (2005) Bioaccessibility of minerals in school meals: comparison between dialysis and solubility methods. Food Chem 92:481–489

Chen J, Gaikwad V, Holmes M et al (2011) Development of a simple model device for in vitro gastric digestion investigation. Food Funct 2:174–182

Dupont D, Mandalari G, Molle D et al (2010) Comparative resistance of food proteins to adult and infant in vitro digestion models. Mol Nutr Food Res 54:767–780

Ekmekcioglu C (2002) A physiological approach for preparing and conducting intestinal bioavailability studies using experimental systems. Food Chem 76:225–230

Etcheverry P, Grusak MA, Fleige LE (2012) Application of in vitro bioaccessibility and bioavailability methods for calcium, carotenoids, folate, iron, magnesium, polyphenols, zinc, and vitamins B6, B12, D, and E. Front Physiol 3:1–21

Fernández-García E, Carvajal-Lérida I, Pérez-Gálvez A (2009) In vitro bioaccessibility assessment as a prediction tool of nutrient efficiency. Nutr Res 29:751–760

Granado-Lorencio F, Donoso-Navarro E, Sánchez-Siles LM et al (2011) Bioavailability of β-cryptoxanthin in the presence of phytosterols: in vitro and in vivo studies. J Agric Food Chem 59:11819–11824

Guerra A, Etienne-Mesmin L, Livrelli V et al (2012) Relevance and challenges in modeling human gastric and small intestinal digestion. Trends Biotechnol 30:591–600

Hur SJ, Lim BO, Decker EA et al (2011) In vitro human digestion models for food applications. Food Chem 125:1–12

Lacomba R, Salcedo J, Alegría A et al (2011) Effect of simulated gastrointestinal digestion on sialic acid and gangliosides present in human milk and infant formulas. J Agric Food Chem 59:5755–5762

Moreda-Piñeiro J, Moreda-Piñeiro A, Romarís-Hortas V et al (2011) In-vivo and in-vitro testing to assess the bioaccessibility and the bioavailability of selenium and mercury species in food samples. Trends Anal Chem 30:324–345

Peixoto RRA, Mazon EAM, Cadore S (2013) Estimation of the bioaccessibility of metallic elements in chocolate drink powder using an in vitro digestion method and spectrometric techniques. J Braz Chem Soc 24:884–890

Perales S, Barberá R, Lagarda MJ et al (2006) Bioavailability of zinc from infant foods by in vitro methods (solubility, dialyzability and uptake and transport by Caco-2 cells). J Sci Food Agric 86:971–978

Rodríguez-Roque MJ, Rojas-Grau MA, Elez-Martínez P et al (2013) Changes in vitamin C, phenolic, and carotenoid profiles throughout in vitro gastrointestinal digestion of a blended fruit juice. J Agric Food Chem 65:1859–1867

Van Campen DR, Glahn RP (1999) Micronutrient bioavailability techniques: accuracy, problems and limitations. Field Crop Res 60:93–113

Wickham M, Faulks R, Mills C (2009) In vitro digestion methods for assessing the effect of food structure on allergen breakdown. Mol Nutr Food Res 53:952–958

Wienk KJH, Marx JJM, Beynen AC (1999) The concept of iron bioavailability and its assessment. Eur J Nutr 38:51–75

Chapter 2
InfoGest Consensus Method

Alan Mackie and Neil Rigby

Abstract This section describes the consensus static digestion method developed within the COST Action InfoGest. Simulated gastro-intestinal digestion is widely employed in many fields of food and nutritional research. Various different digestion models have been proposed, which often impedes the possibility of comparing results across research teams. For example, a large variety of enzymes from different sources such as porcine, rabbit or human have been used and these differ in their activity and characterization. Differences in pH, mineral composition and digestion time that alter enzyme activity and other phenomena may also significantly alter results. Other parameters such as the presence of phospholipids, specific enzymes such as gastric lipase and digestive emulsifiers, etc. have also been discussed at length. In this section, a general standardised and practical static digestion method is given, based on physiologically relevant conditions that can be applied for various endpoints. A framework of parameters for the oral, gastric and small intestinal digestion is outlined and their relevance discussed in relation to available in vivo data and enzymes. Detailed, line-by-line guidance recommendations and justifications are given but also limitations of the proposed model. This harmonised static, in vitro digestion method for food should aid the production of more comparable data in the future.

Keywords In vitro • Digestion • Oral • Gastric • Small intestinal

2.1 Introduction

The static protocol for simulating digestion in the upper GI tract published by InfoGest and led by Andre Brodkorb was the result of more than 2 years' work involving extensive discussion among scientists from a wide range of relevant disciplines (Minekus et al. 2014). The final consensus recommendation is relatively simple, based on physiological parameters that have been cited and is widely

A. Mackie (✉) • N. Rigby
Institute of Food Research, Norwich Research Park, Colney Lane, Norwich NR4 7UA, UK
e-mail: alan.mackie@ifr.ac.uk

© The Author(s) 2015
K. Verhoeckx et al. (eds.), *The Impact of Food Bio-Actives on Gut Health*,
DOI 10.1007/978-3-319-16104-4_2

supported by those undertaking in vitro digestions, especially in food research. In keeping with the requirement for simplicity but not oversimplification discussed in the general introduction to this chapter, this is a static model using values of pH, ionic composition endogenous surfactants and enzyme activity that are fixed at the start of the experiment. All aspects of digestion in the upper GI tract were considered in the development of the method and the reasons for the inclusion or exclusion of specific features will be discussed below. The method comprises up to three stages that mimic the oral, gastric and small intestinal phases of digestion in vivo. At each stage the duration and physical and biochemical environment are described and the reasons for their selection given. The enzymes recommended for inclusion are described using their IUBMB Enzyme Nomenclature and the method has been written in such a way as to allow the sourcing of material from any suitable supplier. The method is outlined in the flow diagram given in Fig. 2.1. All enzyme activities and other concentrations are given per mL of digesta as they will finally be used.

2.2 The Oral Phase

The oral phase of digestion is where solid foods are physically broken down through the process of chewing. Residence time is short, especially for liquid or semi-solid foods, and solids are mixed with saliva to form a bolus with a paste-like consistency before swallowing. In addition to processing there is a great deal of sensing, including taste, texture, aroma, etc. However, most of these functions do not affect digestion in any tangible way and so for the purposes of the method they have been ignored. The exception to this is the texture, which in vivo is continually assessed and generally only when particles of food have been reduced to 2 mm or smaller will the bolus be swallowed (Peyron et al. 2004). Before the oral phase is started a decision needs to be made about what kind of processing is to be included as shown in Fig. 2.1. On the face of it this seems simple as liquid samples don't need to be chewed and so can simply be mixed with simulated salivary or gastric fluid and passed to the gastric phase while solid samples go through the full oral phase as outlined below. However, the user needs to decide where the boundary between solid and liquid lies and whether the addition of salivary amylase is important for their sample.

In addition to chewing the other important factor for solid food is the addition of saliva, which contains a broad range of ions, proteins and peptides, only some of which are directly relevant to digestion (Humphrey and Williamson 2001). Saliva also contains the enzyme α-amylase (EC 3.2.1.1) but not lingual lipase as is often quoted. There is general interest in the importance of mucin in saliva (Sarkar et al. 2009) and much debate about whether it is important to add it or not. There are two types of mucin secreted into saliva MUC5B and MUC7 although there is none in parotid saliva. Mucin represents less than 20 % of the total protein in whole saliva, which is normally around 0.7 mg/mL (Lee et al. 2007). At such low levels as 0.15 mg/mL, other surface active proteins are more likely to be important than mucin for the behaviour of saliva. Also the availability of reliable sources of such

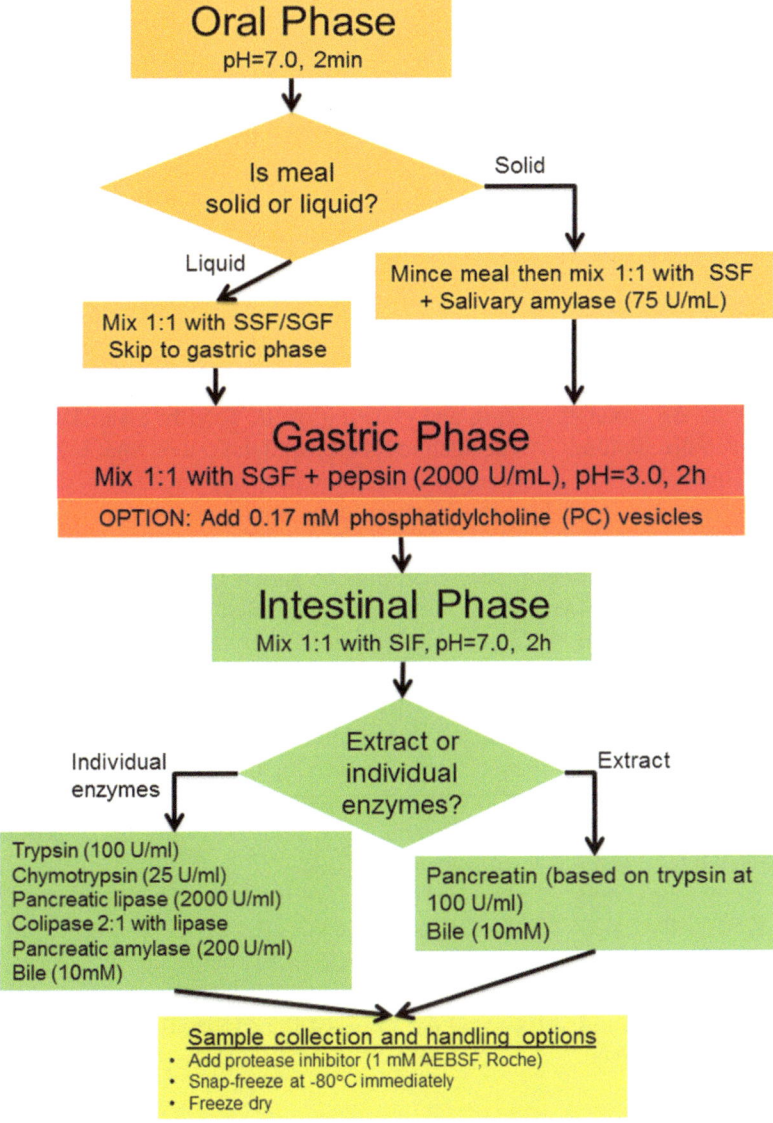

Fig. 2.1 A flow diagram describing the InfoGest digestion method involving simulated salivary fluid (*SSF*), simulated gastric fluid (*SGF*) and simulated intestinal fluid (*SIF*)

salivary mucin would make inclusion difficult under normal circumstances. Thus the method uses a simulated salivary fluid (SSF) containing the ion composition given in Table 2.1 at pH 7.0 and α-amylase at 150 units per mL of SSF (Bornhorst et al. 2014; Hoebler et al. 1998) but no mucin or other proteins. Here, 1 unit is defined as liberating 1.0 mg of maltose from starch in 3 min at pH 6.9 at 20 °C and the activity

Table 2.1 The ionic composition of simulated digestion fluids

Constituent	SSF (pH 7) (mmol/L)	SGF (pH 3) (mmol/L)	SIF (pH 7) (mmol/L)
K^+	18.8	7.8	7.6
Na^+	13.6	72.2	123.4
Cl^-	19.5	70.2	55.5
$H_2PO_4^-$	3.7	0.9	0.8
HCO_3^-, CO_3^{2-}	13.7	25.5	85
Mg^{2+}	0.15	0.1	0.33
NH_4^+	0.12	1.0	–
Ca^{2+}	1.5	0.15	0.6

should be determined using the method of Bernfeld (1955). We now have our saliva but how do we simulate mixing and chewing of the food in a standardised way? After much discussion it was decided to recommend the use of what is known in the UK as a mincer, which is commonly used in kitchens to mince raw or cooked meat. Having chewed the food, how much saliva do we need to add? On average, unstimulated flow rate is 0.3 mL/min but stimulated flow rate is, at maximum, 7 mL/min (Humphrey and Williamson 2001). Stimulated saliva is reported to contribute as much as 80–90 % of the average daily salivary production. Thus based on stimulated flow, the amount of saliva to add is a 1:1 v/w ratio with the food i.e. 5 g of food + 5 mL SSF. The only remaining question is how long should we expose the food to the SSF? Although a value of 0.5 min might be close to the situation in vivo, the practicalities of handling suggest that in order to be confident of reproducing the oral phase in a consistent manner, including mixing of saliva, 2 min would be more appropriate. The temperature at which the amylase containing SSF is mixed with the "chewed" food should of course be 37 °C and the 2 min is the contact time between the food and SSF.

In a typical example: 5 g of solid or 5 mL of liquid food is mixed with 3.5 mL of SSF electrolyte stock solution, either during or after mincing, if necessary. Next, 0.5 mL salivary α-amylase solution of 1,500 U/mL made up in SSF electrolyte stock solution (α-amylase from human saliva Type IX-A, 1,000–3,000 U/mg protein, Sigma) is added followed by 25 μL of 0.3 M $CaCl_2$ and 975 μL of water and thoroughly mixed.

2.3 The Gastric Phase

Following whatever oral processing has been undertaken there needs to be a gastric phase of digestion. Commonly held beliefs about the stomach are that the pH is very low (1–2) and that there is a lot of mixing. Neither of these is a useful idea as the pH is generally only very acidic in the fasted state and there is only mixing in the small region close to the exit of the stomach known as the antrum. The pH in the gastric compartment is rather dynamic and is highly dependent of the buffering capacity of the food (Carrière et al. 1991; Dressman et al. 1990; Kalantzi et al. 2006). However, as this is a static method a specific value needs to be recommended and this was chosen in conjunction with a decision about the duration of the gastric phase. Given that the method needs to be broadly applicable, the method recommends 2 h. This time represents the half emptying of a moderately nutritious and semi-solid meal (Dressman et al. 1990) and in order to match the 2 h recommendation for the dura-tion of the gastric phase the pH we recommend must represent a mean value for a general meal over that time. Thus we recommend the use of a static value of pH 3 combined with the ionic composition outlined in Table 2.1. In this static model the physical environment of the gastric compartment will not be reproduced but some mixing is required and this can be supplied either by shaking or stirring the sample at 37 °C.

The only proteolytic enzyme present in the stomach is pepsin, which is secreted in the form of the precursor pepsinogen. Large variations in pepsin activities are reported in the literature due to the use of different assays and calculations (Ulleberg et al. 2011; Armand et al. 1995). Based on the literature the recommended activity of porcine pepsin (EC 3.4.23.1) is 2,000 U/mL of gastric contents where one unit will produce a ΔA_{280} of 0.001 per minute at pH 2.0 and 37 °C, measured as TCA-soluble products using haemoglobin as a substrate, adapted from Anson (1938) and Anson and Mirsky (1932). The use of lipolytic enzymes is always more difficult and whilst the potential importance of human gastric lipase (HGL) is acknowledged it has not been included for the following reasons. Firstly, because of the relatively low pH, lipid interfaces tend to become saturated and thus gastric lipolysis is gener-ally limited. Secondly, there is no lipase currently widely available, affordable and that has the correct pH and site specificity. The final recommended option is to include phosphatidylcholine (PC) at 0.17 mM in vesicular form (Macierzanka et al. 2009; Mandalari et al. 2009).

In a typical example: 10 mL of liquid sample or oral bolus is mixed with 7.5 mL of SGF electrolyte stock solution, 2.0 mL porcine pepsin solution of 20,000 U/mL made up in SGF electrolyte stock solution (pepsin from porcine gastric mucosa 3,200–4,500 U/mg protein, Sigma), 5 μL of 0.3 M CaCl₂, 0.2 mL of 1 M HCl to reach pH 3.0 and 0.295 mL of water.

2.4   The Small Intestinal Phase

The final and most complex phase is the small intestinal phase in which the pH is again raised to 7 and the gastric effluent is exposed to a broader range of enzymes and surfactants (Kalantzi et al. 2006; Kopf-Bolanz et al. 2012; Versantvoort et al. 2005). As with the gastric compartment the intestinal phase duration is 2 h. This is again a compromise but is based on normal transit times in the human gut but also on the fact that because there is no product removal, inhibition may become a prob-lem at extended times, especially if there is a significant amount of lipid present. The suggested ionic composition for the SIF is again given in Table 2.1. There are two possible approaches that can be used with regard to the enzymes used. Firstly, for reasons of simplicity and cost one can use a pancreatic extract (pancreatin) containing all the relevant enzymes but in a fixed ratio or secondly if preferred one can use the individual enzymes (proteases, lipases and amylase). If pancreatin is used then the amount to add must be based on a specific enzyme activity and we suggest that it is based on the trypsin activity and should be added in sufficient quantity to provide 100 U/mL of intestinal phase content. The activity should be based on the TAME assay. The pancreatin should also be assayed for its chymotrypsin, lipase and amylase activities. Where individual enzymes are to be used the following activities should be added per mL of intestinal content. Bovine trypsin (EC 3.4.21.4) at 100 U/mL where one unit hydrolyses 1 µmole of p-toluene-sulfonyl-L-arginine methyl ester (TAME) per minute at 25 °C, pH 8.1, in the presence of 10 mM calcium ions (Walsh and Wilcox 1970); bovine chymotrypsin (EC 3.4.21.1) at 25 U/mL N-Benzoyl-L-Tyrosine Ethyl Ester (BTEE) units where one unit will hydrolyse 1.0 µmole of BTEE per minute at pH 7.8 at 25 °C (Bergmeyer et al. 1974); porcine pancreatic amylase (EC 3.2.1.1) at 200 U/mL where one unit will liberate 1.0 mg of maltose from corn starch in 3 min at pH 6.9 at 20 °C (Bernfeld 1955); porcine pancreatic lipase (EC 3.1.1.3) at 2,000 U/mL where 1 unit will release 1 µmole of free fatty acid per minute from a substrate of tributyrin at 37 °C, pH 8.0, in the presence of 2 mM calcium ions and 4 mM sodium taurodeoxycholate and excess colipase at a 2:1 molar excess, which is approximately a mass ratio of 1:2 colipase/lipase.

In addition to the enzymes there are also a range of endogenous surfactants including bile salts and phospholipids. These are important in the digestion of both protein and lipid and in the case of the latter they are involved in product removal to the gut epithelium. They can conveniently be added as a bile extract or as frozen porcine or bovine bile. Sufficient should be added to provide 10 mM bile in the final intestinal content (Kalantzi et al. 2006). This can be assayed using a number of different kits such as the bile assay kit 1 2212 99 90 313 from Diagnostic Systems GmbH in Germany (Collins et al. 1984). Of course none of the animal bile will be a really close match for human bile and it is currently unclear what impact that is likely to have on the digestion process. For information, the typical composition of human, bovine and porcine bile are given in Table 2.2. Surprisingly, the bovine bile is a closer match to human than porcine bile, at least in terms of tauro- and glycocholate. The bile will also contain phospholipids and cholesterol in sufficient quantity for the digestion.

In a typical example of intestinal simulation, 20 mL of gastric chyme is mixed with 11 mL of SIF electrolyte stock solution, 5.0 mL of a pancreatin solution 800 U/mL made up in SIF electrolyte stock solution based on trypsin activity, 2.5 mL fresh bile (based on 160 mM fresh bile), 40 µL of 0.3 M $CaCl_2$, 0.15 mL of 1 M NaOH to reach pH 7.0 and 1.31 mL of water. Verification of the pH is recommended to determine the amount of NaOH/HCl required in a test experiment prior to digestion. In this way base/acid can be added more rapidly and followed by final verification of the pH.

Table 2.2 The approximate bile acid composition of human, bovine and porcine bile

Bile acid	Human bile (%)	Bovine bile (%)	Porcine bile (%)
Taurohyodeoxycholate	0	0	37
Glycohyodeoxycholate	0	0	34
Taurocholate	11	31	0
Glycocholate	26	46	0
Taurochenodeoxycholate	13	2	2
Glycochenodeoxycholate	25	3	26
Taurodeoxycholate	5	8	0
Glycodeoxycholate	11	10	0
Other	9	0	0

Table 2.3 Preparation of stock solutions of simulated digestion fluids

			SSF	pH 7	SGF	pH 3	SIF	pH 7
	Stock conc.		Vol of stock	Conc. in SSF	Vol of stock	Conc. in SSF	Vol of stock	Conc. in SSF
Salt	g/L	mol/L	mL	mmol/L	mL	mmol/L	mL	mmol/L
KCl	37.3	0.5	15.1	15.1	6.9	6.9	6.8	6.8
KH$_2$PO$_4$	68	0.5	3.7	3.7	0.9	0.9	0.8	0.8
NaHCO$_3$	84	1	6.8	13.6	12.5	25	42.5	85
NaCl	117	2	–	–	11.8	47.2	9.6	38.4
MgCl$_2$(H$_2$O)$_6$	30.5	0.15	0.5	0.15	0.4	0.12	1.1	0.33
(NH$_4$)$_2$CO$_3$	48	0.5	0.06	0.06	0.5	0.5	–	–
For pH adjustment								
NaOH		1	–	–	–	–	–	–
HCl		6	0.09	1.1	1.3	15.6	0.7	8.4
CaCl$_2$(H$_2$O)$_2$ is not added to the simulated digestion fluids, see legend								
CaCl$_2$(H$_2$O)$_2$	44.1	0.3		1.5 (0.75)[a]		0.15 (0.075)[a]		0.6 (0.3)[a]

[a]In brackets is the corresponding Ca^{2+} concentration in the final digestion mix

2.5   Practicalities

The outline method given above gives the general approach that should be used in terms of enzymes, their activities, ionic composition and endogenous surfactants. However, there are some practicalities that need to be taken into account when executing the method. For example the simulated digestion fluids (SSF, SGF and SIF) are made up using the electrolyte stock solutions given in Table 2.3, enzymes, bile, CaCl$_2$ and water. The volumes are calculated for a final volume of 500 mL for each simulated fluid. However, we recommend making up the stock solution with distilled water to 400 mL, i.e. 1.25 times concentrated, for storage at −20 °C. The addition of enzymes, bile, Ca^{2+} solution etc. and water will result in the correct electrolyte concentration in the final digestion mixture. $CaCl_2$ is not added to the electrolyte stock solutions as precipitation may occur. Instead, it is added to the final mixture of simulated digestion fluid and food.

2.6   Sampling

The way that sampling should be done depends on the nature of the study and should be carefully considered for each study. For example, it may be advisable to have individual sample tubes for each time point rather than withdrawing samples from the reaction vessel. Also, it may be important to sample at multiple time points through both gastric and intestinal phases or it may only be necessary at the end of digestion. Regardless of such questions, the way in which the reactions are stopped will depend on what the samples will be subsequently required for. The following are some recommendations to inhibit further enzyme action in the digesta samples:

- Snap freezing of samples is recommended in liquid nitrogen immediately after the reaction for further analysis. It should be born in mind that enzymes will continue to act, even in frozen samples albeit slowly. Therefore the colder the sample is stored the better.
- If samples are sent to other labs, i.e. by courier or by post, the digestion should be stopped completely and for this, the following procedures are recommended:

  – Neutralize the pH in the gastric phase by adding 0.5 M sodium bicarbonate. This will inactivate the pepsin before snap-freezing in liquid nitrogen and subsequent storage and/or freeze drying.
  – Addition of protease inhibitor (e.g. 1 mM 4-(2-aminoethyl) benzenesulfonyl fluoride hydrochloride [AEBSF], Roche or similar), snap freezing in liquid nitrogen and subsequent freeze drying of samples.

2.7   Conclusions

The InfoGest harmonised static digestion simulation method has been endorsed by a wide range of international experts. We have endeavoured to make it as physiologically relevant as a static model can be but readers should bear in mind that it is still only a simplified model. The main limitations of such a simplified model are the fixed pH and duration of the various phases. However, this can also be seen as an advantage in terms of consistency and comparability. The other potentially problematic issue is the lack of an adsorption step but there are now a number of cellular systems where this aspect can be investigated. It is hoped that this simple model will be widely adopted, allowing faster progress when developing healthier foods and more effective bioactives.

References

Anson ML (1938) The estimation of pepsin, trypsin, papain, and cathepsin with hemoglobin. J Gen Physiol 22(1):79–89

Anson M, Mirsky A (1932) The estimation of pepsin with hemoglobin. J Gen Physiol 16(1):59–63

Armand M, Hamosh M, DiPalma JS, Gallagher J, Benjamin SB, Philpott JR, Lairon D, Hamosh P (1995) Dietary fat modulates gastric lipase activity in healthy humans. Am J Clin Nutr 62(1):74–80

Bergmeyer HU, Gawehn K, Williamson DH, Lund P (1974) Methods of enzymatic analysis, vol 1. Academic, New York, USA

Bernfeld P (1955) Amylases, α and β. In: Methods in enzymology, vol 1. Academic, New York, USA, pp 149–158

Bornhorst GM, Hivert H, Singh RP (2014) Rice bolus texture changes due to α-amylase. LWT-Food Sci Technol 55(1):27–33

Carrière F, Moreau H, Raphel V, Laugier R, Benicourt C, Junien JL, Verger R (1991) Purification and biochemical characterization of dog gastric lipase. Eur J Biochem 202(1):75–83

Collins B, Watt P, O'Reilly T, McFarland R, Love A (1984) Measurement of total bile acids in gastric juice. J Clin Pathol 37(3):313–316

Dressman JB, Berardi RR, Dermentzoglou LC, Russell TL, Schmaltz SP, Barnett JL, Jarvenpaa KM (1990) Upper gastrointestinal (GI) pH in young, healthy men and women. Pharm Res 7(7):756–761

Hoebler C, Karinthi A, Devaux M, Guillon F, Gallant D, Bouchet B, Melegari C, Barry J (1998) Physical and chemical transformations of cereal food during oral digestion in human subjects. Br J Nutr 80:429–436

Humphrey SP, Williamson RT (2001) A review of saliva: normal composition, flow, and function. J Prosthet Dent 85(2):162–169

Kalantzi L, Goumas K, Kalioras V, Abrahamsson B, Dressman JB, Reppas C (2006) Characterization of the human upper gastrointestinal contents under conditions simulating bio-availability/bioequivalence studies. Pharm Res 23(1):165–176

Kopf-Bolanz KA, Schwander F, Gijs M, Vergères G, Portmann R, Egger L (2012) Validation of an in vitro digestive system for studying macronutrient decomposition in humans. J Nutr 142(2):245–250

Lee JY, Chung JW, Kim YK, Chung SC, Kho HS (2007) Comparison of the composition of oral mucosal residual saliva with whole saliva. Oral Dis 13(6):550–554. doi:10.1111/j.1601-0825.2006.01332.x

Macierzanka A, Sancho AI, Mills ENC, Rigby NM, Mackie AR (2009) Emulsification alters simulated gastrointestinal proteolysis of b-casein and b-lactoglobulin. Soft Matter 5(3):538–550

Mandalari G, Mackie AM, Rigby NM, Wickham MS, Mills EN (2009) Physiological phosphatidylcholine protects bovine beta-lactoglobulin from simulated gastrointestinal proteolysis. Mol Nutr Food Res 53(Suppl 1):S131–S139

Minekus M, Alminger M, Alvito P, Ballance S, Bohn T, Bourlieu C, Carrière F, Boutrou R, Corredig M, Dupont D, Dufour C, Egger L, Golding M, Karakaya S, Kirkhus B, Le Feunteun S, Lesmes U, Macierzanka A, Mackie AR, McClements DJ, Ménard O, Recio I, Santos CN, Singh RP, Vegarud GE, Wickham MSJ, Weitschies W, Brodkorb A (2014) A standardised static in-vitro digestion method suitable for food – an international consensus. Food Funct 5:1113–1124

Peyron MA, Mishellany A, Woda A (2004) Particle size distribution of food boluses after mastication of six natural foods. J Dent Res 83(7):578–582

Sarkar A, Goh KK, Singh H (2009) Colloidal stability and interactions of milk-protein-stabilized emulsions in an artificial saliva. Food Hydrocolloids 23(5):1270–1278

Ulleberg EK, Comi I, Holm H, Herud EB, Jacobsen M, Vegarud GE (2011) Human gastrointestinal juices intended for use in in vitro digestion models. Food Digestion 2(1–3):52–61

Versantvoort CH, Oomen AG, Van de Kamp E, Rompelberg CJ, Sips AJ (2005) Applicability of an in vitro digestion model in assessing the bioaccessibility of mycotoxins from food. Food Chem Toxicol 43(1):31–40

Walsh KA, Wilcox PE (1970) Serine proteases. In: Gertrude E, Perlmann LL (eds) Methods in enzymology, vol 19. Academic, New York, USA, pp 31–41

Chapter 3
Approaches to Static Digestion Models

Alan Mackie, Neil Rigby, Adam Macierzanka, and Balazs Bajka

Abstract It is not possible to look in detail at the wide range of static digestion methods that have been used to date. However, this section looks at some of the general approaches that have been used to look at the digestion of various nutrients and bioactives. I have focussed on the two main nutrients that undergo digestion in the upper GI tract, namely protein and lipid. In the case of protein, the research has largely been driven by the need to assess allergenic potential and the parameters used in such an assessment are given along with the justification provided by the authors for their choice. For the lipid digestion, we have drawn heavily upon the work of Julian McClemments and colleagues who have been prolific in generating data in this area. The information provided highlights the fact that a wide range of methods are in use leading to a need for a single method, a role that can be filled by the Infogest method.

Keywords Infogest • Protein • Lipid • Allergy • Bioactive • Delivery

3.1 Introduction

Since the increase in interest in the health implications of specific foods or diets, there has also been an interest in how foods are digested and this has led to the development of a wide range of digestion methods and upper GI tract simulations. Of course the methods have been developed to address specific questions such as the allergenic potential of a protein or the delivery of fat soluble bioactives. Whilst many would argue that specific nutrients should not be considered in isolation, a reductionist approach can sometimes prove helpful. However, this does beg the question of how relevant some of the model digestion systems that have been used are to what happens in vivo after consumption of real foods. The macronutrients that are digested in the upper GI tract are protein, lipid and starch. For many reasons the digestion of both proteins and lipids have often been considered independent of

A. Mackie (✉) • N. Rigby • A. Macierzanka • B. Bajka
Institute of Food Research, Norwich Research Park, Colney Lane, Norwich NR4 7UA, UK
e-mail: alan.mackie@ifr.ac.uk

© The Author(s) 2015
K. Verhoeckx et al. (eds.), *The Impact of Food Bio-Actives on Gut Health*,
DOI 10.1007/978-3-319-16104-4_3

one another and the sections below will outline some of the approaches that have been used when considering the digestion of the micronutrients in isolation. In general, we would recommend that people using static simulations of digestion should use the Infogest model as described in Chap. 2 of this book. However, it is not always possible and so some of the circumstances under which other approaches may be appropriate are given below.

3.2 Static Models for Protein Hydrolysis

The ability of proteins to interact with the immune system in the gut causing intolerance such as coeliac disease and food allergy has led to a significant number of studies. One of the most highly cited articles using in vitro digestion is in the field of allergy. In the article by Astwood et al. (1996) a method is given for determining protein stability. In the article they used a single (gastric) phase of digestion involving simulated gastric fluid (SGF) containing pepsin at 3.2 mg/mL with an activity of 20,100 units in 30 mM NaCl at pH 1.2. The reason given for using these values was that they were "in line with recommendations from the US pharmacopeia". When compared to the value of 2,000 U/mL at pH 3 that is recommended in the Infogest protocol, this seems very high, even if the units are not identical. Surprisingly perhaps, the Astwood article shows that proteins that were food allergens were generally not digested under these conditions. Indeed, this idea led to the use of pepsin resistance as a measure of the allergenic potential of a food protein (Eisenbrand et al. 2002). Under these circumstances it is argued that the method is merely an indicator of structural robustness rather than a precise simulation of how the protein would behave when it is consumed in vivo. It should perhaps be highlighted at this point that the allergenic proteins that were pepsin resistant tended to be those that were thought to sensitise via the oral route. There are a great many allergenic proteins, such as Ara h 1 (Vicillin-type 7S globulin from peanut) or Bos d 8 (casein from cows' milk) that are very susceptible to hydrolysis by pepsin.

In studies undertaken to look at the digestion of allergenic proteins, it is common practice to add protease in a specific proportion relative to the amount of protein being digested. Certainly from the perspective of comparison it is useful to use a consistent activity of enzyme such as the 2,000 U/mL given above and a consistent protein concentration. For example in a ring trial comparing the digestion of milk proteins β-lactoglobulin and β-casein in different laboratories (Mandalari et al. 2009a), two regimes were used, a high and a low protease activity. The high protease used a pepsin activity of 10,560 U/mL based on haemoglobin as a substrate and a substrate concentration of 0.25 mg/mL, equivalent to a pepsin activity of 42,240 U/mg substrate. The low protease part used 165 U/mg of substrate. The pH used in the high protease phase was 1.2 whereas that used in the low protease phase was 2.5. The data from this study are very revealing in terms of comparison of data from different groups. As already stated elsewhere, if a comparison is to be made then the in vitro digestion methods employed must be standardised or at least comparable.

In this study, the methods used were nominally the same in all groups. However, the results varied significantly. For example, under the lower protease condition, β-casein was persistent for 10 min in 62 % of cases but 20 min in 26 % of cases and the remainder showed the protein persistent until either 5 or 40 min. After the simulated gastric phase the study also used a "duodenal" phase lasting 1 h at pH 7.5 or 6.5 for the high and low protease conditions respectively. For the high protease condition pancreatin was used at 12.8 mg per mg substrate and for the low protease condition, trypsin and chymotrypsin were used at 35.4 BAEE U/mg of substrate and 0.4 U/mg of substrate respectively. This is in comparison to the Infogest recommendations of 100 TAME U/mL and 25 U/mL for trypsin and chymotrypsin respectively. There is about a 100-fold difference between TAME and BAEE as a substrate with BAEE giving the higher values. The pancreatin concentration used is likely to have yielded trypsin activities around 8,000 BAEE units per mg of substrate.

In addition to different protease conditions, the study also looked at the effect of 3 mM phospholipid addition to the gastric stage of digestion. The results showed that the addition increased the resistance of β-lactoglobulin to simulated duodenal hydrolysis over 60 min. The mechanism by which this occurs was investigated in more detail in a related paper (Mandalari et al. 2009b). The authors also showed that thermal processing significantly decreased the effect. Such interactions highlight the importance of considering both the protein of interest and other components that may be present during and post consumption in vivo.

The safety assessment of genetically modified products requires consideration of various parameters including assessment of homology with known allergens using various in silico databases, IgE binding studies and resistance of the protein to digestion with simulated gastric fluid (Foster et al. 2013). In all such studies the standard approach has become the use of pepsin at 3.2 mg/mL in 0.03 M NaCl and pH 1.2 (Selgrade et al. 2009). Such amounts of pepsin will typically yield an activity of 10,560 U/mL, as indicated above. In a recent study investigating the safety of the protein osmotin, expressed in transgenic crops to enhance abiotic stress tolerance, the protein was shown to be resistant to pepsin digestion under standard conditions (Sharma et al. 2011). As result, osmotin was regarded as being a potential allergen. In addition to studying proteins for their potential detrimental effects, there has been significant study with regard to the release of bioactive peptides. For example, the group at the Institute of Food Research (CIAL) in Madrid have studied this extensively and in a recent publication they have shown the resistance of casein derived bioactive peptides. The method that they use to simulate adult digestion comprises two phases, gastric and duodenal. The gastric phase uses pepsin at 114 U/mL (11.4 U/mg of substrate) at pH 2.0 for 90 min. The small intestinal phase used Corolase (a pancreatic extract similar to pancreatin) at an enzyme to substrate ratio of 1:25. Given the pepsin activity recommended by Infogest of 2,000 U/mL this seem a little on the low side but it should be born in mind that the pH is also lower (2 rather than 3) and thus the activity of the pepsin in the actual experiment will be slightly higher (Okoniewska et al. 2000).

There has been a significant amount of study of the digestion of protein using the simulated adult gut. However, there have also been many studies of the breakdown of milk proteins in the infant gut. For a good review of the conditions pertaining to the infant gut there is a recent article by the Bourlieu et al. (2014). This review gives a good idea of the physiological environment of the infant gut, both of premature and term infants. In a study where the digestion of protein was compared using infant and adult simulations (Dupont et al. 2010). The adult model used was similar to those given above with a gastric phase at pH 2.5, phospholipid and 182 U/mL pepsin followed by a duodenal phase at pH 6.5, containing 8 mM bile and chymotrypsin and trypsin at 0.4 and 34.5 U/mg of substrate respectively. For the infant model the following changes were made: The pH of the gastric digestion mix was adjusted at 3.0 instead of 2.5; the pepsin concentration in the gastric digestion mix was decreased by a factor of 8 and the duodenal digestion mix was altered by reducing the bile salt concentration by a factor of 4, while the PC, trypsin and chymotrypsin concentrations were reduced by a factor of 10. The proteins used for this comparison were β-lactoglobulin, β-casein and ovalbumin. One might expect that the lower concentrations of proteases used in the infant model would result in less extensive degradation of the three proteins used. Although this was found to be the case for β-casein and ovalbumin, the β-lactoglobulin was more extensively degraded by the infant than the adult digestion simulations. This was thought to be a result of the reduction in the protective effect that gastric phospholipid has on native β-lactoglobulin retarding digestion by trypsin and chymotrypsin. Surprisingly, no information is provided about the justification of the values chosen for the infant model. In a similar, more recent study of simulated gastric digestion of β-lactoglobulin and lactoferrin by a group in Israel, (Shani-Levi et al. 2013) the comparison between adult and infant used gastric pepsin activity of 240 and 210 U/mg of substrate respectively. The main difference between the two models was the way that the pH was lowered going from 6.5 to 3.5 over 4 h in the infant model as opposed to 4.5 to 1.5 over 2 h in the adult model. Needless to say there was little difference in the digestion of β-lactoglobulin but very significant differences in the persistence of lactoferrin, which is a much more labile protein.

Enzyme activity should be measured under the standard conditions recommended by the assay in order to be comparable with other measurements in the literature. However, it should be kept in mind that the activity of the enzyme on the substrate used in the simulation and under the conditions of the simulation is likely to be rather different. For this reason, the simulation should NOT aim to deliver a specific protease activity but rather to deliver a specified amount of active enzyme. This may be a subtle distinction but it has important consequences.

3.3 Static Models for Lipid Hydrolysis

In a similar way that in vitro digestion has been used in some cases to investigate protein digestion in isolation, a number of studies have concentrated on lipid digestion. For a review of this topic, there is an excellent article by Julian McClemments (McClements and Li 2010) in which a large number of different study conditions are given. Perhaps the main message for us from this review is that there is no consistency of approach and everyone uses their own model based on various different requirements and assumptions. Indeed, this very issue was the reason that Infogest was set up. The first stage of digestion may be considered the mouth but whilst there have been a number of studies looking at the behaviour of fat in the mouth (van Aken et al. 2007), as there is no lingual lipase produced in man there is no digestion of fat by endogenous enzymes in the oral cavity. Essentially all the work undertaken on lingual lipase has been done in rodents (Hamosh and Scow 1973) and this has led to the misconception that the same physiology applies to humans. Many studies do not include an oral phase for liquid systems containing fat (Borel et al. 1994; Fernandez et al. 2009; Hedren et al. 2002) or they include an oral phase that merely represents a resting phase after sample preparation (Beysseriat et al. 2006).

The next phase of digestion is the gastric phase containing human gastric lipase (HGL). However, this step is also often excluded from a static digestion focussed on lipolysis for a combination of reasons (Mun et al. 2007; Bonnaire et al. 2008). The most obvious reason is the pH that is being used in simulating the gastric phase is often too low for the HGL to be active as the activity drops rapidly below pH 2 (Ville et al. 2002). There is also the issue of what type of lipase to use as a substitute that has the same pH sensitivity and site specificity as HGL. Also lipolysis under gastric conditions may be considered difficult to follow as the fatty acids (FA) released are not fully dissociated and so not amenable to titration using the normal pH-stat methods. This can be corrected for at the end of the simulation by raising the pH to 9.0, assuring full dissociation of the fatty acids (Helbig et al. 2012). Those that do include a substitute HGL in their gastric simulations often opt for a fungal lipase such as that from *Rhizopus oryzae* (Day et al. 2014; Wooster et al. 2014). This lipase has been well characterised (Hiol et al. 2000), exhibits similar site-specific hydrolysis of triglycerides to that of HGL and is acid stable. However the 'optimal' pH of hydrolysis by *R. oryzae* lipase is 7.5 and the enzyme is only stable in the range pH 4.5–7.5. These values are different from the sensitivity of HGL which is said to have an apparent optimum at pH 4.5 and is still stable at pH 2 (Aloulou and Carriere 2008).

Regardless of the debate as to whether HGL or a substitute should be included in a gastric simulation, there is still the consideration of how much should be added. Recent work has used 0.2 mg/mL fungal lipase at pH 1.9 (Wooster et al. 2014), which given the activity determined by Hiol et al. (2000) of 8,800 U/mg is equivalent to 1,760 U/mL of SGF. As always the method used for the assay is important, and in this case it was against long-chain triacylglycerol plant oils and was determined with 20 mL of substrate emulsion prepared from 40 mL of oil in 400 mL of a 2 % solution of gum acacia prepared in distilled water. One lipase unit corresponded to the release one millimole of fatty acid per minute under assay conditions. This type of assay is difficult to repeat and is thus not comparable with the preferred standard method using tributyrin as a substrate (Carriere et al. 1993). The tributyrin method is preferred because the hydrolysis takes place mainly in solution and is thus not dependant on the surface area of substrate available in an emulsion. This makes it much more reproducible, at least in principle. The activity of HGL has been measured in humans using the tributyrin assay and is around 1,000 U/mg with the activity in gastric fluid of about 100 U/mL in the fed state (Carriere et al. 1993).

In another recent study (van Aken et al. 2011), HGL was substituted by Amano Lipase A, a fungal lipase from *Aspergillus niger* that is quoted by Sigma as having an activity of 120 U/mg but the assay used is not quoted. In the article describing the study the authors go into some detail about the reason for their choice of this enzyme. The main reason for the choice was the broad pH stability meaning that the enzyme remains active at the low gastric pH. However such attention to detail is rare as the small intestine is quite correctly seen as the main site of fat digestion. Despite the lack of importance given to gastric lipolysis, it has been shown that in infants, HGL plays an important role in lipid digestion (Hamosh 1996). This is because in the neonate the production of pancreatic lipases is not fully developed. In a recent study of emulsion digestion using an infant simulation (Lueamsaisuk et al. 2014), fungal (*R. oryzae*) lipase was added at 16 U/mL and a range of pH was assessed (2, 3.5, 4.5 and 5.5). Despite the interesting results confirming the need for a full spectrum of enzymes, one of the main conclusions was the recommendation that outcomes based on in vitro digestion with fungal lipases should always be validated with at least one mammalian gastric lipase. As a final comment, we want to highlight the problems of pH sensitivity in the case of HGL substitutes. If sufficient lipase is added to a gastric simulation to give the relevant activity at the low gastric pH then when the pH is raised for the small intestinal simulation the lipase activity is likely to increase dramatically perhaps dominating the pancreatic lipases. This situation should be avoided.

As already stated the main site of fat digestion is the upper small intestine, duodenum and jejunum. This has led to simulation of this phase of digestion being the focus of most studies. There are three main factors that have been taken into account, enzymes (pancreatic lipase, colipase, etc.), bile (extract or specific composition) and pH. Starting with the simplest parameter, pH, the range of different values used is relatively narrow falling between 6.5 and 7.5 with the occasional study using values as low as pH 5.3 (Beysseriat et al. 2006), which is clearly of no physiological relevance. With this exception, values are generally physiologically relevant to the small intestine as a whole and allow the production of free fatty acids to be reasonably accurately followed by the pH-stat method (Helbig et al. 2012). In the pH-stat method the pH is monitored and any decrease caused by the formation of fatty acids is countered by addition of hydroxide. By monitoring the amount of hydroxide added, the amount of free fatty acid produced can be calculated. If the fatty acid is fully dissociated then the amount of hydroxide added is the same as the amount of fatty acid produced. Endogenous surfactants such as phospholipids and bile acids play a vital role in the hydrolysis and transport of lipids. Bile salts can adsorb onto fat droplets and can remove other materials such as proteins, emulsifiers and lipolysis products from the lipid surface (Maldonado-Valderrama et al. 2011). As a result they should be used in intestinal simulations. The question then arises as to what bile to use and the answer is not clear cut as can be seen from the recommendations in the Infogest method (Chap. 2). Table 2.2 in Chap. 2 shows an analysis of bovine, porcine and human bile using the method of Rossi et al. (1987). It is clear from the table that whilst neither is a perfect match, the bovine bile is closer to human in composition.

3.4 Other Static Models

In addition to studying the digestion of proteins and lipids, static models of gastro-intestinal digestion have been used for a range of other things. In particular, starch resistance has been studied using such models for many years (Ring et al. 1988; Wolf et al. 1999). However, the key to the functionality of resistant starch is its lack of digestion in the upper GI tract, thus the models have tended to focus on colonic fermentation. Despite this focus there have been some more recent articles that look at the digestion of starch in the upper GI tract. In an article by Wooster et al. (2014), an emulsion was combined with a range of different polymers including starch and the in vitro digestion simulated the small intestine with the use of pancreatin at 125 mg/mL but the activity was not assayed.

In addition to the three groups of macronutrients (protein, lipid and carbohydrate) food provides a wide range of other bioactive molecules and many of these have been studied using static simulations of the upper GI tract. For example the release of polyphenols from orange juice was assed using a static digestion model (Gil-Izquierdo et al. 2001) in which the gastric phase was simulated for 2 h at pH 2 with 315 U of pepsin per mL of juice. The small intestinal phase was incubated for ~2.5 h at pH ~5 with 1 g pancreatin in 250 mL digesta and 6.25 g of bile extract. The conclusion was that although orange juice is a very rich source of flavanones, the concentration of compounds that are in a soluble bioaccessible form under the conditions of the small intestine, is probably much smaller but again the conditions were not those recommended.

Our final example looks at a GI simulation used to assess the iron availability from meals (Miller et al. 1981). The conditions used in this simulation are essentially identical to those used in the above simulation used to follow polyphenol release. It includes an interesting way of raising the pH between the gastric and small intestinal phases. Segments of dialysis tubing containing 25 mL water and an amount of $NaHCO_3$ equivalent to the titratable acidity measured previously were placed in the gastric sample. This was then sealed incubated in a 37 °C shaking water bath until the pH reached about 5 (approximately 30 min). This method provides the gentle rise in pH necessary for working with minerals. However the final pH is rather low compared to what might be expected in vivo. This method also highlights approaches that tend to be used out of context and one could argue that this might not be the most appropriate simulation for following the release of poly-phenols from orange juice. When using a static model of digestion the parameters used should be appropriate to the question and physiologically relevant.

References

Aloulou A, Carriere F (2008) Gastric lipase: an extremophilic interfacial enzyme with medical applications. Cell Mol Life Sci 65(6):851–854

Astwood JD, Leach JN, Fuchs RL (1996) Stability of food allergens to digestion in vitro. Nat Biotechnol 14(10):1269–1273

Beysseriat M, Decker EA, McClements DJ (2006) Preliminary study of the influence of dietary fiber on the properties of oil-in-water emulsions passing through an in vitro human digestion model. Food Hydrocolloids 20(6):800–809

Bonnaire L, Sandra S, Helgason T, Decker EA, Weiss J, McClements DJ (2008) Influence of lipid physical state on the in vitro digestibility of emulsified lipids. J Agric Food Chem 56(10):3791–3797

Borel P, Armand M, Ythier P, Dutot G, Melin C, Senft M, Lafont H, Lairon D (1994) Hydrolysis of emulsions with different triglycerides and droplet sizes by gastric lipase in-vitro – effect on pancreatic lipase activity. J Nutr Biochem 5(3):124–133

Bourlieu C, Ménard O, Bouzerzour K, Mandalari G, Macierzanka A, Mackie AR, Dupont D (2014) Specificity of infant digestive conditions: some clues for developing relevant in vitro models. Crit Rev Food Sci Nutr 54(11):1427–1457

Carriere F, Barrowman J, Verger R, Laugier R (1993) Secretion and contribution to lipolysis of gastric and pancreatic lipases during a test meal in humans. Gastroenterology 105(3):876–888

Day L, Golding M, Xu M, Keogh J, Clifton P, Wooster TJ (2014) Tailoring the digestion of structured emulsions using mixed monoglyceride–caseinate interfaces. Food Hydrocolloids 36:151–161

Dupont D, Mandalari G, Molle D, Jardin J, Leonil J, Faulks RM, Wickham MS, Mills EN, Mackie AR (2010) Comparative resistance of food proteins to adult and infant in vitro digestion models. Mol Nutr Food Res 54(6):767–780

Eisenbrand G, Pool-Zobel B, Baker V, Balls M, Blaauboer BJ, Boobis A, Carere A, Kevekordes S, Lhuguenot JC, Pieters R, Kleiner J (2002) Methods of in vitro toxicology. Food Chem Toxicol 40(2–3):193–236

Fernandez S, Chevrier S, Ritter N, Mahler B, Demarne F, Carriere F, Jannin V (2009) In vitro gastrointestinal lipolysis of four formulations of piroxicam and cinnarizine with the self emulsifying excipients Labrasol (R) and Gelucire (R) 44/14. Pharm Res 26(8):1901–1910

Foster ES, Kimber I, Dearman RJ (2013) Relationship between protein digestibility and allergenicity: comparisons of pepsin and cathepsin. Toxicology 309:30–38

Gil-Izquierdo A, Gil MI, Ferreres F, Tomas-Barberan FA (2001) In vitro availability of flavonoids and other phenolics in orange juice. J Agric Food Chem 49(2):1035–1041

Hamosh M (1996) Digestion in the newborn. Clin Perinatol 23(2):191

Hamosh M, Scow RO (1973) Lingual lipase and its role in digestion of dietary lipid. J Clin Invest 52(1):88–95

Hedren E, Diaz V, Svanberg U (2002) Estimation of carotenoid accessibility from carrots determined by an in vitro digestion method. Eur J Clin Nutr 56(5):425–430

Helbig A, Silletti E, Timmerman E, Hamer RJ, Gruppen H (2012) In vitro study of intestinal lipolysis using pH-stat and gas chromatography. Food Hydrocolloids 28(1):10–19. doi:10.1016/j.foodhyd.2011.11.007

Hiol A, Jonzo MD, Rugani N, Druet D, Sarda L, Comeau LC (2000) Purification and characterization of an extracellular lipase from a thermophilic *Rhizopus oryzae* strain isolated from palm fruit. Enzyme Microb Technol 26(5–6):421–430

Lueamsaisuk C, Lentle RG, MacGibbon AKH, Matia-Merino L, Golding M (2014) Factors influencing the dynamics of emulsion structure during neonatal gastric digestion in an in vitro model. Food Hydrocolloids 36:162–172

Maldonado-Valderrama J, Wilde PJ, Macierzanka A, Mackie AR (2011) The role of bile salts in digestion. Adv Colloid Interface Sci 165(1):36–46

Mandalari G, Adel-Patient K, Barkholt V, Baro C, Bennett L, Bublin M, Gaier S, Graser G, Ladics GS, Mierzejewska D, Vassilopoulou E, Vissers YM, Zuidmeer L, Rigby NM, Salt LJ, Defernez M, Mulholland F, Mackie AR, Wickham MSJ, Mills ENC (2009a) In vitro digestibility of beta-casein and beta-lactoglobulin under simulated human gastric and duodenal conditions: a multi-laboratory evaluation. Regul Toxicol Pharmacol 55(3):372–381

Mandalari G, Mackie AM, Rigby NM, Wickham MS, Mills EN (2009b) Physiological phosphatidylcholine protects bovine beta-lactoglobulin from simulated gastrointestinal proteolysis. Mol Nutr Food Res 53(Suppl 1):S131–S139

McClements DJ, Li Y (2010) Review of in vitro digestion models for rapid screening of emulsion-based systems. Food Funct 1(1):32–59

Miller DD, Schricker BR, Rasmussen RR, Vancampen D (1981) An in vitro method for estimation of iron availability from meals. Am J Clin Nutr 34(10):2248–2256

Mun S, Decker EA, McClements DJ (2007) Influence of emulsifier type on in vitro digestibility of lipid droplets by pancreatic lipase. Food Res Int 40(6):770–781

Okoniewska M, Tanaka T, Yada RY (2000) The pepsin residue glycine-76 contributes to active-site loop flexibility and participates in catalysis. Biochem J 349:169–177

Ring SG, Gee JM, Whittam M, Orford P, Johnson IT (1988) Resistant starch – its chemical form in foodstuffs and effect on digestibility in vitro. Food Chem 28(2):97–109

Rossi SS, Converse JL, Hofmann AF (1987) High-pressure liquid-chromatographic analysis of conjugated bile-acids in human bile – simultaneous resolution of sulfated and unsulfated lithocholyl amidates and the common conjugated bile-acids. J Lipid Res 28(5):589–595

Selgrade MK, Bowman CC, Ladics GS, Privalle L, Laessig SA (2009) Safety assessment of biotechnology products for potential risk of food allergy: implications of new research. Toxicol Sci 110(1):31–39

Shani-Levi C, Levi-Tal S, Lesmes U (2013) Comparative performance of milk proteins and their emulsions under dynamic in vitro adult and infant gastric digestion. Food Hydrocolloids 32(2):349–357

Sharma P, Singh AK, Singh BP, Gaur SN, Arora N (2011) Allergenicity assessment of osmotin, a pathogenesis-related protein, used for transgenic crops. J Agric Food Chem 59(18):9990–9995

van Aken GA, Vingerhoeds MH, de Hoog EHA (2007) Food colloids under oral conditions. Curr Opin Colloid Interface Sci 12(4–5):251–262

van Aken GA, Bomhof E, Zoet FD, Verbeek M, Oosterveld A (2011) Differences in in vitro gastric behaviour between homogenized milk and emulsions stabilised by Tween 80, whey protein, or whey protein and caseinate. Food Hydrocolloids 25(4):781–788

Ville E, Carriere F, Renou C, Laugier R (2002) Physiological study of pH stability and sensitivity to pepsin of human gastric lipase. Digestion 65(2):73–81

Wolf BW, Bauer LL, Fahey GC (1999) Effects of chemical modification on in vitro rate and extent of food starch digestion: an attempt to discover a slowly digested starch. J Agric Food Chem 47(10):4178–4183

Wooster TJ, Day L, Xu M, Golding M, Oiseth S, Keogh J, Clifton P (2014) Impact of different biopolymer networks on the digestion of gastric structured emulsions. Food Hydrocolloids 36:102–114

Chapter 4
Dynamic Digestion Models: General Introduction

Eva C. Thuenemann

Abstract The first section of this chapter has focused on static digestion models and their specific applications. Whilst these static models have many advantages, they mainly function to mimic the biochemical processes in the gastrointestinal (GI) tract and usually use a single set of initial conditions (pH, concentration of enzymes, bile salts, etc.) for each part of the GI tract. However, this simplistic approach is often not a realistic simulation of the more complex in vivo conditions, where the biochemical environment encountered is constantly changing and physical parameters such as shear and grinding forces can have a large impact on the breakdown of larger food particles and the release of nutrients. Several dynamic digestion models have been developed in recent years to address these complex aspects of digestions, and four of these dynamic models will be presented in more detail in the following subchapters. This introduction will provide a brief overview of how the aspects of geometry, biochemistry and physical forces have been addressed in these and other dynamic digestion models.

Keywords Dynamic model • Digestion • GI tract • In vitro

4.1 Geometry

The human gastrointestinal tract consists of distinct compartments of differing shapes, sizes and orientations. These need to be considered when designing a realistic dynamic model. The stomach has a shape of an expanded J, with food entering from the esophageal sphincter at the top and eventually being released through the pylorus at the bottom. During digestion, body position may have an influence on some aspects of gastric digestion, especially gastric sieving of larger particles and pharmaceuticals. Three main approaches have been followed in the design of the models' gastric compartments, each with their own advantages and disadvantages: vertical alignment, horizontal alignment and beaker. Vertical alignment of the

E.C. Thuenemann (✉)
John Innes Centre, Colney Lane, Norwich, Norfolk NR4 7UH, UK
e-mail: eva.thuenemann@jic.ac.uk

© The Author(s) 2015
K. Verhoeckx et al. (eds.), *The Impact of Food Bio-Actives on Gut Health*,
DOI 10.1007/978-3-319-16104-4_4

gastric model allows phase separation to occur during digestion, as in vivo, but has the disadvantage of gravity influencing the sedimentation of larger particles towards the bottom opening (examples include the Dynamic Gastric Model [DGM, Chap. 6] and the Human Gastric Simulator [HGS; Chap. 7]). Horizontal alignment may be more suitable for the simulation of gastric sieving, but it does not provide a realistic representation of the low mixing environment of the gastric fundus (for example: TNO Intestinal Model [TIM-1, Chap. 5]). A stirred beaker is used in some models and is reminiscent of most static models (e.g. the DIDGI system [Chap. 8], and the in vitro Digestion System [IViDiS; Tompkins et al. 2011]). In a newer version of the IViDiS, Campbell et al. (2011) report the use of a molded, elastic stomach construct which more closely resembles the shape and size of the human stomach and is designed to mimic gastric peristalsis through the use of external rollers.

In contrast to the stomach, the peristaltic movements and tube-like structure of the small intestine may reduce the impact of body position on its function. Several dynamic digestion models use a horizontal, tube-like alignment to represent the small intestine. In these models, the peristaltic movements of the small intestine are simulated either by alternating pressure on the flexible wall of the compartment (TIM-1) or constrictions within the wall of the compartment (IViDiS). Other models use one or more thermostated beakers to simulate the small intestine (e.g. DIDGI-system; SHIME model, Chap. 27).

The design of the stomach and intestinal parts of digestion models has a direct impact on the physical forces exerted on the chyme, and how realistically the model simulates in vivo shear forces.

4.2 Physical Forces

Whilst passing through the stomach and small intestine, food particles and drugs are subjected to physical shear and grinding forces as well as pressure exerted by peristaltic movements. This is particularly true in the fed state within the stomach: During a meal, a complex mixture of masticated food bolus enters the fundus of the stomach where it may reside for several hours, depending on meal volume and calorific content. Within the fundus only gentle mixing occurs, whereas closer to the antrum peristaltic waves strengthen and the bolus is subjected to strong mixing and shear. The complex physical forces exerted by the GI tract are not well simulated by a stirred beaker approach; therefore the geometry of some dynamic models (see above) has been designed in such a way as to simulate these physical forces. Three notable examples are the HGS, the DGM and the TIM-agc (TNO's advanced gastric compartment), each of which has undergone validation of the physical forces exerted on the food bolus.

The Human Gastric Simulator's (Chap. 7) vertically aligned, cylindrical gastric compartment is periodically squeezed by the action of Teflon rollers on its flexible wall. These rollers impinge the compartment successively more towards the bottom, thereby simulating stronger forces nearer the antrum of the stomach. Kong and

Singh (2010) validated the model by measuring the pressure exerted on a rubber bulb within the compartment, and these pressures were found to fall within the rage of mechanical stresses reported in literature for the human stomach.

The Dynamic Gastric Model (Chap. 6) consists of two connected compartments, simulating the fundus/main body and the antrum. Within the antral part, the food bolus is repeatedly passed through a flexible disc (annulus) to simulate the mixing and shear stresses encountered during antral contraction waves (ACWs) in vivo. Vardakou et al. (2011) validated these forces by comparing the mean breaking times of agar beads in high- and low-viscosity meals within the DGM, to the mean breaking times of beads in an equivalent in vivo study (Marciani et al. 2001a).

The TIM advanced gastric model (Chap. 5) simulates ACWs by modulating the pressure within a water jacket surrounding the antral compartment. Pressure profiles within this compartment were measured using a pressure-measuring capsule and compared to in vivo gastric pressure profiles (see Sect. 5.5).

4.3 Biochemistry

Dynamic digestion models, like static models, are built to mimic the biochemical environment of the compartments of the GI tract. Many of the considerations high-lighted in the earlier sections of this chapter therefore apply to dynamic models as well: What concentrations of which enzymes, bile salts and phospholipids should be used? Can porcine, bovine or fungal versions of enzymes be used? Is it better to use complex mixtures (e.g. pancreatin) or individual, purified enzymes? At what pH should digestion take place? How long should a given meal reside in the stomach?

In contrast to static models, however, the exact conditions within the different com-partments of a dynamic model will change over time to simulate the in vivo digestion processes. Dynamic digestion models generally have a number of different digestive secretions which are added to the compartments of the model over time. This addition can either follow a steady secretion rate (e.g. the simulated gastric juice of HGS is added at a rate of 2.5 mL/min), or it can follow a pre-programmed pattern allowing the rate to change over time (e.g. in the TIM-1 model), or it can be programmed to change in response to other parameters, such as the fill volume of the model (e.g. gastric secretion in the DGM). The pH is often monitored in real-time within dynamic mod-els and is used to control the rate of addition of hydrochloric acid, allowing the acidi-fication of the meal within the gastric compartment to follow a pre-determined curve. In dynamic models which incorporate a duodenal step, at this stage the pH of the chyme is neutralized by controlled addition of sodium bicarbonate solution, and secretions of bile and pancreatic enzymes (or pancreatin) are added.

Whilst the concentrations of enzymes, bile, electrolytes and phospholipids are set for the various secretions used in dynamic models, the concentrations of these components within the digesta cannot be readily determined. This is in part due to the dynamic nature of the models, allowing secretion rates to change throughout the digestion process. However, it is also due to the inhomogeneity of the bolus within the models. In vivo, a solid meal will be ingested over a period of time as small balls of chewed food. Within the fundus of the stomach, the bolus is only subjected to gentle contractions and therefore is not well mixed. Whilst the outside of the bolus is acted upon by gastric secretions, it can take over 1 h for these secretions to penetrate to the center of the bolus (Marciani et al. 2001b). In the antral part of the stomach, strong peristaltic waves mix the bolus more readily, producing a more homogeneous chyme.

Some of the more advanced dynamic digestion models have a geometry designed to represent the fundus and antrum of the stomach, and/or the duodenum. These designs allow for the simulation of the physical forces exerted on the digesta during transit through the GI tract, which in turn allows for simulation of the inhomogeneous nature of digesta and localized biochemical environments, as in vivo.

References

Campbell G, Arcand Y, Mainville I (2011) Development of a tissue mimicking stomach construct for *in-vitro* digestive system. Paper presented at the plant bioactives research in Canada. Canada-United Kingdom Gut and Health Workshop, Saint-Hyacinthe, QC, Canada, 24–25 February 2011

Kong FB, Singh RP (2010) A human gastric simulator (HGS) to study food digestion in human stomach. J Food Sci 75(9):E627–E635. doi:10.1111/j.1750-3841.2010.01856.x

Marciani L, Gowland PA, Fillery-Travis A, Manoj P, Wright J, Smith A, Young P, Moore R, Spiller RC (2001a) Assessment of antral grinding of a model solid meal with echo-planar imaging. Am J Physiol Gastrointest Liver Physiol 280(5):G844–G849

Marciani L, Gowland PA, Spiller RC, Manoj P, Moore RJ, Young P, Fillery-Travis AJ (2001b) Effect of meal viscosity and nutrients on satiety, intragastric dilution, and emptying assessed by MRI. Am J Physiol Gastrointest Liver Physiol 280(6):G1227–G1233

Tompkins TA, Mainville I, Arcand Y (2011) The impact of meals on a probiotic during transit through a model of the human upper gastrointestinal tract. Benef Microbes 2(4):295–303. doi:10.3920/Bm2011.0022

Vardakou M, Mercuri A, Barker SA, Craig DQM, Faulks RM, Wickham MSJ (2011) Achieving antral grinding forces in biorelevant in vitro models: comparing the USP dissolution apparatus II and the dynamic gastric model with human in vivo data. AAPS PharmSciTech 12(2):620–626. doi:10.1208/s12249-011-9616-z

Chapter 5
The TNO Gastro-Intestinal Model (TIM)

Mans Minekus

Abstract The TNO Gastro-Intestinal Model (TIM) is a multi-compartmental model, designed to realistically simulate conditions in the lumen of the gastro-intestinal tract. TIM is successfully used to study the gastro-intestinal behavior of a wide variety of feed, food and pharmaceutical products. Experiments in TIM are based on a computer simulation of the digestive conditions in the lumen of the gut during transit and digestion of a meal in vivo. These conditions include controlled parameters such as gastric and small intestinal transit, flow rates and composition of digestive fluids, pH values, and removal of water and metabolites. Simulation protocols have been developed for young, adult and elderly humans, dogs, pigs and calves after ingestion of various meals. The typical end point from results obtained with TIM is the availability of a compound for absorption through the gut wall (bio-accessibility). Results from TIM—with or without additional intestinal cell assays and in silico modeling—show a high predictability as compared to in vivo data (Marteau et al., J Dairy Sci 80:1031–1037, 1997; Verwei et al., J Nutr 136:3074–2078, 2006; Bellmann et al., TIM-carbo: a rapid, cost-efficient and reliable in vitro method for glycaemic response after carbohydrate ingestion. In: van der Kamp J-W, Jones JM, McCleary BV (eds) Dietary fibre: new frontiers for food and health. Wageningen Academic Publishers, Wageningen, p 467–473, 2010; Van Loo-Bouwman et al., J Agric Food Chem 62(4):950–955, 2014).

Keywords Multi-compartmental dynamic gastric intestinal model • Physiological • Gastric • In vitro • Digestion • Bio-accessibility • Nutrient • Digestion

5.1 Introduction

The TNO Gastro-Intestinal Model (TIM) is a multi-compartmental dynamic model that was developed in the early 1990s in response to industrial demand to study food products under more physiologically relevant conditions as compared to contemporary digestion models (Minekus et al. 1995). During the past years TIM has

M. Minekus (✉)
TNO Triskelion, Utrechtseweg 48, 3704 HE Zeist, The Netherlands
e-mail: mans.minekus@tno.triskelion.nl

© The Author(s) 2015
K. Verhoeckx et al. (eds.), *The Impact of Food Bio-Actives on Gut Health*,
DOI 10.1007/978-3-319-16104-4_5

developed from an experimental lab setup—controlled by an 8 MHz PC—into a platform of cabinet systems that are successfully used for a broad range of studies, serving the feed, food and pharmaceutical industries. This chapter describes the concept of TIM, the TIM gastro-intestinal systems and some examples of methods to study the digestion of nutrients.

5.2 Concept of TIM

The gastro-intestinal tract is a tube like organ with different compartments (stomach, small intestine, large intestine) for each step of digestion. During the gradual transit of the meal through the compartments different fractions of the meal are exposed to changing conditions due to gradual secretion of digestive fluids and absorption of water and nutrients. TIM intends to simulate the dynamic conditions in the lumen of the gastrointestinal tract. It is designed to combine the controllability and reproducibility of a model system with physiological parameters such as mixing, meal transit, variable pH values in place and time, realistic secretion and composition of digestive fluids, and removal of digested compounds and water. These parameters are combined in a protocol as an input for a computer simulation of a specific digestive setting. Such settings includes species (human, dog, pig, calve), age (infant, adult, elderly), pathology and meal-related parameters, obtained from in vivo data (Marteau et al. 1997; Minekus 1998; Smeets-Peeters 2000; Havenaar et al. 2013). Based on the computer simulation, the physical model is controlled to reproduce the underlying in vivo settings.

5.3 TIM-1

TIM-1 (Fig. 5.1) is the most frequently used configuration of the TIM platform. It comprises four compartments, representing the stomach, duodenum, jejunum and ileum. Compartments are connected by peristaltic valve pumps (PVP) that allow the transfer of controlled amounts of chyme. The PVPs are designed to have low dead volume in the closed position. They are not blocked by particles and able to handle complete meals. Mixing for each compartment is achieved by alternating the pressure on flexible walls. Temperature is maintained by controlling the temperature of the water circulating outside the flexible walls. Prior to introduction into the gastric compartment, the meal is masticated with a food processor (Solostar II, Tribest) and mixed with artificial saliva containing electrolytes and α-amylase. Gastric secretion contains electrolytes, pepsin and a fungal lipase (F-AP 15, Amano) as an alternative to gastric lipase. The pH is measured and controlled with hydrochloric acid, to follow a predetermined curve or at a variable rate in time. Duodenal secretion consists of, electrolytes, bile and pancreatin. The pH is controlled at pre-set values for each compartment with sodium bicarbonate. All flows of secretion are programmable in time. Digestion products are removed by two different systems. Water soluble products are removed by dialysis through membranes with a molecular weight cutoff of app.

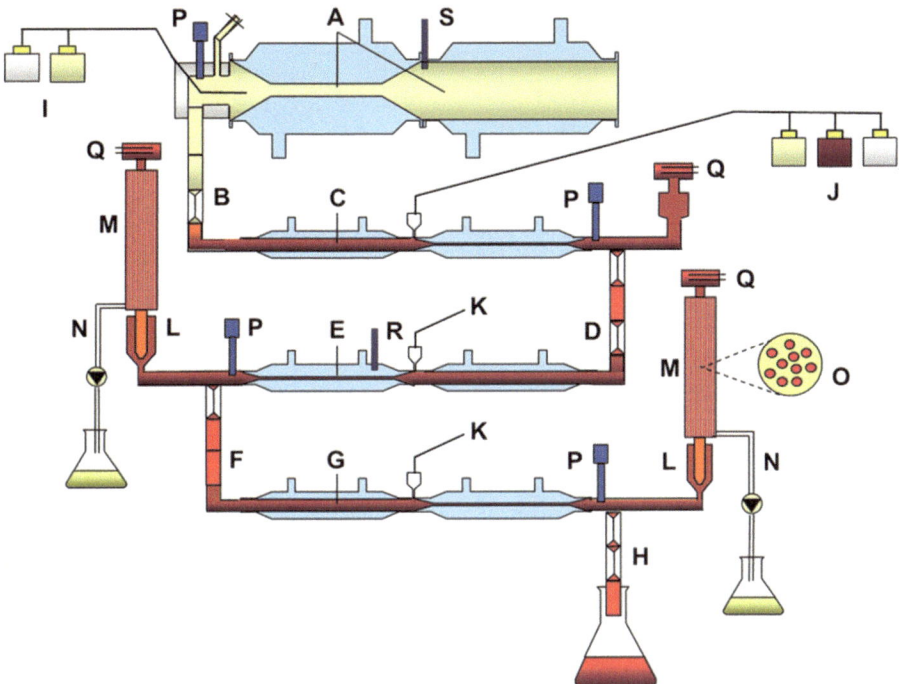

Fig. 5.1 Schematic presentation of TIM-1, equipped with filters to study the bio-accessibility of lipids. A. gastric compartment; B. pyloric sphincter; C. duodenal compartment; D. peristaltic valve; E. jejunal compartment; F. peristaltic valve; G. ileal compartment; H. ileal-cecal valve; I. gastric secretion; J. duodenal secretion; K. bicarbonate secretion; L. pre-filter; M. filtration system; N. filtrate with bio-accessible fraction; O. hollow fiber system (cross section); P. pH electrodes; Q. level sensors; R. temperature sensors; S. pressure sensor

10 kDa, connected to the jejunal and ileal compartments. Lipophilic products cannot be removed efficiently by these membranes since they are incorporated in micelles that are too big to pass the membrane. Lipophilic products are removed through a 50 nm filter that passes micelles but retains fat droplets. Meal transit is controlled by dictating the gastric- and ileal-emptying according to the formula (Fig. 5.2) described by Elashoff et al. (1982). A typical protocol for the simulation of the digestion of a high fat meal in a human adult is presented in Table 5.1. An overview of the composition of physiological relevant secreted fluids for a human adult is given in Chap. 2 on the Infogest consensus method for static digestion.

5.4 TinyTIM

TinyTIM (Fig. 5.3) is a simplified version of the TIM-1, designed to increase the throughput as compared to TIM-1, with focus on studies that do not need separate intestinal steps. The TinyTIM is used with the same gastric compartment as TIM-1

Fig. 5.2 Gastric emptying (*circles*, $t_{1/2}$=80 min, β=2) and ileal emptying curve (*squares*, $t_{1/2}$=220 min, β=2.2) to control transit of a solid meal in TIM-1

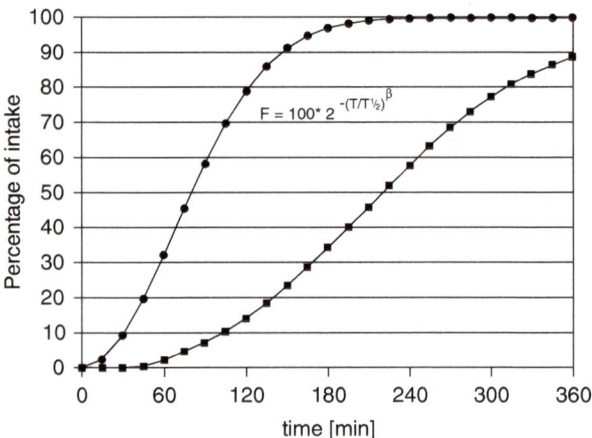

$$F = 100 \cdot 2^{-(T/T_{1/2})^{\beta}}$$

Table 5.1 Typical parameter settings for digestion of a high fat meal in TIM-1 with filters

Volume (ml)	Stomach: 300, duodenum: 55, jejunum: 130, ileum: 130
Meal size (g)	300
Gastric secretion (ml/min)	1
Gastric emptying curve	$t_{1/2}$=80 min, β=2
Gastric pH curve (time, pH)	(0, 5.2) (30, 3.2) (60, 2.2) (120, 1.7)
Bile secretion (ml/min)	0.5
Pancreatin/electrolytes (ml/min)	0.5
Ileal emptying curve	$t_{1/2}$=220 min, β=2.2
Small intestinal pH	Duodenum: 6.2, jejunum: 6.5, ileum: 7.4
Filtration rate (ml/min)	Jejunum: 4.5, ileum: 4.5

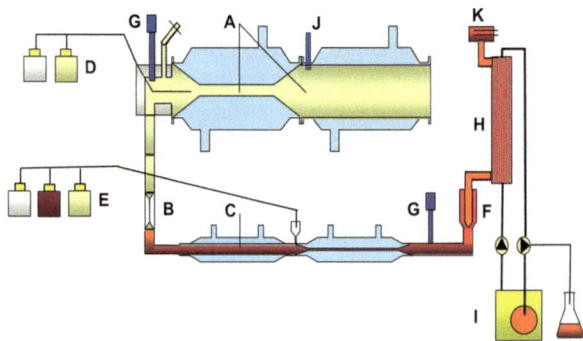

Fig. 5.3 Schematic presentation of TinyTIM, equipped with a dialysis membrane to study the bio-accessibility of water soluble compounds. A. gastric compartment; B. pyloric sphincter; C. duodenal compartment; D. gastric secretion; E. duodenal secretion; F. pre-filter; G. pH electrodes; H. dialysis membrane; I. dialysis system; J. pressure sensor; K. level sensor

when the ratio between amount of food and ingested material, such as pharmaceutical formulations, is important. For other experiments, a half size gastric compartment is used. All functions of the gastric compartment are similar to the gastric compartment of TIM-1. TinyTIM has a single small intestinal compartment instead of three and no ileal efflux. All fluids entering the small intestinal compartment are removed through the filtration- or dialysis-membrane. This implies that small intestinal transit is simulated by assuming a plug of chyme traveling through the small intestine, instead of a "flow through" compartment such as in TIM-1.

5.5 Advanced Gastric Compartment (TIM-agc)

In the standard gastric compartment, the meal is mixed to obtain a homogenized gastric content and a consequent predictable gastric emptying of compounds. This is particularly important to compare the digestion of compounds under exactly controlled conditions. In order to include the effect of gastric motility on the gastric behavior of food components and pharmaceuticals, a gastric compartment is designed that mimics the shape and motility of the stomach in a more realistic manner (Fig. 5.4). The system consists of a body part with a flexible wall that gradually contracts to simulate gastric tone and consequent reduction of gastric volume during emptying. Two antral units can be moved to simulate mixing by an antral wave. A valve is synchronized with an antral wave to simulate the opening of the pyloric sphincter during gastric emptying. Similar to other TIM models, the contractions are achieved by modulating the pressure on water that is circulated in the space between a glass jacket and a flexible membrane. All contractile movements and the resulting mixing and pressure profiles are accurately controlled and synchronized. Motility patterns as well as gastric emptying and secretion of digestive fluids are dictated by a predetermined protocol that describes a specific condition (e.g. fed or fasting) in time. A study has been performed in both the TIM-agc and human volunteers to compare gastric pressure profiles, using a smartpill® (Given Imaging GmbH, Hamburg, Germany) (Fig. 5.5, Minekus et al. 2013).

Fig. 5.4 Schematic presentation of the TIM advanced gastric compartment (TIM-agc). The *left* and *right* pictures show a filled and completely empty gastric compartment, respectively. A. body; B. proximal antrum; C. distal antrum; D. pyloric sphincter

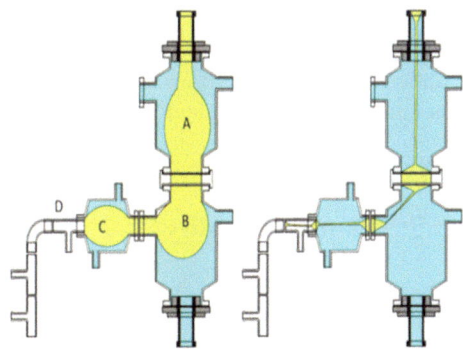

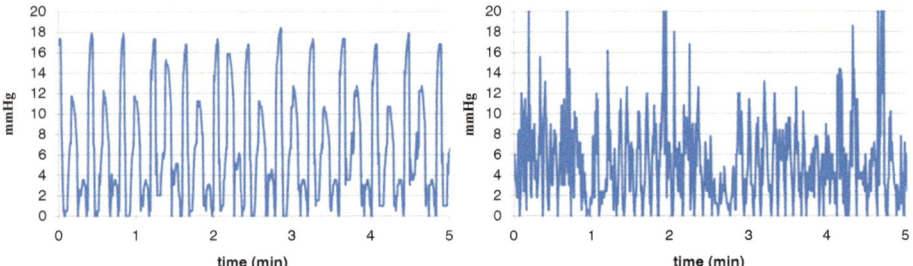

Fig. 5.5 Gastric pressure profile measured with a smartpill® in the TIM-agc (*left*) as compared to an in vivo gastric pressure profile measured with a smartpill® during the digestive phase

5.6 The Use of TIM to Study the Bio-accessibility of Nutrients

The TIM has been successfully used to study the bio-accessibility of macro nutrients, minerals, fat- and water soluble vitamins, and bioactive compounds (Larsson et al. 1999; Verwei et al. 2006; Mateo Anson et al. 2009). Bio-accessibility is defined as the fraction of a compound that is available for absorption through the gut wall. In TIM, this is determined by measuring the fraction of a compound that has passed the dialysis or filtration membrane. When validating TIM with in vivo data, it is important to realize that a valid correlation between bio-accessibility and bio-availability can only be achieved when transport through the gut wall (mucus layer and enterocytes) is not a limiting step. This gap can be bridged by using TIM data in combination with transport data obtained with intestinal cells and/or in silico modeling. The digestion of a nutrient is determined by the characteristics of the nutrient, the composition and structure of the meal matrix and the individual's physical response to the digesting nutrient and meal while travelling through the gastro-intestinal tract. As is the case in all digestive models, TIM does not include feedback on the characteristics of the meal. Therefore, the approach is taken to assume a set of conditions that is based on in vivo data and normal for the type of meal and target group. The effect of variability of a specific condition on digestion within a population can be tested by changing only this condition in the digestive protocol. The reproducible conditions allow comparison of different compounds under the same conditions and do not need as many replicates as are necessary to obtain sufficient statistical power for in vivo studies.

5.7 Protein Quality

The basic method to study protein digestion in TIM is to determine the bio-accessibility by expressing the amount of protein nitrogen dialyzed as a percentage of the amount of protein nitrogen in the meal. The bio-accessibility data are corrected with the bio-accessibility of protein nitrogen from the secreted protein, thus presenting the true digestibility of the protein.

To optimize the nutritional quality of food and feed, it is important for the food and feed industries to have insight in the nutritional quality of the protein in their ingredients. The nutritional quality of proteins depends on the amino acid composition profile and the bio-accessibility of essential amino acids, while digestion of proteins may be affected by processing steps during manufacturing. Essential amino acids cannot be produced by the body and need to be supplied in sufficient quantities in the diet. A protein of high biological value contains all essential amino acid in proportion to the need. The amino acid that is in shortest supply in relation to the need is referred to as the limiting amino acid. Traditionally protein quality is evaluated by determination of the Protein Efficiency Ratio (PER). The PER method reflects the amino acid requirements of young animals such as broiler chickens and rats, as determined with growth experiments on protein sources (Skinner et al. 1991). However, these experiments are relatively slow and do not give insight into the availability of the relevant amino acids. Also, this method determines the requirements of rats and broilers, not humans. Moreover, such experiments can result in strong growth retardation due to amino acid deficiencies and have therefore ethical drawbacks. A method was developed with TinyTIM as an alternative to the time consuming PER test that uses young animals. In this method the bio-accessible amount of the limiting amount of amino acid is determined after digesting the feed in TinyTIM (Minekus and Van der Klis 2001; Minekus et al. 2006). The FAO/WHO adopted the Protein Digestibility Corrected Amino Acid Score (PDCAAS) and later the Digestible Indispensable Amino Acid Score (DIAAS) as best method to determine the protein quality (Schaafsma 2005; FAO 2013). The method with TinyTIM to determine the true digestibility of protein and (limiting) amino acids offers an alternative to the use of rats for the determination of the PDCAAS and DIAAS, respectively.

5.8 Prediction of Glycemic Response

Studies on the digestion of carbohydrates and consequent glucose plasma levels are important for diabetic patients, obesity control and designing sport foods. As an alternative to expensive and time consuming human trials, a rapid in vitro method has been developed to predict the glycemic response after intake of carbohydrates (Bellmann et al. 2010). In this method the carbohydrates are digested in TinyTIM and a successive step with brush border enzymes. The released glucose and fructose are analyzed as a function of time and processed with in silico modeling based on the homeostatic model assessment (HOMA; Matthews et al. 1985) to predict the glycemic response. The method was validated against in vivo plasma glucose curves of 21 different food products (R = 0.91).

5.9 Lipids

The uptake of fat soluble compounds needs a realistic simulation of the digestion of the compound and the food matrix, and the formation of mixed micelles. In TIM, fresh porcine bile is used to supply adequate quantities of bile salts, phospholipids and cholesterol for mixed micelle formation. Filters with a pore size of 50 nm are used to differentiate between undigested fat and micelles, and to remove lipolytic products to avoid product inhibition. It is assumed that the products in micelles are available for absorption through the gut wall. Figure 5.6 shows the cumulative appearance of total fatty acids in the jejunal and ileal filtrates during digestion of intra lipid in TIM-1. This method has been used to study the bio-accessibility of carotenoids (Southon 2001; Van Loo-Bouwman et al. 2014), the study of Partially Hydrolysed Guar Gum (PHGG) on lipid digestion (Minekus et al. 2005) and the bio-accessibility of blueberry anthocyanins (Ribnicky et al. 2014).

5.10 Conclusions

TIM is designed to reproduce the conditions in the lumen of the gastro-intestinal tract by realistic mixing, transit of the meal, rate and composition of secretions and removal of digested products and water. It is designed to predict the bio-accessibility of a wide variety of ingested compounds present in a broad range of foods and pharmaceutical matrices. Accurate simulation and control of the multi-compartmental

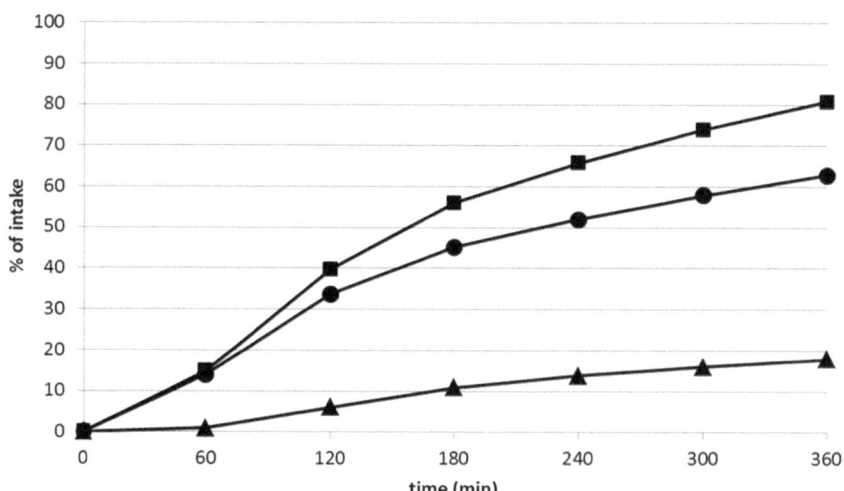

Fig. 5.6 Bio-accessibility profiles of fatty acids from the jejunal compartment (*circles*), ileal compartment (*triangles*) and total from both compartments (*squares*), during digestion of intra-lipid in TIM-1

processes in the lumen of the GIT allows the testing of compounds under exactly the same conditions or specifically modified conditions for "what if" studies. Bio-accessibility profiles directly from TIM or after a separate transport assay with intestinal cells can be processed further with in silico modeling to predict the bio-availability of compounds. In contrast with static methods, that are only useful for specific studies within a narrow range of products (Minekus et al. 2014), TIM is intended to simulate the dynamic conditions in the lumen of the GIT to predict the bio-accessibility of a variety of nutrients in a wide range of meals. The more complex system with a lower throughput and higher costs is well compensated by the high predictability and broad applicability. As a high-end digestion model system, it may offer a faster and a more ethical alternative to studies in animals and humans.

Contract research on gastro-intestinal behavior of nutrients and pharmaceuticals, as well as sales and lease of TIM equipment is located at TNO-Triskelion, Zeist, The Netherlands (http://www.tnotriskelion.com/).

References

Bellmann S, Minekus M, Zeijdner E, Verwei M, Sanders P, Basten W, Havenaar R (2010) TIM-carbo: a rapid, cost-efficient and reliable in vitro method for glycaemic response after carbohydrate ingestion. In: van der Kamp J-W, Jones JM, McCleary BV (eds) Dietary fibre: new frontiers for food and health. Wageningen Academic Publishers, Wageningen, pp 467–473

Elashoff JD, Reedy TJ, Meyer JH (1982) Analysis of gastric emptying data. Gastroenterology 83(6):1306–1312

FAO Food and Nutrition Paper 92. Dietary protein quality evaluation in human nutrition. Report of an FAO Expert Consultation', Rome 2013

Havenaar R, Anneveld B, Hanff LM, de Wildt SN, de Koning BAE, Mooij MG, Lelieveld JPA, Minekus M (2013) In vitro gastrointestinal model (TIM) with predictive power, even for infants and children? Int J Pharm http://dx.doi.org/10.1016/j.ijpharm.2013.07.053

Larsson M, Minekus M, Havenaar R (1999) Estimation of the bioavailability of iron and phosphorus in cereals using a dynamic in vitro gastrointestinal model. J Sci Food Agric 74(1):99–106

Marteau P, Minekus M, Havenaar R, Huis in't Veld JHJ (1997) Survival of lactic acid bacteria in a dynamic model of the stomach and small intestine: validation and the effects of bile. J Dairy Sci 80:1031–1037

Mateo Anson N, Havenaar R, Bast A, Haenen GRMM (2009) Antioxidant and anti-inflammatory capacity of bioaccessible compounds from wheat fractions after gastrointestinal digestion. J Cereal Sci 51(1):110–114

Matthews DR, Hosker JP, Rudenski AS, Naylor BA, Treacher DF, Turner RC (1985) Homeostasis model assessment: insulin resistance and β-cell function from fasting plasma glucose and insulin concentrations in man. Diabetologia 28(7):412–419

Minekus M (1998) Development and validation of a dynamic model of the gastrointestinal tract. PhD thesis, University of Utrecht, Elinkwijk b.v., Utrecht, Netherlands

Minekus M, van der Klis JD (2001) Development and validation of a dynamic in vitro method as alternative to PER studies in chickens. TNO report V4326

Minekus M, Marteau P, Havenaar R, Huis in't Veld J (1995) A multi-compartmental dynamic computer-controlled model simulating the stomach and small intestine. ATLA 23:197–209

Minekus M, Jelier M, Xiao JZ, Kondo S, Iwatsuki K, Kokubo S, Bos M, Dunnewind B, Havenaar R (2005) Effect of partially hydrolyzed guar gum (PHGG) on the bioaccessibility of fat and cholesterol. Biosci Biotechnol Biochem 69(5):932–938

Minekus M, Lelieveld J, Havenaar R (2006) Rapid, reliable and cost-effective methodology for measuring feed quality. In: Proceeding VIV Europe, Utrecht, Netherlands

Minekus M, Lelieveld J, Anneveld B, Zeijdner E, Barker R, Banks S, Miriam Verwei M (2013) Pressure forces in the TIM-agc, an advanced gastric compartment that simulates shape and motility of the stomach. Poster 5th WCDATD/OrBiTo Meeting Uppsala, Sweden, 24–27 June

Minekus M, Alminger M, Alvito P, Ballance S, Bohn T, Bourlieu C, Carriere F, Boutrou R, Corredig M, Dupont D, Dufour C, Egger L, Golding M, Karakaya S, Kirkhus B, Le Feunteun S, Lesmes U, Macierzanka A, Mackie A, Marze S, McClements DJ, Menard O, Recio I, Santos CN, Singh RP, Vegarud GE, Wickham MSJ, Weitschies W, Brodkorb A (2014) A standardised static in vitro digestion method suitable for food – an international consensus. Food Funct. http://dx.doi.org/10.1039/C3FO60702J

Ribnicky DM, Roopchand DE, Oren A, Grace M, Poulev A, Lila MA, Havenaar R, Raskin I (2014) Effects of a high fat meal matrix and protein complexation on the bioaccessibility of blueberry anthocyanins using the TNO gastrointestinal model (TIM-1). Food Chem 142:349–357

Schaafsma G (2005) The protein digestibility-corrected amino acid score (PDCAAS) – a concept for describing protein quality in foods and food ingredients: a critical review. J AOAC Int 88(3):988–994

Skinner JT, Izat AL, Waldroup PW (1991) Effects of dietary amino acid levels on performance and carcass composition of broilers 42 to 49 days of age. Poult Sci 70(5):1223–1230

Smeets-Peeters MJE (2000) Feeding FIDO: development, validation and application of a dynamic in vitro model of the gastrointestinal tract of the dog. PhD thesis, Wageningen University, Universal Press, Veenendaal, The Netherlands

Southon S (2001) Model systems in vitro and in vivo, for predicting the bioavailability of lipid soluble components in food. FAIR CT97-3100 (Final Report)

Van Loo-Bouwman CA, Naber THJ, Minekus M, van Breemen RB, Hulshof PJM, Schaafsma G (2014) Matrix effects on bioaccessibility of β-carotene can be measured in an in vitro gastrointestinal model. J Agric Food Chem 62(4):950–955

Verwei M, Freidig AP, Havenaar R, Groten JP (2006) Predicted plasma folate levels based on in vitro studies and kinetic modeling consistent with measured plasma profiles in humans. J Nutr 136:3074–3078

Chapter 6
Dynamic Gastric Model (DGM)

Eva C. Thuenemann, Giuseppina Mandalari,
Gillian T. Rich, and Richard M. Faulks

Abstract The Dynamic Gastric Model (DGM) was developed at the Institute of Food Research (Norwich, UK) to address the need for an in vitro model which could simulate both the biochemical and mechanical aspects of gastric digestion in a realistic time-dependent manner. As in the human stomach, masticated material is processed in functionally distinct zones: Within the fundus/main body of the DGM, gastric acid and enzyme secretions are introduced around the outside of the food bolus which is subjected to gentle, rhythmic massaging. Secretion rates adapt dynamically to the changing conditions within this compartment (acidification, fill state). Portions of gastric contents are then moved into the DGM antrum where they are subjected to physiological shear and grinding forces before ejection from the machine (and subsequent separate duodenal processing).

The DGM has been used extensively for both food and pharmaceutical applications, to study, for example, release and bioaccessiblity of nutrients and drugs. The system allows the use of complex food matrices (as used in in vivo studies) and processes these under physiological conditions in real-time, thereby providing a realistic tool for the simulation of human gastric digestion.

Keywords Dynamic gastric model • Physiological • Gastric • In vitro • Digestion • Bioaccessiblity • Nutrient • Dissolution • Pharmaceutical • Real-time

6.1   Origins and Design of the DGM

The Dynamic Gastric Model was developed at the Institute of Food Research (Norwich, UK) to address the need for an in vitro model which could simulate both the biochemical and mechanical aspects of human gastric digestion in a realistic

E.C. Thuenemann (✉)
John Innes Centre, Colney Lane, Norwich, Norfolk NR4 7UH, UK
e-mail: eva.thuenemann@jic.ac.uk

G. Mandalari • G.T. Rich • R.M. Faulks
The Model Gut, Institute of Food Research, Norwich Research Park,
Colney Lane, NR4 7UA Norwich, UK

© The Author(s) 2015   47
K. Verhoeckx et al. (eds.), *The Impact of Food Bio-Actives on Gut Health*,
DOI 10.1007/978-3-319-16104-4_6

time-dependent manner. This was initially done in the interest of food-based research to enable the study of parameters such as nutrient bioaccessibility, effect of food structure on nutrient delivery, nutrient interactions, survival and delivery of functional foods. Over the years, the DGM has also increasingly been used by pharmaceutical industry as an in vitro tool to study the effect of food matrices on the disintegration and dissolution of drug formulations and delivery profile to the duodenum. This is in part due to its ability to realistically process any complex food matrix for direct comparison with the results of in vivo/clinical studies.

The design of the DGM is based on extensive research into the process of digestion and the physiology of the human stomach, both biochemical and mechanical (reviewed by Wickham et al. 2012). It builds on these findings as well as literature data to closely mimic the conditions encountered by food particles and drug formulations as they move through the upper gastrointestinal tract. To this end, digestions using the DGM are performed in real-time, and the length of each experiment is designed around the estimated gastric residence time of the particular meal used. Experiments typically last between 25 min (glass of water) and 4.5 h (high-fat FDA breakfast) depending on meal size, composition and calorific content.

The following paragraphs provide an introduction to the parts of the DGM and how these are used to simulate the natural physiology of the healthy adult human stomach (from ingestion to release into the duodenum). Schematic drawings and further descriptions are provided in Fig. 6.1 and Table 6.1 respectively.

Masticated food, a complex inhomogeneous mixture of accessible and inaccessible protein, carbohydrate and fat (in particles of varying sizes) as well as water and saliva, enters the stomach in portions from the esophagus. It initially encounters an acidic environment of resting gastric fluid (24±5 ml, Dubois et al. 1977), whose pH is subsequently altered by the buffering capacity of the meal. In the DGM, masticated food can be introduced in real-time or as a bulk from the top into the fundus and main body (Fig. 6.1), where it encounters a previously added 20 ml volume of gastric priming acid (Table 6.2).

Initiation of gastric digestion of a food bolus in the stomach is via secretions from the mucosal gastric surface and a change from resting to rhythmic phase 2 contractions. The secretion rates of acid and enzyme are dependent on, amongst others, the composition of the food bolus and fill volume of the stomach, and are therefore not constant throughout the digestion process but change in response to factors such as the acidification of the bolus and emptying of the stomach contents into the duodenum (Konturek et al. 1974; Schubert and Peura 2008). Within the DGM, gastric acid and enzyme solutions (Table 6.2) are added through a perforated hoop situated around the wall of the fundus (Fig. 6.1 and Table 6.1). The flow rates of these secretions is controlled dynamically: The rate of gastric acid addition slows gradually in response to the acidification of the meal as detected by the pH electrode inserted within the fundus; The rate of gastric enzyme addition slows in response to the gradual decrease in food bolus volume as recorded in response to ejection of samples from the antrum.

The human stomach has distinct zones which differ markedly in the physical forces applied to the meal bolus: the fundus/main body (proximal) and the antrum

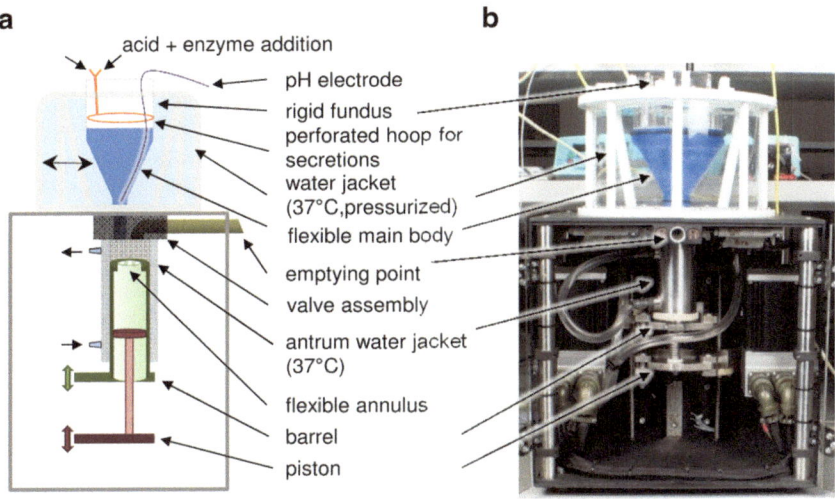

Fig. 6.1 The dynamic gastric model (DGM). (**a**) Schematic representation of the main components of the DGM (*side view*) (**b**) Photographic image of the DGM (*front view*)

(distal) (Bornhorst and Singh 2014). Within the fundus, the meal is subjected to low physical forces exerted by gentle rhythmic peristaltic contractions. Upon ingestion, the bulk of the meal resides within this part of the stomach and it was found that the penetration of this bolus by acidic gastric secretions occurs very slowly (Marciani et al. 2001b). Additionally, it has been shown that gastric secretions can form an acidic pool on top of a dense meal (Holloway and Sifrim 2008). This not only influences conditions such as gastric reflux disease, but also has a bearing on the microenvironment encountered by drug formulations when they are taken after a meal. The DGM also models the two distinct regions of the stomach. Within the fundus/main body, the food bolus is subjected to rhythmic squeezing brought about by cyclical pressurization of the 37 °C water jacket surrounding the main body. Depending on the meal viscosity, the gastric secretions applied to the outside surface of the fundus can take a considerable amount of time to fully acidify the meal bolus. Also, a pool of gastric secretions forms on the surface, mimicking the phase separation of the meal within the stomach.

Within the antral part of the human stomach, the food bolus is subjected to higher shear forces and turbulence, resulting in greater mixing as well as gradual size reduction of particles. Peristaltic contractions push the food towards the pyloric sphincter which provides resistance resulting in retrograde movement of the chyme which is pushed forwards again by the next contraction (Ferrua and Singh 2010). Within a meal of high viscosity, the shear and grinding forces are higher than in a meal of low viscosity, as shown in an in vivo study of the gastric residence time of agar beads of different strengths within the two meal types (Marciani et al. 2001a). A similar study using the same agar beads and meal types was conducted in the DGM to calibrate the physical forces within the DGM antrum

Table 6.1 Functional parts of the DGM

Schematic[a]	Name[a]	Functions	Relevance in vivo
	pH electrode	Records pH within DGM main body, enabling automatic adjustments to gastric acid flow rate (within physiological production rates) in response to acidification of meal. Can be forced outside of "normal" range if necessary	Gastric acid production is controlled through a pH-regulated feedback loop
	Perforated ring for gastric secretions	Distribution of gastric secretions around outside of DGM main body, fed by computer-controlled pumps. Secretion rates respond dynamically to pH and volume changes of food bolus	Acid and enzymes are secreted from the walls of the stomach and can form an acid pool on top of the meal
	Water jacket	Maintains DGM main body at 37 °C allowing heat transfer to food bolus. Enables gentle mixing within DGM main body by cyclical pressurization	Normal body temperature is ca. 37 °C
	Rigid fundus and flexible main body	Holds up to 800 ml of masticated real food and drink. Gentle mixing due to cyclical pressurization of water jacket. Heat transfer rates from 37 °C water jacket similar to those in vivo	Typical meal sizes are less than 1 l Main body of stomach is characterized by gentle movements and slow mixing
	Valve assembly	Inlet valve allows movement of portions of food bolus from main body into antral part of DGM. Inlet valve allows retrograde movement from antral part to main body during processing outlet valve allows ejection of samples from DGM	Phase II contractions periodically empty digesta from antrum through pyloric sphincter into the duodenum
	Antrum water jacket	Maintains antral temperature at 37 °C	Normal body temperature is ca. 37 °C
	Barrel and flexible membrane	Barrel moves flexible membrane rhythmically through food bolus contained within antral part of DGM, creating an environment of high shear and mixing	Contractions of the proximal stomach strengthen towards the antrum, creating potentially high shear forces dependent on meal viscosity and particle sizes
	Piston	Allows food bolus to move from main body through valves into antral part of DGM Compensates volume changes within antrum due to barrel movement to modulate reflux Maintains a dead volume within barrel to simulate gastric sieving	Large, dense food particles and/or pharmaceuticals can sink to the greater curvature of the stomach, thereby delaying their emptying from the stomach (gastric sieving)

The main functions of each part along with their relevance to the human stomach are given
[a]Schematic representations and names of parts refer to Fig. 6.1

Table 6.2 Composition of solutions used in the DGM

Solution	Component	Source	Concentration[a]
Artificial saliva (pH 6.9)	Salt (NaCl)	Sigma[b]	150 mM
	Urea	Sigma[b]	3 mM
	Salivary amylase (human)	Sigma[b]	36 U/ml
Gastric priming acid	Salts (NaCl, KCl, CaCl$_2$, NaH$_2$PO$_4$)	Sigma[b]	89 mM (total)
	HCl	Sigma[b]	10 mM
Gastric acid	Salts (NaCl, KCl, CaCl$_2$, NaH$_2$PO$_4$)	Sigma[b]	89 mM (total)
	HCl	Sigma[b]	200 mM
Gastric enzyme	Salts (NaCl, KCl, CaCl$_2$, NaH$_2$PO$_4$)	Sigma[b]	89 mM (total)
	Egg lecithin	Lipid P.[c]	0.38 mM
	Lipase (fungal, DF15)	Amano[d]	60 U/ml
	Gastric pepsin (porcine)	Sigma[b]	8.9 kU/ml
Duodenal hepatic	Salts (NaCl, KCl, CaCl$_2$)	Sigma[b]	154 mM (total)
	Egg lecithin	Lipid P.[c]	6.5 mM
	Cholesterol	Sigma[b]	3 mM
	Bile salts (Na-taurocholate, Na-glycodeoxycholate)	Sigma[b]	25 mM (total)
Duodenal pancreatic	Salts (NaCl, KCl, CaCl$_2$, MgCl$_2$, ZnSO$_4$)	Sigma[b]	154 mM (total)
	Pancreatic lipase (porcine, Type VI-S)	Sigma[b]	590 U/ml
	Colipase (porcine)	Roche[e]	0.2 mg/ml
	Trypsin (porcine, Type IX-S)	Sigma[b]	11 U/ml
	α-Chymotrypsin (bovine, Type II)	Sigma[b]	24 U/ml
	α-Amylase (porcine, Type VI-B)	Sigma[b]	300 U/ml

[a]Stated concentrations are within the stock solutions used. Final concentrations within gastric/duodenal compartment will be significantly lower due to dilution with food bolus and other solutions, bringing them within physiological ranges presented in literature
[b]Sigma Aldrich, Gillingham, Dorset, UK
[c]Lipid Products Ltd, Redhill, UK
[d]Amano Enzyme Inc., Nagoya, Japan
[e]Roche Diagnostics GmbH, Mannheim, Germany

(Vardakou et al. 2011a). The two studies show a good correlation between the physical forces of the human stomach and the DGM antrum, using both low viscosity and high viscosity meals.

The DGM antrum consists of a barrel and a piston, which move within a 37 °C water jacket (Fig. 6.1). While the piston draws portions of food bolus through an inlet valve from the fundus into the antrum, it is the upward and downward movement of the barrel during processing which exerts shear stresses on the antral contents. This is due to a flexible annulus mounted within the top part of the barrel through which food (and formulations) passes during every stroke, thereby simulating the rhythmic peristaltic contractions of the human stomach. While the speed of movements has been calibrated to provide physiological shear forces (Vardakou et al. 2011a), the actual volume of food bolus processed within the antrum at any one time, as well as duration of processing are tailored to the specific meal used (volume, composition, calorific content). At pre-defined intervals, the inlet valve closes and the outlet valve opens, allowing the processed chyme to be ejected from the DGM.

A phenomenon observed in the human stomach is that of gastric sieving, whereby larger, denser particles/formulations can be retained within the greater curvature of the stomach longer than smaller particles, therefore subjecting them to extended processing (Meyer 1980). Gastric sieving is simulated within the DGM by definition of a "dead volume," i.e. a defined space between barrel and piston whose volume is maintained during ejection thereby allowing large, dense particles to remain in the antrum and undergo repeated processing cycles. At the end of a simulated digestion, any material remaining in this dead volume is ejected to simulate the phase III contraction (housekeeper wave) which fully empties the human stomach at the end of gastric digestion (Meyer 1987).

Following ejection from the DGM, samples can be subjected to further digestion using a static duodenal model. To this end, the pH of the samples is elevated and a physiological mix of bile salts with lecithin and cholesterol and pancreatic enzymes, is added to simulate conditions found within the duodenum.

6.2   General Protocol for DGM Experiments

Planning First, the following information about the test meal (and/or drink) is gathered: mass, energy content, composition (carbohydrate, fat, protein content). This information is used to estimate total gastric residence time of the meal as well as maximum rate of gastric secretion.

Using parameters such as the volume and frequency of ejection from the antrum, a program is designed to allow the DGM to empty the test meal fully within the calculated gastric residence time. The chosen sample volume and frequency are dictated by the gastric residence time and volume of the meal, as well as any downstream processing and analysis of the samples which may require certain minimum volumes to be ejected.

Preparation Any enzyme solutions needed during the experiment (salivary, gastric, duodenal) are prepared immediately prior to the experiment. The components of solutions and enzymes used in the experiments are detailed in Table 6.2. Solutions have been designed to provide biochemical conditions (e.g. concentrations of salts, enzymes, etc., and/or secretion rates) within the "normal" physiological range of healthy subjects (Lentner 1981). For example, the contents of the DGM at any one time will be a complex inhomogeneous mixture of food matrix and the components of gastric priming acid, gastric acid and gastric enzyme solutions.

Mastication Depending on the requirements of the project and food type, several methods of mastication can be used.

- No mastication (e.g. for liquid meals, drinks and meals where mastication is not required)
- Simulated mastication using a food processor, mincer or grinder, with or without addition of artificial or human saliva

- Human chew, whereby the test meal is chewed to the naturally perceived point of swallowing and collected in a beaker before addition to the DGM
- Real-time human chew, similar to the above but with real-time transfer of each mouthful to the DGM

Dynamic Gastric Processing Before the start of gastric processing, the DGM pH electrode is allowed to equilibrate in the test meal/drink. This sets the range of pH during processing of this specific meal (from start reading to pH 2) and allows the DGM to dynamically adjust gastric acid addition as the pH drops. The DGM is primed with 20 ml priming acid to simulate the residual gastric fluid normally found in the resting human stomach.

Upon start of the program, the meal is added to the fundus of the DGM. Fluid portions of the meal are poured slow over a spoon and allowed to trickle down the outside edge of the fundus and main body. This simulates how fluid would enter the stomach from the esophagus and minimizes any artificial turbulence and mixing of the gastric contents. Masticated foods are usually added slowly over the course of several minutes to mimic the swallowing of food.

Any food/drink added at the start of a run is immediately in contact with the walls of the main body, allowing heat transfer to occur. The contents are also subjected to gentle squeezing in the main body (three contractions per minute). Gastric acid and enzyme secretions are added through the perforated ring at physiological rates dependent on meal size and buffering capacity. These rates slow down progressively during the experiment in response to reducing bolus volume and change in pH. Shortly after start of the run, the lower part of the DGM is activated and pulls a portion of the food bolus into the antrum for processing (high shear, mixing).

At programmed intervals, the inlet valve of the DGM closes, the outlet valve opens and the defined proportion of bolus from the antrum is ejected by an upward movement of the barrel and piston. Any pre-defined dead volumes between piston, barrel and valves are maintained to allow for large dense particles to remain in the machine for further processing (gastric sieving). The outlet valve then closes, the inlet valve opens and the next portion of partially digested food enters the antrum from the main body.

At the end of the DGM run, the final sample is ejected by complete upward movement of barrel and piston, thereby ejecting any remaining dead volumes that were present in the antrum region and simulating the housekeeper wave. The DGM is disassembled to recover any residues.

Static Duodenal Processing Where required, samples from the DGM can be subjected to further static digestion in a duodenal model. First, the pH of the sample is adjusted to pH 6.8 to reduce further activity of gastric enzymes and to simulate the change of pH in the duodenum. One of two main methods is then employed:

- From each DGM gastric sample, a subsample is transferred to a separate vessel, and pancreatic enzymes as well as bile salts, lecithin and cholesterol are added at physiological levels depending on food. These separate duodenal incubations (3–4 h each in an orbital shaker, 170 rpm, at 37 °C) can then be sampled at defined intervals to establish separate duodenal nutrient/pharmaceutical release profiles for each gastric ejection. This approach is often followed for nutrient bioaccessibility studies and pharmaceutical studies using formulations that disperse in the stomach.

- Subsamples of each DGM gastric sample are neutralized and pooled in a single vessel and kept on ice until the end of the gastric phase. Pancreatic enzymes and bile salts, lecithin and cholesterol are added at physiological levels depending on food, and the vessel is incubated in an orbital shaker for 3–4 h at 170 rpm, 37 °C. Samples are taken at defined intervals to establish a single duodenal release profile. This method is often used for gastro-resistant pharmaceutical formulations which are recovered from the DGM intact and transferred to this duodenal pool to monitor dispersal and dissolution in the duodenal phase.

Controls Control experiments are designed for any project involving DGM runs. In the case of food applications such as bioavailability studies, these control runs essentially follow the same protocol as the actual experiments, but do not include any digestive enzymes. The starting material before gastric digestion is also analysed. This enables a distinction to be made between the effects of mechanical processing in the DGM alone, and full mechanical and biochemical digestion. For pharmaceutical applications, control experiments generally follow the full experimental protocol including enzymes but without the drug.

DGM experiments are generally carried out in triplicate in the first instance.

6.3   Uses of the DGM

The DGM has a wide variety of applications and has so far been used to study nutrient bioaccessibility and structural changes of food matrices during digestion, as well as the disintegration and dissolution of various drug formulations. Normal experimental readouts include a photographic/video record of digesta appearance, acidification profile and temperature profile. Some previously studied parameters are summarized Table 6.3.

6.3.1   Food-Based Research

The DGM has extensively been used to evaluate the rate and extent of nutrient and phytochemical release from plant foods and the effect of food matrix on their release in the upper GI tract (Mandalari et al. 2010, 2013). The effect of mastication and processing on the lipid release from almond seeds in the upper GI tract has recently been investigated (Mandalari et al. 2014a).

The bioaccessibility of pistachio polyphenols, xanthophylls and carotenoids during simulated human digestion was recently assessed using the DGM followed by duodenal incubation: results demonstrated that the presence of a food matrix, such as muffin, decreased the bioaccessibility of certain polyphenols in the upper GI tract (Mandalari et al. 2013).

Table 6.3 Applications of DGM samples (this list is not exhaustive)

Parameter	Method of analysis	Output	References
Viscosity	Rheological analysis	Plot of changes in viscosity	
Cellular structure	Microscopy	Changes in cell/surface integrity	Mandalari et al. (2014b)
Phytochemicals bioaccessibility	GC, HPLC	Effect of food matrix on the bioaccessibility of bioactives from pistachios during simulated digestion	Mandalari et al. (2013)
Starch digestion	Starch analysis	Changes in ratio of glucose:starch, total starch Effect of particle size on starch degradation from *Durum* wheat	Ballance et al. (2013) IFR, unpublished data
Lipid separation (digestion?)	Total lipid release and fatty acid analysis	Effect of mastication and processing on lipid release from almond seeds	Mandalari et al. (2014a)
Protein digestion	1D and 2D SDS-PAGE, RP-HPLC, MALDI-ToF, immunoblotting	Effect of food matrix on protein digestion from almond seeds	Mandalari et al. (2014b)
Peptide production	1D and 2D SDS-PAGE, RP-HPLC, MALDI-ToF, immunoblotting	Persistence of allergens from cow's milk and peanut flour	IFR, unpublished data
Probiotic survival	Culturing on selective media	Effect of food matrix on probiotic survival in the upper GI tract	Lo Curto et al. (2011) and Pitino et al. (2010, 2012)
Prebiotic delivery	Culturing on selective media, genetic analysis	Delivery of potential prebiotics in the distal GI tract	Mandalari et al. (2010)

Ongoing research aims to establish the key parameters involved in starch digestion from cereal (Ballance et al. 2013) and *Durum* wheat (IFR, unpublished data). The data obtained using the DGM compared well with in vivo data of glycaemic response, indicating that the DGM was predictive of the kinetics of digestible starch hydrolysis (Ballance et al. 2013).The effect of particle size on starch degradation from *Durum* wheat is currently being investigated using the DGM coupled with the static duodenal model and compared with an in vivo ileostomy study (IFR, unpublished data).

The DGM has recently been used to assess digestibility of almond protein in the upper GI tract, evaluate the effects of food matrix on protein release and assess the persistence of immunoreactive polypeptides generated during simulated digestion (Mandalari et al. 2014b). The results obtained are useful to investigate the relationship between food matrix and almond allergy.

The persistence of allergens present in cow's milk and peanut flour as measured by gastric and duodenal aspirates from human volunteers has been compared with data sets obtained from similar meals processed by the DGM. The comparison suggests that the DGM was predictive, not only of the persistence of the original allergens, but also of the profile of peptide fragments generated during digestion (IFR, unpublished data).

Probiotic strains of *Lactobacillus* spp. were investigated for their ability to survive in the upper GI tract using a number of vehicles and different growth phases: the results obtained showed that probiotic survival using dynamic models was affected by the buffering capacity of the matrix in relation to the pH decrease in the stomach (Lo Curto et al. 2011; Pitino et al. 2010). Cheese was also found to be a good vehicle for passage of probiotic bacteria in the upper GI tract and scanning electron microscopy indicated production of extracellular polysaccharides by *Lactobacillus rhamnosus* strains as a response to acid stress in the gastric compartment (Pitino et al. 2012).

A full model of the gastrointestinal tract, including in vitro gastric and duodenal digestion, followed by colonic fermentation using mixed faecal bacterial cultures, was used to investigate the prebiotic potential of natural (NS) and blanched (BS) almond skins, which are rich in dietary fiber (Mandalari et al. 2010). Both NS and BS significantly increased the population of bifidobacteria and *Clostridium coccoides*/*Eubacterium rectale* group, which are known for their beneficial effect in relation to health.

6.3.2 *Pharmaceutical-Based Research*

In recent years, the DGM group has seen a marked increase in projects from pharmaceutical industry. Current in vitro methods used in the study of disintegration and dissolution of oral solid dosage forms do not provide a physiological representation of the dynamic biochemical and physical environment of the human stomach (Vardakou et al. 2011a) and are therefore sometimes not predictive of the in vivo behaviour of dosage forms, particularly in the case of dosing with or after a typical meal. The DGM is well placed to bridge the gap between these simpler dissolution tests and in vivo studies (animal or human) and can be used in either to explain unexpected in vivo results, or as a predictive tool (Mann and Pygall 2012; Wickham et al. 2009).

Past studies in the DGM have involved a wide variety of dosage forms (capsule, tablet, powder, liquid) and types e.g. immediate release, modified release, gastro-retentive, self-emulsifying drug delivery system (Vardakou et al. 2011b; Mercuri et al. 2009, 2011). Particularly in the case of gastro-retentives, the ability to introduce sequential meal cycles (e.g. breakfast, lunch, dinner) within the same experiment allows for the real-time, realistic simulation of the range of conditions (pH, viscosity, shear forces) that these formulations are likely to encounter in vivo. The DGM may also find future use in the assessment of alcohol-drug interactions (dose-dumping) as well as the modelling of pediatric and/or geriatric physiology, all of which provide challenges (ethical and otherwise) in the justification of in vivo tests.

6.4 Advantages, Disadvantages and Limitations

Some advantages and limitations of the DGM are provided in (Table 6.4).

Table 6.4 Advantages and limitations of DGM

Advantages	Limitations
Capacity: Full meals up to 800 ml	*Not transparent*: Visual observations not possible during antral processing
Meals: Any masticated food/drink matrix	*Orientation*: Vertical alignment of main body and antrum
In vivo correlation: Use of exact meal used in a clinical study	*Open top*: Fundus always exposed to air
Temporal simulation: Real-time digestion and monitoring of pH and temperature	*Satiety*: no in vivo satiety signals controlling rate of digestion
Temporal simulation: allows time dependent processes to be studied	
Sequential meals: A full day's feeding regimen can be followed in real-time	

6.5 Availability of the System

At the time of writing, two DGM machines are in operations (Mark I and Mark II), with a third machine in development. The machines can be used for food- and pharmaceutical-based research, by both industry and academia. The Dynamic Gastric Model is protected by granted patents and pending patent applications owned by Plant Bioscience Limited (PBL). Enquiries about purchasing a DGM unit for academic or commercial use should be directed to PBL (Plant Bioscience Limited, Norwich, UK; martin@pbltechnology.com; http://pbltechnology.com). Access to the DGM as an outsourced contract research facility is available exclusively through Bioneer:FARMA (Bioneer:FARMA, Copenhagen, Denmark; bioneer@bioneer.dk; http://www.bioneer.dk/DGM/).

References

Ballance S, Sahlstrom S, Lea P, Nagy NE, Andersen PV, Dessev T, Hull S, Vardakou M, Faulks R (2013) Evaluation of gastric processing and duodenal digestion of starch in six cereal meals on the associated glycaemic response using an adult fasted dynamic gastric model. Eur J Nutr 52(2):799–812. doi:10.1007/s00394-012-0386-5

Bornhorst GM, Singh RP (2014) Gastric digestion in vivo and in vitro: how the structural aspects of food influence the digestive process. Ann Rev Food Sci Technol. doi:10.1146/annurev-food-030713-092346

Dubois A, Vaneerdewegh P, Gardner JD (1977) Gastric-emptying and secretion in Zollinger–Ellison syndrome. J Clin Invest 59(2):255–263. doi:10.1172/Jci108636

Ferrua MJ, Singh RP (2010) Modeling the fluid dynamics in a human stomach to gain insight of food digestion. J Food Sci 75(7):R151–R162. doi:10.1111/j.1750-3841.2010.01748.x

Holloway RH, Sifrim DA (2008) The acid pocket and its relevance to reflux disease. Gut 57(3):285–286. doi:10.1136/Gut.2006.118414

Konturek SJ, Biernat J, Oleksy J (1974) Serum gastrin and gastric-acid responses to meals at various Ph levels in man. Gut 15(7):526–530. doi:10.1136/Gut.15.7.526

Lentner C (ed) (1981) Units of measurement, body fluids, composition of the body, nutrition, Volume 1 of Geigy scientific tables, 8th edn. Ciba-Geigy Limited, Basle

Lo Curto A, Pitino L, Mandalari G, Dainty JR, Faulks RM, Wickham MSJ (2011) Survival of probiotic lactobacilli in the upper gastrointestinal tract using an in vitro gastric model of digestion. Food Microbiol 28(7):1359–1366. doi:10.1016/J.Fm.2011.06.007

Mandalari G, Faulks RM, Bisignano C, Waldron KW, Narbad A, Wickham MSJ (2010) In vitro evaluation of the prebiotic properties of almond skins (*Amygdalus communis* L.). FEMS Microbiol Lett 304(2):116–122. doi:10.1111/j.1574-6968.2010.01898.x

Mandalari G, Bisignano C, Filocamo A, Chessa S, Saro M, Torre G, Faulks RM, Dugo P (2013) Bioaccessibility of pistachio polyphenols, xanthophylls, and tocopherols during simulated human digestion. Nutrition 29(1):338–344. doi:10.1016/j.nut.2012.08.004

Mandalari G, Grundy MM, Grassby T, Parker ML, Cross KL, Chessa S, Bisignano C, Barreca D, Bellocco E, Laganà G, Butterworth PJ, Faulks RM, Wilde PJ, Ellis PR, Waldron KW (2014a) The effects of processing and mastication on almond lipid bioaccessibility using novel methods of in vitro digestion modelling and micro-structural analysis. Br J Nutr 112(9):1521–1529. doi:10.1017/S0007114514002414

Mandalari G, Rigby NM, Bisignano C, Lo Curto RB, Mulholland F, Su M, Venkatachalam M, Robotham JM, Willison LN, Lapsley K, Roux KH, Sathe SK (2014b) Effect of food matrix and processing on release of almond protein during simulated digestion. LWT – Food Sci Technol 59(1):439–447

Mann JC, Pygall SR (2012) A formulation case study comparing the dynamic gastric model with conventional dissolution methods. Dissolution Technol 19(4):14–19

Marciani L, Gowland PA, Fillery-Travis A, Manoj P, Wright J, Smith A, Young P, Moore R, Spiller RC (2001a) Assessment of antral grinding of a model solid meal with echo-planar imaging. Am J Physiol Gastrointest Liver Physiol 280(5):G844–G849

Marciani L, Gowland PA, Spiller RC, Manoj P, Moore RJ, Young P, Fillery-Travis AJ (2001b) Effect of meal viscosity and nutrients on satiety, intragastric dilution, and emptying assessed by MRI. Am J Physiol Gastrointest Liver Physiol 280(6):G1227–G1233

Mercuri A, Faulks R, Craig D, Barker S, Wickham M (2009) Assessing drug release and dissolution in the stomach by means of dynamic gastric model: a biorelevant approach. J Pharm Pharmacol 61:A5

Mercuri A, Passalacqua A, Wickham MSJ, Faulks RM, Craig DQM, Barker SA (2011) The effect of composition and gastric conditions on the self-emulsification process of ibuprofen-loaded self-emulsifying drug delivery systems: a microscopic and dynamic gastric model study. Pharm Res 28(7):1540–1551. doi:10.1007/s11095-011-0387-8

Meyer JH (1980) Gastric-emptying of ordinary food - effect of antrum on particle-size. Am J Physiol 239(3):G133–G135

Meyer JH (1987) Motility of the stomach and gastroduodenal junction. In: Johnson LR (ed) Physiology of the gastrointestinal tract, vol 1, 2nd edn. Raven Press, New York, pp 613–629

Pitino L, Randazzo CL, Mandalari G, Lo Curto A, Faulks RM, Le Marc Y, Bisignano C, Caggia C, Wickham MSJ (2010) Survival of *Lactobacillus rhamnosus* strains in the upper gastrointestinal tract. Food Microbiol 27(8):1121–1127. doi:10.1016/J.Fm.2010.07.019

Pitino I, Randazzo CL, Cross KL, Parker ML, Bisignano C, Wickham MSJ, Mandalari G, Caggia C (2012) Survival of *Lactobacillus rhamnosus* strains inoculated in cheese matrix during simulated human digestion. Food Microbiol 31(1):57–63. doi:10.1016/J.Fm.2012.02.013

Schubert ML, Peura DA (2008) Control of gastric acid secretion in health and disease. Gastroenterology 134(7):1842–1860. doi:10.1053/j.gastro.2008.05.021

Vardakou M, Mercuri A, Barker SA, Craig DQM, Faulks RM, Wickham MSJ (2011a) Achieving antral grinding forces in biorelevant in vitro models: comparing the USP dissolution apparatus II and the dynamic gastric model with human in vivo data. AAPS PharmSciTech 12(2):620–626. doi:10.1208/s12249-011-9616-z

Vardakou M, Mercuri A, Naylor TA, Rizzo D, Butler JM, Connolly PC, Wickham MSJ, Faulks RM (2011b) Predicting the human in vivo performance of different oral capsule shell types using a novel in vitro dynamic gastric model. Int J Pharm 419(1–2):192–199. doi:10.1016/j. ijpharm.2011.07.046

Wickham M, Faulks R, Mills C (2009) In vitro digestion methods for assessing the effect of food structure on allergen breakdown. Mol Nutr Food Res 53(8):952–958. doi:10.1002/ mnfr.200800193

Wickham MJS, Faulks RM, Mann J, Mandalari G (2012) The design, operation, and application of a dynamic gastric model. Dissolution Technol 19(3):15–22

Chapter 7
Human Gastric Simulator (Riddet Model)

Maria J. Ferrua and R. Paul Singh

Abstract An in vitro 'dynamic' model for food digestion diagnosis, the Human Gastric Simulator (HGS), has been designed to reproduce the fluid mechanical conditions driving the disintegration and mixing of gastric contents during digestion. The HGS simulates the stomach as a flexible compartment, and mimics its contractive motility by a series of rollers that continuously impinge and compress the compartment wall with increasing amplitude. Operated at 37 °C, the HGS facilitates a precise control of the mechanical forces to which foods are exposed during the process, as well as of the rate of simulated gastric secretions and emptying patterns.

Applications of the HGS have illustrated the need to better understand, and mimic, the fluid mechanic conditions that develop during digestion to improve the performance and reliability of novel in vitro models. To date, the HGS has been used to analyse the digestion behaviour of different foods, and the role of their materials properties on the physicochemical changes that they experience during the process. While the ability of the HGS to reproduce the gastric forces that develop in vivo has been proved, further studies are needed to achieve a thorough validation of its digestive capabilities.

Keywords Human gastric simulator • In vitro model • Digestion • Gastric motility • Digesta fluid mechanics

7.1 Origins of the HGS

Central to the delivery of optimal nutrition, the stomach is, after the mouth, the main site for food disintegration during digestion (Wickham et al. 2012). Once in the stomach, products are stored, digested and progressively emptied into the

M.J. Ferrua (✉)
Riddet Institute, Massey University, Palmerston North, New Zealand
e-mail: m.j.ferrua@massey.ac.nz

R.P. Singh
Department of Biological and Agricultural Engineering, University of California, Davis, CA, USA
e-mail: rpsingh@ucdavis.edu

© The Author(s) 2015
K. Verhoeckx et al. (eds.), *The Impact of Food Bio-Actives on Gut Health*,
DOI 10.1007/978-3-319-16104-4_7

61

duodenum by a synergy of physicochemical processes triggered and regulated by the motor and secretory activities of the gastric wall (Barrett and Raybould 2010a; Mayer 1994).

From a functional point of view, the stomach is divided into two main regions. Within the proximal region (upper half), changes in the compliance and secretory activity of the gastric wall allow the stomach to accommodate the ingested meal and provide the biochemical environment needed for its conditioning (Schwizer et al. 2002; Wickham et al. 2012). The distal region, on the other hand, is expected to play a major role in the structural disintegration of the meal. It is within this region where a series of peristaltic antral contraction waves (ACWs) continuously mix, compress and shear gastric contents during the process (Schwizer et al. 2006; Schulze 2006). As a result, food is converted into a semi-liquid mass of partially digested food, whose emptying from the stomach is feedback-regulated by a series at physicochemical receptors within the intestine (Barrett and Raybould 2010b).

Despite the complexities of gastric processes, increasing evidence indicates that the hydrodynamic conditions that develop during digestion have a central role on the material response and subsequent bioavailability of nutrients and bioactive compounds (Dikeman et al. 2006; Lentle and Janssen 2010). In particular, the poor in vitro–in vivo performance of many of the in vitro models currently used for digestion diagnosis has been largely attributed to their inability to reproduce the in vivo mechanics of the gastrointestinal (GI) tract (Yoo and Chen 2006).

Significant efforts have been made during the last decade to better understand the overall functioning of the human stomach and to develop a new generation of in vitro models of enhanced biochemical and mechanical relevance (Boulby et al. 1999; Faas et al. 2002; Kunz et al. 2005; Goetze et al. 2007, 2009; Kwiatek et al. 2006; Marciani et al. 2001a, 2007, 2012; Marciani 2011; Schwizer et al. 2002, 2006; Steingoetter et al. 2005; Treier et al. 2006; Mackie et al. 2013). More notably among those models are the TNO and DGM systems discussed in the previous sections. However, it is noteworthy that there is still no consensus agreement on the way in which these models reproduce the hydrodynamic conditions that develop in vivo, with none of them being able to replicate the actual motility of the gastric wall during digestion.

The Human Gastric Simulator (HGS) was specifically designed and developed by Kong and Singh (2010) to mimic the peristaltic activity of ACWs as reported in vivo (Kwiatek et al. 2006; Schwizer et al. 2006). Aimed at reproducing one of the main features driving the dynamics of gastric contents, this model is expected to better simulate the fluid mechanical forces driving food disintegration during digestion. Since its development, the HGS has been used to investigate not only the physicochemical changes experienced by different food products during digestion, but also the role of gastric motility on the outcomes of the process.

7.2 Model Description

The HGS consists of a cylindrical latex chamber that simulates the stomach compartment, and four conveyor belts that periodically impinge a series of Teflon rollers upon its wall to mimic the antral contraction wave activity of the stomach wall (Fig. 7.1). The system operates inside an insulated chamber maintained at 37 °C, while facilitating the delivery of gastric juices and emptying of simulated digesta in a continuous and controlled manner.

7.2.1 Gastric Compartment

The stomach is represented by a 5.7 L cylindrical vessel (20 cm high and 10.2 cm diameter) that ends in the form of a conic frustum (13 cm height and 2.5 cm final diameter).

Made of latex rubber, the gastric compartment sits straight up, wrapped onto a stainless steel ring (10.2 cm high and 15.2 cm diameter) that is supported by four

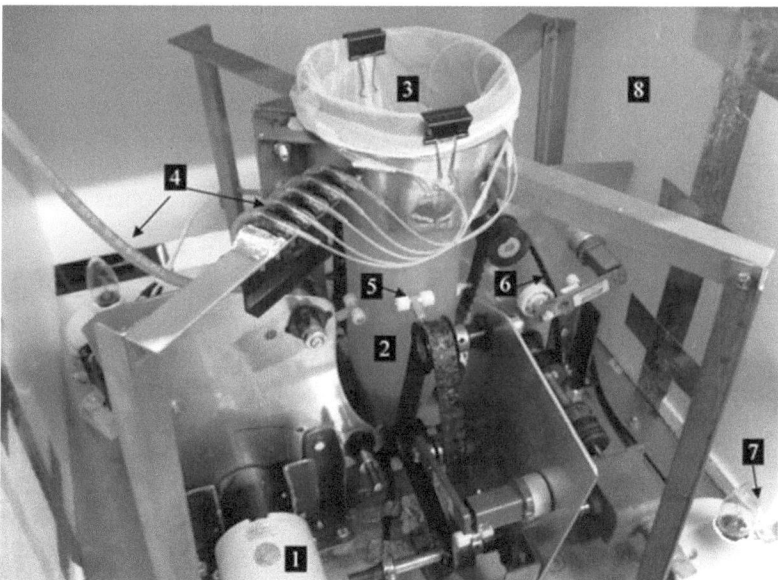

Fig. 7.1 Human gastric simulator. (1) Motor (2) Gastric compartment (3) Mesh bag (4) Simulating secretion tubes (5) Teflon roller set (6) Conveying belt (7) Insulated chamber. From Kong and Singh (2010)

diametrically opposed legs welded to the base. The open end of the container provides a simple way for loading food materials into the unit and for sampling of the simulated digesta during the process.

7.2.2 Gastric Motility

With the primary goal of mimicking the physical processes driving food disintegration, the HGS was designed to reproduce the motor activity of the antral contraction waves (ACWs) along the distal region of the stomach.

The dynamics of the ACWs along the lower part of the cylindrical vessel is mimicked by a mechanical drive system. Four conveyor rubber belts move along the height of the gastric compartment at 90° from each other.

Each belt is supported by four 0.95 cm pulleys, which attached to a low-carbon steel plate are moved by a drive shaft (1.27 cm diameter brass rod) connected to a 115 V Stir-Pak Heavy-Duty Mixer head (model R-50002-10, Cole-Parmer) motor. A Stir-Pak controller (model R-50002-02, Cole-Parmer) is used to allow for speed adjustments within the range of 2–180 rpm. Power is transmitted from one drive shaft to another via two bevel gears coupled at a 90° angle.

Each timing belt (0.95 cm wide) carries three sets of equally spaced Teflon rollers located every 20 cm of each other. Each set of rollers consists of two wide Teflon wheels (1.27 cm diameter, 0.9 mm long) placed together through an aluminium rod, that is secured to the belt by a male threaded screw (0.2 cm diameter, 1.5 cm long). As the belt moves, rollers start impinging the compartment wall (at about two-thirds of its total height). As the rollers propagate down, they replicate the increasing compression pattern of the ACWs by getting successively closer to the rollers on the opposite belt. The closer the rollers, the higher the compression forces.

To avoid possible interference between neighbouring rollers as they get further down the bottom of the gastric compartment, the lower pulleys closer to the compartment are placed at two different levels, with one pair of opposite pulleys located 3.0 cm higher than the other pair.

In order to simulate the motility pattern of the ACWs, the drive system is set to impose three propagating contractions per minute (with one finishing while another commences). If desired, this propagation speed can be changed by simply adjusting the rpm of the driving shaft. The force imposed by the rollers on the simulated digesta can be controlled by adjusting the distance between opposite rollers through the screw engagement depth inside the aluminium rod.

7.2.3 Gastric Emptying

To simulate the sieving effect of the pylorus, a polyester mesh bag (pore size of 1.5 mm) is used to line the inner surface of the gastric compartment and prevent larger particles from emptying the compartment. At the end of the trial, the mesh can be easily taken out through the open top section of the compartment, facilitating the removal and analysis of the remaining digesta.

The rate of gastric emptying is controlled by means of a peristaltic pump (Masterflex Pump Controller 7553-50/7090-42 Pump, Cole-Parmer, Chicago, Ill., U.S.A.) connected to the bottom of the gastric compartment through a 0.32 cm plastic tube.

7.2.4 Gastric Secretions

Simulated gastric juices are delivered at about 10–15 cm from the bottom of the compartment through five polyethylene tubes (I.D. 0.86 mm) uniformly distributed between the mesh bag and the latex wall.

The delivery rate of the simulated gastric juice into the compartment is controlled by a mini peristaltic pump (Model 3385, VWR, Scientific, Rochester, N.Y., U.S.A.) and a control valve placed on a 6.4 mm plastic tube that later on divides into five tubes going into the compartment. The flow rate of the simulated secretion can be adjusted between 0.03 and 8.2 mL/min.

It is noteworthy that while it is possible to control the release of gastric juices during the simulated processes, there are no mechanisms in place to automatically adjust this gastric response to the specific composition and volume of simulated digesta.

7.2.5 Temperature Control

The system is placed inside an insulated plastic foam chamber, where two 60 W light bulbs and a mini-fan are installed to maintain the system operating at uniform and constant temperature of 37 °C. The operation of the bulbs is automatically controlled by a thermostat (Model T675A 1516, Honeywell, Honeywell Inc., Minneapolis, Minn., U.S.A.).

7.3 Analysis of HGS Biomechanical Relevance

The ability of the HGS to replicate the biomechanics of the human stomach was confirmed by analysing its ability to simulate the mechanical forces that develop in vivo (Kong and Singh, 2010). Based on the significant variation that exists on the levels of gastric forces and contractive activity of the stomach wall during digestion, the mechanical forces within the HGS were investigated for two different compressions levels: 50 % and 70 % (as determined by a minimum distance between opposite rollers of 1.2 cm and 0.6 cm, respectively). The HGS was filled with water and the forces that develop at the bottom of the compartment were determined by measuring the pressure to which a rubber bulb is exposed due to the contractive activity of the rollers. Details of the experimental methodology and force computations can be found in Kong and Singh (2010). Normalized by the cross-sectional area of the bulb, the maximum stresses recorded within the HGS were 6.7 ± 1.2 kPa and 8.9 ± 2.5 kPa, for a 50 % and a 70 % of compression, respectively. As stated by the authors, these results, as well as the periodic changes in the pressure values inside the HGS, were in reasonably good agreement with in vivo data, that commonly report mechanical stresses varying from 5.1 to 67 kPa (Marciani et al. 2001b; Kamba et al. 2000).

7.4 Operating Protocol

To provide a reference frame for the operation of the HGS, a brief description of the methodologies applied during the use of the HGS is presented in the following.

7.4.1 Preparation of a Food Bolus

Different methods can be used to prepare the simulated bolus. In particular, during current applications of the HGS, food samples are either cut or ground to emulate the particle size distribution observed in human boluses. The particulate sample is then mixed with simulated saliva (100 g food: 20 mL saliva) for 30 s and allowed to stand at 37 °C for 2 min. The exact composition of the artificial saliva can be varied depending on the type of food and scope of the study (Kong and Singh 2010; Guo et al. 2014).

7.4.2 Gastric Processing

Simulated gastric juice is prepared by dissolving pepsin (1 g), gastric mucin (1.5 g), and NaCl (8.775 g) in 1 L distilled water with pH of 1.3 adjusted using 6 N HCl. To simulate the fasting conditions of the stomach, 50–70 mL of simulated gastric juice is first loaded into the HGS and equilibrated at 37 °C. The release of gastric juice within the gastric compartment starts immediately after the bolus is loaded, and continues at a rate of 2.5 mL/min during the entire processes (Hoebler et al. 2002). Gastric digesta is removed at a rate of 3 mL/min and subject to different chemical and physical analysis. Depending on study, after 3–5 h of simulated process the digesta remaining inside the HGS is removed for further analysis.

7.5   Uses of the HGS

The HGS has been used to investigate the role that ACW dynamics and food material properties have on the structural changes and disintegration profile of different foods during digestion.

7.5.1   Role of ACW Activity on Food Digestion

To investigate the relevance of ACW motility on the digestion behaviour of foods, Kong and Singh (2010) compared the performance of the HGS against the more traditional shaking bath method.

Apple cubes and extra-long white rice kernels were mixed with simulated saliva and exposed to 2 h of digestion in both, the HGS and a shaking bath. It is noteworthy that a batch approach was employed in both cases, with gastric juices being added to the systems only at the beginning of the process. The results clearly illustrated the significant effect that the crushing and squeezing forces generated within the HGS have on the breakdown of both food models during the process. In the case of the apples, 61 % of the total dry matter from the shaking bath was still in particles larger than 6.3 mm, and only 20 % in particles smaller than 2.8 mm. In comparison, only 16 % of the total dry matter from the HGS remained in particles larger than 6.3 mm, with a 69 % of it distributed in particles smaller than 2.8 mm. A similar result was found in the case of rice. Most kernels were intact after digestion in the shaking bath, while 52 % of the dry matter from the HGS was associated with particles smaller 0.8 mm.

Considering that the structural breakdown of the diet will have a significant impact on the rate of nutrient release during digestion, this study confirmed the need to better emulate the fluid mechanical conditions that develop during digestion. To further investigate the role of gastric motility on the disintegration kinetics of foods, Kong and Singh (2010) investigated the disintegration profile of white rice when exposed to two different levels of compression. Unlike the previous study, the HGS was operated under dynamic conditions, where a continuous release of 2.5 mL/min of gastric juice was imposed. In agreement with in vivo data, an exponential decay of the digesta's pH from an initial value of 4.27 to a final constant value of 1.35 was observed within the first 2 h of process. Simulated digesta was continuously removed from the HGS at a rate of 3 mL/min, leading to a 60 % of the total dry mass being emptied after 3 h of process. The HGS was operated under two levels of compression (50 % and 70 %). The higher the compression, the higher the disintegration of the food particles. In particular, a 75 % compression was able to break down 75 % of the rice kernels into particles much smaller than 1.2 mm in size. This study showed once again the need to better understand and mimic the biomechanical functions of the human stomach in order to improve the performance and reliability of in vitro digestive systems.

7.5.2   Role of Food Material Properties

Kong et al. (2011) used the HGS to investigate the physical changes that white and brown rice experience during digestion. The authors cooked the rice, mixed it with simulated saliva, and placed it in the HGS (previously loaded with 50 mL of simulated gastric juice). During the 3 h of process, gastric juice was added to the system at a rate of 2.5 mL/min. The simulated digesta was emptied at a rate of 3 mL/min and exposed to a maximum compression level of 50 %. The solid composition of the emptied digesta clearly illustrated the greater level of disintegration and dissolution experienced by white rice. After 3 h of digestion, 55 % of the white rice solids were emptied from the HGS, as compared with 45 % for the brown rice. Sieving of the digesta allowed the authors to associate the slower emptying rate of brown kernels to its slower rate of disintegration. By the end of the process, 80 % of the particles within the white rice digesta were smaller than 10 mm^2 compared to only 40 % for the brown rice. The differences observed in the physical changes of both types of rice were largely associated with the bran layer surrounding the brown kernels. As illustrated by the authors, this bran layer not only delayed the diffusion of gastric juice into the kernels, but also protected them from the mechanical forces that develop during the process. In addition, while both digesta samples behave as weak-gels, the brown rice digesta was found to have an enhanced elastic component, which could further slow down its mechanical disintegration.

Guo et al. (2014) used the HGS to investigate the effect of different emulsion gels' structures on their disintegration profile during digestion. Homogeneous 'soft' gels and heterogeneous 'hard' gels were mechanically grounded and mixed with artificial saliva to specifically simulate in vivo masticated gel boluses. The simulated boluses were loaded into the HGS already containing 70 mL of simulated gastric juice and exposed to 5 h of digestion process. Gastric juice was delivered at 2.5 mL/min and digesta samples removed at a rate of 3 mL/min (starting after 30 min). Rollers were set to impose a maximum compression level of 60 %. Despite the differences in the initial strength and size of the gel particles, similar amounts of solids were emptied from the HGS ($\approx$74 %) at the end of the process, and a similar distribution of particles sizes was found in the digesta remaining inside the HGS. Despite this similarities, the authors did find significant differences in the emptying profile and size distribution of the digesta leaving the HGS during the process. Initially, the amount of solids leaving the HGS was higher in the case of the 'hard' gel, but this trend reversed after 3 h of process. This initial trend was explained by the smaller size of hard gel particles in the simulated masticated bolus. In addition, they also found that while the diameter ($d_{4,3}$) of oil droplets in the empited digesta of the 'hard' gel did not change during the process, in the case of the soft gel it remained unchanged only for the first hour. After that time, the $d_{4,3}$ of emptied oil droplets from the 'soft' gel increased to reach a maximum at about 2.5 hours of process. These differences in the emptying profile of solids and size distribution of oil droplets at later stages of the process were related to the way in which the structure of gels ingluences their chemical digestion by pepsin. Within the first hour of process, a combination of chemical and mechanical effects gradually broke down both gels to particles of about 10 µm. During this process, the microstructure of the gels was largely maintained and only very small quantities of oil were released. As time evolves, the fine-stranded structure of the 'soft' gel allowed pepsin to further disintegrate the gel particles down to a size of 0.45 µm (a process not observed in the case of the 'hard' gel). This further disintegration of the 'soft' gel enhanced its rate of emptying from the HGS and the release of oil droplets from the matrix. The transitional increase in the size of the liberated oil droplets after 1 h of process was associated with their flocculation, as the digesta passes through the isoelectric point of the denatured whey proteins.

7.6   Advantages and Limitations

Specifically designed to mimic the motor activity of the antral contractions waves during digestion, the HGS has been proved to reproduce the fluid mechanical forces that develop in vivo. Preliminary applications of the HGS have demonstrated the need to better understand and mimic the physical processes underlying digestion. The possibility to control the motor activity of the ACW offer new opportunities to investigate the impact gastric motility dysfunctions on food digestion. Further efforts needs to be done to automate the secretory and emptying patterns of the HGS in response to digesta properties during the process, and to pursue a thorough validation of its digestive capabilities.

7.7   Availability of the System

Two HGS models are in operation. One in the Department of Biological and Agricultural engineering at the University of California (Davis), where it was first created. A second replicate was made and currently used at the Riddet Institute, Massey University (New Zealand).

References

Barrett KE, Raybould HE (2010a) The gastric phase of the integrated response to a meal. In: Koeppen BM, Stanton BA (eds) Berne & Levy physiology, 6th edn. Elsevier, Philadelphia, pp 504–516

Barrett KE, Raybould HE (2010b) The small intestinal phase of the integrated response to a meal. In: Koeppen BM, Stanton BA (eds) Berne & Levy physiology, 6th edn. Elsevier, Philadelphia, pp 516–532

Boulby P, Moore R, Gowland P, Spiller RC (1999) Fat delays emptying but increases forward and backward antral flow as assessed by flow-sensitive magnetic resonance imaging. Neurogastroenterol Motil 11:27–36

Dikeman CL, Murphy MR, Fahey GC Jr (2006) Dietary fibers affect viscosity of solutions and simulated human gastric and small intestinal digesta. J Nutr 136:913–919

Faas H, Steingoetter A, Feinle C, Rades T, Lengsfeld H, Boesiger P, Fried M, Schwizer W (2002) Effects of meal consistency and ingested fluid volume on the intragastric distribution of a drug model in humans—a magnetic resonance imaging study. Aliment Pharmacol Ther 16:217–224

Goetze O, Steingoetter A, Menne D, van der Voort IR, Kwiatek MA, Bösiger P et al (2007) The effect of macronutrients on gastric volume responses and gastric emptying in humans: a magnetic resonance imaging study. Am J Physiol Gastrointest Liver Physiol 292:G11–G17

Goetze O, Treier R, Fox M, Steingoetter A, Fried M, Boesiger P, Schwizer W (2009) The effect of gastric secretion on gastric physiology and emptying in the fasted and fed state assessed by magnetic resonance imaging. Neurogastroenterol Motil 21:725–742

Guo Q, Ye A, Lad M, Dalgleish D, Singh H (2014) Effect of gel structure on the gastric digestion of whey protein emulsion gels. Soft Matter 10:1214–1223

Hoebler C, Lecannu G, Belleville C, Devaux MF, Popineau Y, Barry JL (2002) Development of an in vitro system simulating bucco-gastric digestion to assess the physical and chemical changes of food. Int J Food Sci Nutr 53(5):389–402

Kamba M, Seta Y, Kusai A, Ikeda M, Nishimura K (2000) A unique dosage form to evaluate the mechanical destructive force in the gastrointestinal tract. Int J Pharm 208:61–70

Kong F, Singh RP (2010) A human gastric simulator (HGS) to study food digestion in human stomach. J Food Sci 75(9):E627–E635

Kong F, Oztop MH, Singh RP, McCarthy MJ (2011) Physical changes in white and brown rice during simulated gastric digestion. J Food Sci 76(6):E450–E457

Kunz P, Feinle-Bisset C, Faas H, Boesiger P, Fried M, Steingotter A et al (2005) Effect of ingestion order of the fat component of a solid meal on intragastric fat distribution and gastric emptying assessed by MRI. J Magn Reson Imaging 21:383–390

Kwiatek MA, Steingoetter A, Pal A, Menne D, Brasseur JG, Hebbard GS et al (2006) Quantification of distal antral contractile motility in healthy human stomach with magnetic resonance imaging. J Magn Reson Imaging 24:1101–1109

Lentle RG, Janssen PWM (2010) Manipulating digestion with foods designed to change the physical characteristics of digesta. Crit Rev Food Sci Nutr 50:130–145

Mackie AR, Rafiee H, Malcolm P, Salt L, van Aken G (2013) Specific food structures supress appetite through reduced gastric emptying rate. Am J Physiol Gastrointest Liver Physiol 304:G1038–G1043

Marciani L, Gowland PA, Spiller RC, Manoj P, Moorel RJ, Young P, Fillery-Travis AJ (2001a) Effect of meal viscosity and nutrients on satiety, intragastric dilution, and emptying assessed by MRI. Am J Physiol Gastrointest Liver Physiol 280(6):G1227–G1233

Marciani L, Gowland PA, Fillery-Travis A, Manoj P, Wright J, Smith A, Young P, Moore R, Spiller RC (2001b) Assessment of antral grinding of a model solid meal with echo-planar imaging. Am J Physiol Gastrointest Liver Physiol 280:G844–G849

Marciani L, Wickham M, Singh G, Bush D, Pick B, Cox E, Fillery-Travis A, Faulks R, Marsden C, Gowland PA, Spiller RC (2007) Enhancement of intragastric acid stability of a fat emulsion meal delays gastric emptying and increases cholecystokinin release and gallbladder contraction. Am J Physiol Gastrointest Liver Physiol 292(6):G1607–G1613

Marciani L (2011) Assessment of gastrointestinal motor functions by MRI: a comprehensive review. Neurogastroenterol Motil 23:399–407

Marciani L, Hall N, Pritchard SE, Cox EF, Totman JJ, Lad M, Hoad CL, Foster TJ, Gowland PA, Spiller RC (2012) Preventing gastric sieving by blending a solid/water meal enhances satiation in healthy humans. J Nutr 142(7):1253–1258

Mayer EA (1994) The physiology of gastric storage and emptying. In: Johnson L (ed) Physiology of the gastrointestinal tract, 3rd edn. Raven Press, New York, pp 929–976

Schulze K (2006) Imaging and modeling of digestion in the stomach and the duodenum. Neurogastroenterol Motil 18(3):172–183

Schwizer W, Steingotter A, Fox M, Zur T, Thumshirn M, Bösiger P, Fried M (2002) Noninvasive measurement of gastric accommodation in humans. Gut 51(Suppl 1):i59–i62

Schwizer W, Steingoetter A, Fox M (2006) Magnetic resonance imaging for the assessment of gastrointestinal function. Scand J Gastroenterol 41:1245–1260

Steingoetter A, Kwiatek MA, Pal A, Hebbard G, Thumshirn M, Fried M et al (2005) MRI to assess the contribution of gastric peristaltic activity and tone to the rate of liquid gastric emptying in health. Proc Int Soc Magn Reson Med 13:426

Treier R, Steinetter A, Weishaupt D, Goetze O, Bösiger P, Fried M et al (2006) Gastric motor function and emptying in the right decubitus and seated body position as assessed by magnetic resonance imaging. J Magn Reson Imaging 23:331–338

Wickham MJS, Faulks RM, Mann J, Mandalari G (2012) The design, operation, and application of a dynamic gastric model. Dissolution Technol 19(3):15–22

Yoo JY, Chen XD (2006) GIT physicochemical modeling - a critical review. Int J Food Eng 2:1–7

Chapter 8
The DIDGI® System

Olivia Ménard, Daniel Picque, and Didier Dupont

Abstract A simple two-compartment in vitro dynamic gastrointestinal digestion system allowing the study of the disintegration of food during digestion has been recently developed at INRA. As a first application, it has been used for understanding the mechanisms of infant formula disintegration in the infant gastrointestinal tract. The developed system was validated by comparing the kinetics of proteolysis obtained in vitro towards in vivo data collected on piglets. Results showed a good correlation between in vitro and in vivo data and prove the physiological relevance of the newly developed system.

Keywords Two-compartment in vitro dynamic gastrointestinal digestion system • Disintegration of food • In vitro • Digestion • Infant gastrointestinal tract

8.1 Origins and Design of the DIDGI® System

At the French National Institute for Agricultural Research (INRA), several groups are trying to improve our understanding of the fate of different foods (dairy, egg, meat, bakery products, etc.) in the gastrointestinal tract. Our first objective is to unravel the mechanisms of food disintegration in the gastrointestinal tract and identify the molecules (nutrients, bioactive compounds, contaminants etc.) that are released during digestion (Barbé et al. 2013). A second objective is to determine how the structure of food matrices affects food digestion and nutrient bioaccessibility and bioavailability (Barbé et al. 2014). Finally, we model digestion and translate this cascade of events into mathematical models (Le Feunteun et al. 2014) in order to design new foods through a reverse engineering approach i.e. starting

O. Ménard • D. Dupont (✉)
UMR 1253 INRA – Agrocampus Ouest, STLO, 65 rue de Saint-Brieuc,
35042 Rennes Cedex, France
e-mail: didier.dupont@rennes.inra.fr

D. Picque
UMR 782 INRA-AgroParisTech, GMPA, 78850 Thiverval Grignon, France

© The Author(s) 2015
K. Verhoeckx et al. (eds.), *The Impact of Food Bio-Actives on Gut Health*,
DOI 10.1007/978-3-319-16104-4_8

from the bioactivity we want to deliver to the body and going back to the most adapted structure.

Since dynamic digestion devices are not very available on the market or quite expensive, we decided to build our own system. The system developed had to be cheap, simple, robust and applicable to all kind of foods INRA is working on. This model was built in order to monitor the disintegration and the kinetics of hydrolysis of the food occurring during a simulated digestion. It focuses on the upper parts of the digestive tract, i.e. the stomach and the small intestine. To be physiologically realistic, the computer-controlled system reproduces the gastric and intestinal transit times, the kinetics of gastric and intestinal pH, the sequential addition of digestive secretions and the stirring of the stomach and small intestine contents.

The gastrointestinal digestion system (Fig. 8.1) consists of two consecutive compartments simulating the stomach and the small intestine. Each compartment is surrounded by a glass jacket filled with water pumped using a temperature-controlled water bath. The system is equipped with temperature, pH and redox sensors and variable speed pumps to control the flow of meal, HCl, Na_2CO_3, bile, enzymes and the emptying of each compartment. Flow rates are regulated by specific computer-controlled peristaltic pumps. Anaerobic conditions can be simulated by purging air with nitrogen. A Teflon membrane with 2 mm holes is placed before the transfer pump between the gastric and the intestinal compartment to mimic the sieving effect of the pylorus in human, as described previously (Kong and Singh 2008).

The computer program was designed to accept parameters and data obtained from in vivo studies in animals or human volunteers, such as the quantity and duration of a meal, the pH curves for the stomach and small intestine, the secretion rates into the different compartments and the gastric and small intestine emptying rates. The system is controlled by software named StoRM for Stomach regulation and monitoring (Guillemin et al. 2010). To control the transit time of the chyme in each compartment, a power exponential equation for gastric and intestinal delivery is used $f = 2^{-(t/t1/2)\beta}$ where f represents the fraction of the chyme remaining in the stomach, t is the time of delivery, $t_{1/2}$ is the half time of delivery and β is the coefficient describing the shape of the curve, as described previously (Elashoff et al. 1982).

Fig. 8.1 Presentation of the gastro-intestinal dynamic digestion system

8.2   Validation of DIDGI® for the Digestion of Infant Formula

The DIDGI® system is a very recent one. Although several matrices (dairy, meat, fruits and vegetables, emulsions) have been submitted to digestion using the DIDGI® system, only data obtained on the digestion of infant formulas have been published so far (Ménard et al. 2014). In order to demonstrate that this system was physiologically-relevant, a comparison of the in vitro and in vivo digestion of an infant formula was performed. The in vivo trial was conducted on 18 piglets that were fed the infant formula for which the concentration in lipids and proteins was increased compared to a standard one, but the ratio lipids/proteins was kept constant. In parallel, in vitro gastro-intestinal digestion was performed on this enriched infant formula using the newly developed system and the extent of milk proteolysis was monitored and compared to the one obtained in vivo.

8.2.1   Protocol for the In Vitro Dynamic Digestion of Infant Formula Using the DIDGI® System

Infant formula (150 ml) was introduced into the gastric compartment within a pre-set period of time, i.e. 10 min. The pH values were controlled via the computer by secreting either 0.5 M HCl to decrease the pH in the stomach or 0.1 M Na_2CO_3 to neutralize the pH in the small intestine.

The dynamic digestive system was set up using parameters taken from the literature that are listed in Table 8.1. The pH curve in the stomach was obtained by combining data from different in vivo experiments on piglets (Moughan et al. 1991; Chiang et al. 2008; Bouzerzour et al. 2012) whereas intestinal pH was kept constant at 6.5. Other parameters like the transit time of the formula in the stomach (determination of $t_{1/2}$ and β) was obtained from an exhaustive review on human infant gastrointestinal physiological conditions (Bourlieu et al. 2014) and fixed at $t_{1/2} = 70$ min or 200 min and β = 1.23 or 2.2 for gastric or intestinal transit time, respectively. Volumes, flow rates of secretions, nature and quantity of enzymes in the different stages of the gastro-intestinal model have been described previously (Minekus et al. 1995; Blanquet et al. 2004; Bouzerzour et al. 2012) based on results of in vivo experiments. Digestive enzymes and bile were diluted in 150 mM NaCl, pH 6.5. After rehydration, all the digestive enzymes were kept on ice throughout the experiment in order to avoid autolysis. Digestion experiments were performed in triplicates. Samples were collected during the digestion in each compartment at 30, 60, 90, 120 and 210 min after ingestion. Before being frozen at −20 °C, phenyl-methanesulfonyl fluoride was added at 0.37 g/kg of digested sample in order to inhibit proteolysis.

Table 8.1 Parameters of in vitro gastro-intestinal conditions

Gastric conditions		
Volume ingested (ml)		150
Fasted volume (ml)		30
		Pepsin 625 U/mL diluted in NaCl 150 mM pH = 3
pH (acidification curve)		Y = −0.011*t + 5.4 [t: time after ingestion(min)]
Secretions		
	Pepsin flow	1,250 U/mL
		0.25 ml/min
	Lipase flow	60 U/mL
		0.25 mL/min
Transit time	$t_{1/2}$(min)	70 min
	β	1.23
Intestinal conditions		
Fasted volume		5 ml of bile solution (1 %)
		+5 ml of pancreatin solution (10 %)
pH		Maintained at 6.5
Secretions		
	Bile flow	1 %
		0.5 mL/min
	Pancreatin flow	10 %
		0.25 mL/min
Transit time	$t_{1/2}$(min)	200
	β	2.2

8.2.2   In Vivo Digestion of Infant Formula on Piglets

An in vivo study on piglets was conducted in parallel in accordance with the guide-lines formulated by the European Community for the use of experimental animals (EU Directive 2010/63/EU). Eighteen piglets [Pietrain × (Large White × Landrace)] were separated from their mothers after 2 days and fed for 26 days with the formula using an automatic milk feeder, as described previously (Bouzerzour et al. 2012). The experimental design was a complete block design with 1 × 3 factorial arrangement of 1 diet and 3 slaughter times after the last meal (30 min, 90 min and 210 min). The daily net energy ration of 1,450 kJ/body weight $(BW)^{0.75}$ was partitioned into ten meals automatically distributed during the day. Body weights were recorded weekly, and feeding schedules were adjusted accordingly. At the age of 28 days, piglets were allotted in three groups according to their slaughter times after the last meal: 30 min, 90 min and 210 min. They were slaughtered by electro-narcosis immediately followed by exsanguination. Immediately afterwards, the digestive tract was removed, dissected and samples were collected from the stomach to the ileum.

The total content of each segment was collected, dispersed and pH was measured. Sodium benzoate and phenylmethanesulfonyl fluoride (10 and 0.37 g/kg of content, respectively) were added to each digested sample in order to avoid further protein breakdown. All effluents were stored at −20 °C until further analyses.

8.2.3   Comparison In Vitro/In Vivo Data

The evolution of the volume emptied from the stomach into the small intestine, the volume remaining in the stomach and the total volume of gastric content (emptied + remaining) monitored during three independent digestion experiments is represented on Fig. 8.2. The rapid increase in the total volume during the first 10 min of digestion corresponds to the ingestion of the infant formula. Subsequently, its increase is due to the injection of gastric secretions.

The emptied volume follows the gastric transit described by the Elashoff's equation (Elashoff et al. 1982). Figure 8.2 also shows an excellent reproducibility between experiments since the curves of the replicates are similar in the gastric compartment. In the intestinal compartment, the reproducibility was also excellent (data not shown). Measuring all the volumes coming in and out of the two compartments allowed an assessment of the reliability of the system by comparing the values obtained by the system with those predicted by the software. The initial pH of the formula was 6.5. The pH regulation started only 5 min after the beginning of the experiment. Indeed, a minimum volume of c.a. 15 ml is needed for the pH electrode to give reliable measurements. Then pH followed the equation of the gastric

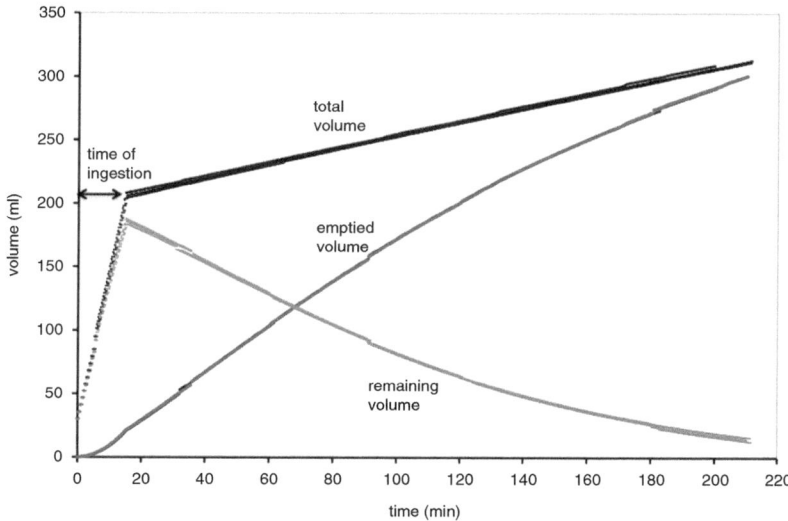

Fig. 8.2 Evolution of the volume emptied from the stomach into the small intestine, the volume remaining in the stomach and the total volume of gastric content (emptied + remaining) monitored during three independent in vitro digestion experiments

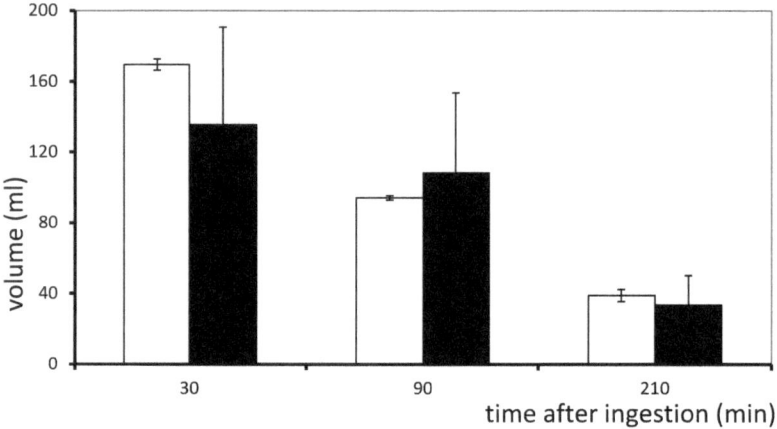

Fig. 8.3 Comparison of the volumes remaining in the gastric compartment between in vivo (*filled square*) and in vitro (*open square*) experiments

acidification curve. The decrease in pH throughout the experiment is due to the addition of hydrochloric acid to the gastric compartment. After 210 min of digestion, gastric pH reached 3.10; at this stage, only 40 mL were present in the stomach mainly consisting of simulated gastric fluid.

Volumes of the stomach content observed in vitro with the dynamic digestion system were compared to the ones observed in vivo in piglets (Fig. 8.3). No significant differences were observed 30, 90 and 210 min after ingestion confirming that the parameters chosen for mimicking the gastric transit of infant formula in vitro were physiologically relevant.

Evolution of caseins (Fig. 8.4a) and β-lactoglobulin (Fig. 8.4b) throughout in vitro and in vivo digestion, as determined by ELISA, was compared. Results showed that the kinetics of hydrolysis of both proteins during in vitro and in vivo digestion were similar. The proportion of immunoreactive caseins appeared not to be significantly different between both experiments for samples collected in the stomach as well as in the small intestine after 30, 90 and 210 min of digestion (Fig. 8.4a). Similarly, the percentage of immunoreactive β-lactoglobulin showed no significant differences for samples collected in vivo and in vitro in the stomach after 30, 90 and 210 min. However, the percentage of immunoreactive β-lactoglobulin in the small intestine was significantly higher in vitro than in vivo. The correlation coefficient, between in vitro and in vivo ELISA determination for caseins and β-lactoglobulin was 0.987 ($p < 0.001$), proving a good agreement between in vitro and in vivo proteolysis during digestion.

8.3   Advantages, Disadvantages and Limitations

Some advantages and limitations of the DIDGI® are provided in Table 8.2.

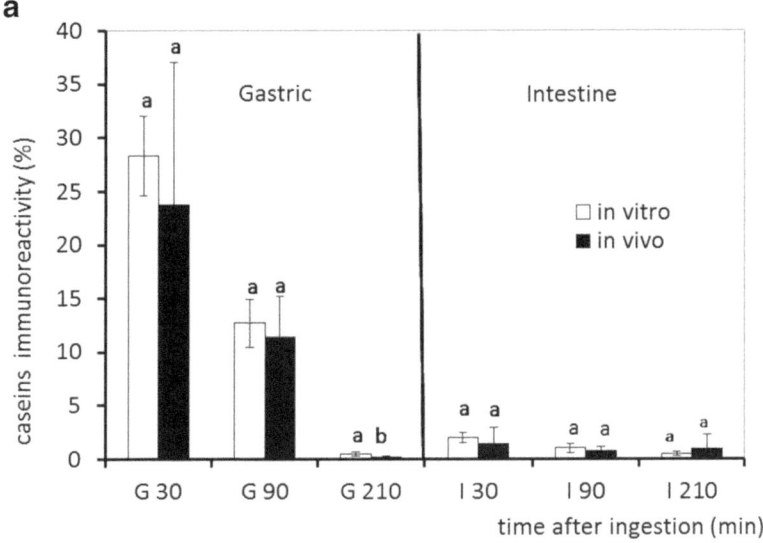

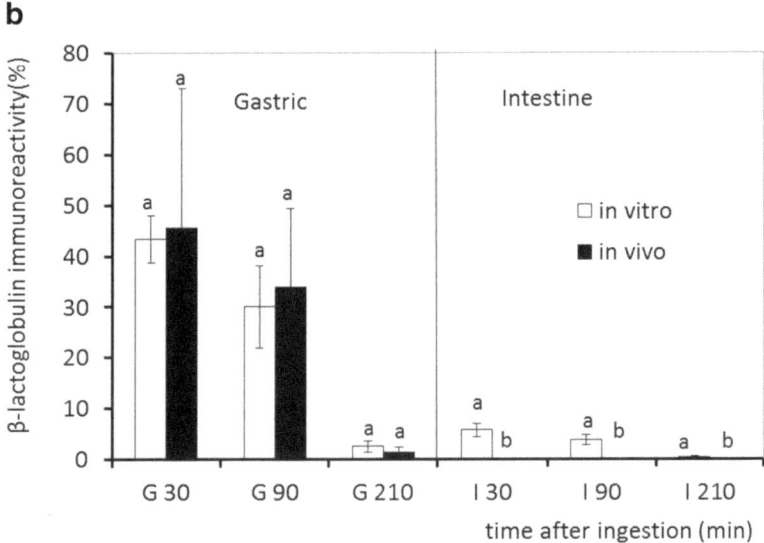

Fig. 8.4 Casein (**a**) and β-lactoglobulin (**b**) concentrations determined by ELISA

8.4   Conclusion and Prospects

We present here the development of a simple and new dynamic digestion system, DIDGI® that has been validated against in vivo (porcine) data for the digestion of infant formula. The system is currently being applied to study (1) the digestion of

Table 8.2 Advantages and limitations of DIDGI®

Advantages	Limitations
Capacity: Full meals up to 200 mL	No simulation of nutrient absorption
Temporal simulation: Real-time digestion and monitoring of pH and temperature	No simulation of the biomechanical aspects of food digestion (stomach contractions, peristalsis…)
Possibility of modifying the atmosphere and work in controlled gas conditions	Bioreactors do not mimic the anatomy of the gut compartments
Meals: any masticated food/drink matrix	Satiety: no in vivo satiety signals controlling rate of digestion
Transparent: visual observations possible during gastric processing	Orientation: Vertical alignment of main body and antrum
Model validated against in vivo data	

human milk, dairy gels and emulsions, cheese, and (2) the survival of food microorganisms in the gastro-intestinal tract. The results obtained so far are quite promising. Some improvements have recently been applied to the system: it now has three compartments (stomach, duodenum and jejunum + ileum) instead of two; dialysis membranes will soon be added to the system to mimic absorption; the pump allowing the bolus to enter the stomach has been modified to allow the introduction of solid matrices (meat, bread, fruits and vegetables).

References

Barbé F, Ménard O, Le Gouar Y, Buffière C, Famelart MH, Laroche B, Le Feunteun S, Dupont D, Rémond D (2013) The heat treatment and the gelation are strong determinants of the kinetics of milk proteins digestion and of the peripheral availability of amino acids. Food Chem 136(3–4):1203–1212

Barbé F, Ménard O, Le Gouar Y, Buffière C, Famelart M-H, Laroche B, Le Feunteun S, Rémond D, Dupont D (2014) Acid and rennet gels exhibit strong differences in the kinetics of milk protein digestion and amino acid bioavailability. Food Chem 143:1–8

Blanquet S, Zeijdner E, Beyssac E, Meunier JP, Denis S, Havenaar R, Alric M (2004) A dynamic artificial gastrointestinal system for studying the behavior of orally administered drug dosage forms under various physiological conditions. Pharm Res 21(4):585–591

Bourlieu C, Menard O, Bouzerzour K, Mandalari G, Macierzanka A, Mackie A, Dupont D (2014) Specificity of infant digestive conditions: some clues for developing relevant in vitro models. Crit Rev Food Sci Nutr 54:1427–1457

Bouzerzour K, Morgan F, Cuinet I, Bonhomme C, Jardin J, Le Huerou-Luron I, Dupont D (2012) In vivo digestion of infant formula in piglets: protein digestion kinetics and release of bioactive peptides. Br J Nutr 108(12):1–10. doi:10.1017/S000711451200027X

Chiang C, Croom J, Chuang S, Chiou P, Yu B (2008) Development of a dynamic system simulating pig gastric digestion. Asian-Australasian J Anim Sci 21(10):1522–1528

Elashoff JD, Reedy TJ, Meyer JH (1982) Analysis of gastric-emptying data. Gastroenterology 83(6):1306–1312

Guillemin H, Perret B, Picque D, Menard O, Cattenoz T (2010) Logiciel StoRM - stomach and duodenum regulation and monitoring. IDDNFR001230009000RP201000031235:290

Kong F, Singh RP (2008) Disintegration of solid foods in human stomach. J Food Sci 73(5):R67–R80. doi:10.1111/j.1750-3841.2008.00766.x

Le Feunteun S, Barbé F, Rémond D, Ménard O, Le Gouar Y, Dupont D, Laroche B (2014) Impact of the dairy matrix structure on milk protein digestion kinetics: mechanistic modelling based on mini-pig in vivo data. Food Bioprocess Technol 7(4):1099–1113

Ménard O, Cattenoz T, Guillemin H, Souchon I, Deglaire A, Dupont D, Picque D (2014) Validation of a new in vitro dynamic system to simulate infant digestion. Food Chem 145:1039–1045

Minekus M, Marteau P, Havenaar R, Huisintveld JHJ (1995) A multicompartmental dynamic computer-controlled model simulating the stomach and small-intestine. Altern Lab Anim 23(2):197–209

Moughan PJ, Cranwell PD, Smith WC (1991) An evaluation with piglets of bovine-milk, hydrolyzed bovine-milk, and isolated soybean proteins included in infant milk formulas.2. Stomach-emptying rate and the postprandial change in gastric Ph and milk-clotting enzyme-activity. J Pediatr Gastroenterol Nutr 12(2):253–259

Part II
General Introduction to Cells, Cell Lines and Cell Culture

Introduction

Cell and tissue culture technology holds a central position in modern biomedical research. Cells in culture are useful model systems for studying normal physiological and biochemical processes in cells and can even be used for diagnostic purposes. The main advantage of using cell culture systems is the consistency and reproducibility obtained.

In most situations cells or tissues have to be cultivated for longer periods to obtain sufficient numbers of cells for analysis. Alternatively, the same cells can be continuously propagated in the laboratory for use in repeated experiments. Normal cells (primary cells) usually divide only a limited number of times before entering into senescence. However, cells can be immortalized through transformation, a process that can occur spontaneously or by treating the cells with chemicals or viruses. Long-term culture of cells requires well established aseptic techniques to avoid contamination with bacteria, yeasts, molds or viruses. Contamination may cause erroneous results and eventually the loss of valuable cell material.

Usually, there are two different ways of growing cells in culture, either as cells attaching to the surface of the culture vessel, adherent cells, or as cells growing in suspension. Most cells from vertebrates, either in the form of primary cells or cell lines, adhere to the plastic surface of the culture vessel. Some cells adhere rather loosely and may require that the plastic of the culture tray is coated with a 1 % solution of gelatin, collagen or fibronectin before use. Most cells and cell lines of hemopoietic origin, typically lymphocytes, grow free-floating in the culture medium.

Salt Solutions

In vitro, tissue and cells should only be exposed to isotonic solutions at physiological pH, either in the form of *balanced salt solutions* or *culture media*. The simplest salt solutions are saline (0.9 % NaCl) or phosphate-buffered saline (PBS, 10–15 mM phosphate buffer in saline). Many balanced salt solutions (typically Earle's balanced salt solution (EBSS), Hank's balanced salt solution (HBSS), Dulbecco's phosphate-buffered saline (D-PBS) or similar) have been developed to provide optimal short term support for different cell types in vitro and for different culturing conditions.

Culture Media

Most culture media are based on balanced salt solutions with a variable number of additions in the form of amino acids, vitamins, trace minerals, a pH indicator, and eventually antibiotics. Such media are called *basal media*. Typical examples are RPMI 1640 (Roswell Park Memorial Institute 1640) and DMEM (Dulbecco's Modified Eagle Medium), eventually with the Ham's F12 nutrient mixture (DMEM/ F12). In addition to the basal media special medium formulations that require less serum addition, so called Reduced Serum Media, alternatively Serum-Free Media (SFM), are available for special cell types and special applications.

The culture medium is absolutely crucial for successful cell culture experiments. In general, recommendations are available for each particular cell line or cell type. However, different media may satisfy the required conditions for a particular cell type. Thus, personal experience could be valuable when choosing medium. Such personal experience may also include observed beneficial effects due to the addition of non-essential amino acids, pyruvate, reducing agents etc. It is extremely important to know that any addition to ready-made media requires components that have been tested and certified for cell culture work. Even chemicals of *pro analysi* quality may contain traces of toxic compounds that can be detrimental to cells in culture.

Medium Quality

More important than medium type is perhaps the overall "quality" of the medium. Most researchers today prefer the simplicity of using sterile, ready-to-use media. However, many media contain storage-sensitive components. Glutamine for example tends to decompose upon storage creating pyrrolidone carboxylic acid and ammonia. At 4 °C approximately 10 % of the glutamine will decompose in 9 days. Thus, to be on the safe side many researchers add fresh glutamine to their medium before use. If glutamine addition is repeatedly carried out, the build-up of ammonia may become toxic to the cells affecting protein glycosylation patterns and even cell viability (Heeneman et al. 1993). As an alternative to glutamine, GlutaMAX™-based media may be used. GlutaMAX™ is a stable dipeptide of alanyl-glutamine serving the same functions as glutamine in the culture medium (Roth et al. 1988). Generally, media should be as fresh as possible. Storage in a cold room for more than a few months may provide suboptimal culture conditions. This will, however, depend on the cell type or cell line under study. If working with demanding cells, one can benefit from preparing fresh medium from powder. In this situation water quality is absolutely essential. Distilled or deionized water is usually not of high enough quality for medium preparation. Double or triple glass-distilled water or similar should be used.

pH and CO_2

The optimum pH for most mammalian cells is pH 7.2–7.4, and there is very little variability. This is usually obtained with a sodium bicarbonate buffer that requires a delicate balance of dissolved carbon dioxide (CO_2) and bicarbonate (HCO_3^-). Changes in atmospheric CO_2 may affect the pH of the medium. Therefore, a supply of exogenous CO_2 (5–7 % CO_2 in air) is necessary when using media based on CO_2/bicarbonate as buffering system. Alternatively, the synthetic buffer HEPES (4-(2-hydroxyethyl)-1-piperazineethanesulfonic acid) can be used. HEPES allows experiments to be carried out when a CO_2 incubator is not available. However, HEPES is expensive, and toxic effects of HEPES on some cell lines have been reported.

Serum Addition

Although chemically defined, serum-free media are available, most cell culture media require the addition of serum to provide essential transport proteins, fatty acids, growth factors and hormones. However, serum from different animal species, and even different batches of serum from the same species, may have very different properties in supporting cell growth in vitro. Depending on the cells in question, serum from horse, swine, cattle, newborn calves, calf fetuses and humans can be used. Usually, fetal bovine serum/fetal calf serum (FBS/FCS) is preferred.

The properties of FBS may vary from vendor to vendor and from batch to batch. Large cell culture laboratories often test a number of different batches and screen for select properties depending on the type of cells under study. One particular batch may provide good growth support, while another batch from the same supplier may be inferior in comparison. Thus, to be able to work reproducibly with cell cultures over time will require access to larger volumes of the same serum batch, and selection of new serum batches must be based on experiences with the prior or reference batch used in the laboratory. Serum is usually added to the medium in concentrations between 5 and 20 %. People tend to believe that medium quality improves with increasing serum concentrations. However, 20–25 % of a low quality serum may be inferior in performance to 5–10 % of a good quality serum.

One of the main reasons for choosing fetal serum is the low content of antibodies present. Antibodies together with serum complement could have negative effects on cell viability. Thus, routinely FBS is still inactivated, i.e. incubated at 56 °C for

30 min to destroy heat-labile complement proteins. However, the usefulness of this procedure can be questioned. First, complement protein levels in FBS are only a few percent of the levels found in adult serum. Secondly, heat inactivation may destroy other valuable serum components, as well. Today, it can be considered unnecessary to heat-inactivate FBS from established suppliers.

Most cell culture work today is carried out with continuously growing, transformed cell lines. Primary cells have limited life span. Based on experience with primary cells with proliferating potential, like multipotent stromal cells (MSC) from bone marrow or adipose tissue, it is possible to cultivate primary cells for in excess of 40 population doublings before they reach senescence and die. Among immune cells, T lymphocytes can also be expanded in vitro by stimulation and periodic restimulation with antigen or antibodies against the T-cell receptor/CD3 complex in the presence of recombinant cytokines like interleukin 2 (IL-2). This is currently not possible for B lymphocytes since activated B lymphocytes differentiate into short-lived antibody-producing plasma cells that do not express antigen receptors, and cannot be restimulated. Monocytes, macrophages and dendritic cells do not proliferate in vitro. Epithelial cells have also been very difficult to establish in culture as primary cells. So, for most purposes, we depend on using transformed cell lines that have been established from tumors of different tissues. This means that such cell lines are not necessarily clonal in their origin. Transformed cell lines can be very heterogeneous containing cells with different properties. During propagation, culture conditions can be expected to influence and select for cells with special properties. Thus, during long-term cultivation a particular cell line changes in composition and properties. It is absolutely recommended that when a cell line arrives in a laboratory, its passage number should be known. Then the cells should be expanded and frozen to create a stock cell bank so that new cells can be thawed and brought into culture to preserve the properties of the original cell line. This is also one of the reasons why comparison of results with the "same" cell line from different laboratories often turns out to be very difficult. Cells or cell lines subject to different cultivation conditions, e.g. different media, different serum source, differences in passage number, differences in density before subcultivation, will eventually acquire and express different properties and respond differently to the same stimuli in in vitro experiments.

The Cell Culture Hood

Aseptic working conditions are absolutely essential for a cell culture laboratory. A separate cell culture room is preferred. However, a designated area for cell culture work as part of a large laboratory can also be used. A cell culture hood represents an aseptic work area allowing protection of both cells and laboratory personnel. Class II cell culture hoods will provide the necessary conditions for long term cell culture work. Today's class II biosafety cabinets efficiently protect the working environment from dust and microbial contaminants by means of a HEPA-filtered vertical laminar air flow. All remedies like medium bottles, trays, culture flasks, tube racks, pipettor etc. should be kept in the back or at the side of the cabinet leaving a free working space. Remember that all equipment found inside the cabinet is potentially contaminated, and it is of utmost importance to avoid disturbing the laminated airflow above the open media bottles and culture trays with any contaminated material. As a rule of thumb, never work with more than one cell line at the time inside the cell culture hood. Likewise, never use the same medium bottle to feed different cell lines!

Cell Culture Terminology (Fig. 1)

Most cells and cell lines cultivated in vitro adheres to the tissue culture plastic, and this adherence is often essential for maintaining the cells in a healthy, viable state. A culture of primary cells can be grown to confluence, e.g. in a continuous monolayer, in any tissue culture vessel like a petri dish, a tissue culture flask or plates with 6-, 12- or 24 wells. When reaching confluence they have to be *subcultured* or *passaged*. If not, continued culture of confluent cells may result in reduced mitotic activity, selection of variant cells and even cell death. Subcultivation of cells requires detachment of the cells from the surface of the tissue culture tray, usually by *trypsinization* or scraping with a mechanical device like *a rubber policeman* (see Fig. 2).

The cell suspension is then subdivided and seeded into fresh medium. Such secondary cultures are fed at regular intervals, and may eventually be subcultured in the same way to produce tertiary cultures. The time between feeding and subculturing of the cells may vary from cell type to cell type depending on growth rate.

General Cell Culture Protocols

Trypsinisation and Subculturing of Cells

Remove gently all medium from the culture to be passaged with a sterile Pasteur pipette or by suction from a vacuum pump. The adherent cell layer should then be rinsed one to two times with HBSS without Ca^{2+}/Mg^{2+} to get rid of any FBS that may interfere with the trypsinization process. Add just enough trypsin/EDTA solution to cover the culture plate/tray. Keep at 37 °C for 1–2 min. Tap the culture tray against the working plate of the laminated air flow cabinet, and inspect the cells in an inverted microscope to ascertain that the cells are rounded up and detached from the surface. It may be necessary to increase trypsinization up to 15 min. However, prolonged incubation with trypsin may reduce cell viability.

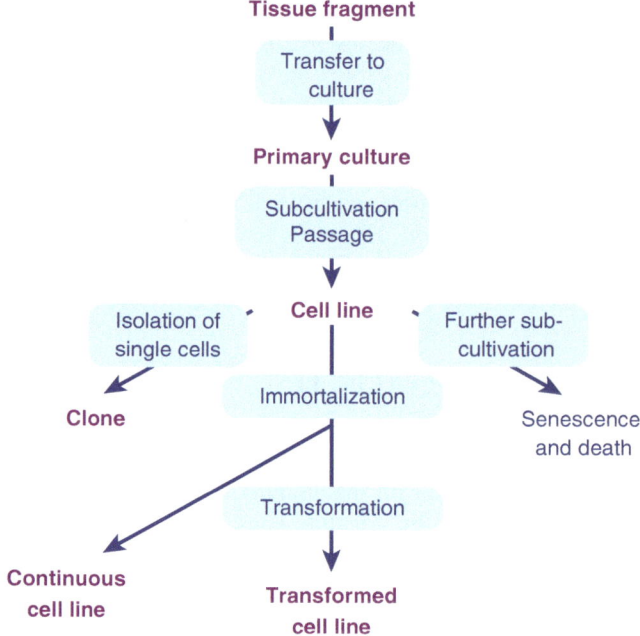

Fig. 1 The figure describes the relationship and differences between primary cells, transformed and immortalized cell lines and clones

Add a sufficient volume of complete medium with FBS and draw the cell suspension into a Pasteur pipet. Rinse the plate carefully with the cell suspension to dislodge any residual cells. When all cells are detached, add more complete medium with FBS to inhibit residual trypsin activity that may damage the cells.

Count the cells in a haemocytometer, alternatively in a Coulter counter or similar device. Dilute the cells in complete medium and seed them at a suitable density in the tissue culture tray of choice.

If necessary, feed subconfluent cell layers after 3–4 days by removing most of the spent medium and replacing it with fresh new medium at 37 °C. Follow the same protocol for subsequent subcultivations. Notify the number of passages for each culture.

Passaging of Cells in Suspension Culture

Although most primary cells and cell lines are plastic adherent, some cell types are non-adherent. Typically, primary immune cells like B and T lymphocytes and transformed lymphoid cells, NK cells and granulocytes grow in suspension. It is easier to passage cells growing in suspension than adherent cells since no trypsinization or detachment from the tissue culture plastic is necessary. Cells in suspension culture

Fig. 2 The drawing shows a rubber policeman (*left*) and a typical cell scraper to collect adherent cells from the surface of the tissue culture plastic ware

do not grow to confluence, and different cells may tolerate growing at different densities. Thus, optimal cell concentration has to be determined in each instance. Also, seeding densities may vary. Some cells are able to grow at very low densities while others need support from accompanying cells and have to be reseeded at higher densities. As a rule of thumb, when the cells reach a density of around 2×10^6/ml, they should be split back to $2-3 \times 10^5$ cells/ml. An optimal cell concentration will usually be around 10^6/ml.

If you are going to expand the cells to obtain large cell numbers, remove the flask from the incubator, swirl the flask gently to resuspend the cells and remove a small sample to measure cell density by counting in a hemocytometer or a Coulter counter. Depending on cell concentration aseptically distribute the contents into new flasks allowing a new cell concentration of $2-3 \times 10^5$ cells/ml.

Add fresh medium and ascertain that the height of the medium never exceeds 2 cm above the bottom of the flask or tray. This is important to allow proper buffering of the medium and physiological pH for the cells sedimenting to the bottom of the flask.

If cells are growing slowly, they might be fed by aseptically removing 1/3 of the old medium to be replenished by fresh medium.

Freezing Cells

Both adherent and suspension culture cells can be frozen and stored in liquid nitrogen for extended periods of time. In general, the same protocol can be used for both cell types.

Harvest the cells and spin them down in a table top centrifuge. Resuspend the cells carefully with a Pasteur pipette in cold medium containing 10 % FBS to a concentration of around 5×10^6 cells/ml. Keep tube or bottle with cells on ice in the sterile hood. Add a similar volume of cold freezing medium dropwise to the tube with cells while gently swirling the tube.

Distribute the cell suspension in 1 ml aliquots in 2 ml cryovials. Tighten caps properly. Place vials in a styrofoam box and place in a −80 °C freezer overnight before transfer to liquid nitrogen storage tank.

The Thawing and Recovery of Cells

Cells can be stored in liquid N_2, or in the gas phase over the liquid N_2 for long periods of time. However, valuable cells should be thawed and brought back into culture once a year to ascertain viability and function. Thawing is as important as the freezing process to maintain cell viability, and thawing should be carried out according to a rigorous protocol.

Frozen vials with cells are placed on ice directly from the liquid N_2 storage tank.

The vials should be immersed in a water bath at 37 °C and thawed leaving a small lump of ice in the vial before transfer of the vial to melting ice again. When fully thawed, transfer the contents of the vial into a tube and add slowly cold medium to dilute the DMSO in the freezing medium. It is very important that the cell suspension is kept cold during this procedure.

Spin the cells carefully down in a table top centrifuge, remove the DMSO containing medium and add a small volume of complete medium to resuspend the cells.

Take a small sample to monitor cell viability by trypan blue exclusion and light microscopy, or similar viability test (see Sect. 9.5). If cell viability is below 70 % it may turn out difficult to obtain proper cell growth. Reduced cell viability is most likely due to improper handling of cells either before freezing, during freezing or thawing of the cells.

Dilute with complete medium to the desired cell concentration, and add the cell suspension to a suitable culture vessel. Place in incubator. Monitor cell morphology and cell growth during the following days.

Cell Viability Testing

Assessment of cell viability is fundamental in all cell culture work. Cell viability will provide information about culture conditions, in general, and may also represent the final read-out system for a cytotoxicity experiment. A large number of viability assays are reported in the literature, ranging from the simple trypan blue exclusion test to complex analysis of individual cells with Raman spectroscopy.

Cell viability assays will often overlap with cell proliferation assays. They are divided into assays that monitors viability at the single cell level and those that analyses viability at the population level. The simplest and cheapest viability test exploits the vital stain trypan blue. Trypan blue is excluded from cells and tissues with intact membranes, and the staining of dead cells is easy to detect in a standard light microscope.

Prepare a 0.8 mM trypan blue solution in phosphate-buffered saline (PBS). Mix cells and trypan blue solution in a 1:1 ratio. Inspect and count dead and alive cells under a light microscope in a haemocytometer. Dead cells will stain blue due to trypan blue uptake, while live cells appear transparent. Cells should not be exposed to trypan blue solution for more than 20 min. Prolonged incubation will increase cell death and reduce viability.

A variety of fluorescent dyes can also be used to monitor cell viability, typical in the combination with flow cytometry. Propidium iodide and 7-actinomycin D are commonly used for this purpose.

Assessing cell viability at the population level has become very popular. These assays are typically carried out by using different tetrazolium compounds. Among the most popular are MTT ((3-(4,5-dimethylthiazol-2-yl)-2, 5-diphenyltetrazolium bromide), MTS, XTT and WST-1. Among these MTT is positively charged and readily penetrates viable cells, while the others are negatively charged and are excluded from viable cells. The MTT assay is perhaps the most popular to assess cell viability and proliferation in a population of cells. It was the first assay to be developed for high trough put screening in a 96-well format. Viable cells convert MTT into a purple colored formazan product. When cells die, they lose ability to convert MTT into formazan. The mechanism behind this process is not well understood, although many publications suggest that the MTT assay reflects changes in mitochondrial activity. Commercial kits containing MTT and a solubilization reagent can be obtained from several vendors like Sigma Aldrich, Promega and Millipore, among others.

Contamination of Cell Cultures

Cell cultures may become contaminated by bacteria, molds, viruses and mycoplasmas. In general, contamination by bacteria and molds are easy to detect by visual inspection. Mycoplasmas, however, are perhaps the most severe and difficult cell culture contaminant to observe and handle. Since its discovery in 1956 (Armstrong et al. 2010), the presence of mycoplasmas in cell cultures has been a great challenge, both in research as well as in industrial laboratories and production facilities for cell-derived biological and pharmaceutical products. Although the risks and consequences of mycoplasma infection have been known for decades, recent studies (Drexler and Uphoff 2002) indicate that there has been no decline in infection rates. Mycoplasmas are small (0.1–0.6 μm) bacteria that lack cell wall. Thus, they are difficult to observe by conventional light microscopy. Furthermore, they are unaffected by antibiotics that target cell wall biosynthesis like penicillin and β-lactams, so they will continue to grow in infected cell cultures despite the presence of standard antibiotics and antimycotics. Mycoplasma species have been found associated with a range of hosts, including humans and other mammals, birds, reptiles, insects and plants. Despite the fact that more than 100 mycoplasma species have been described, most mycoplasmas that contaminate cell culture laboratories belong to only six species of bovine, swine and human origin. Bovine and swine mycoplasmas typically derive from contaminated sera or other animal-derived products used in the cell culture laboratory. Currently, many infections are transfered from already infected cultures within the same laboratory due to inadequate aseptic and cell culture techniques or from equipment that has been contaminated through prior handling of infected cells. However, laboratory personnel also represent an important source of mycoplasma infections, and mycoplasma species from humans, mainly of oral origin, are responsible for 40–80 % of mycoplasma infections in cell cultures. In total it is estimated that 15–35 % of all cultures based on continuous cell lines are mycoplasma infected (Barile et al. 1973), while contaminated cultures based on primary cells are seldom found.

Their flexibility in shape due to the lack of cell wall, allows most mycoplasmas to pass through 0.2 µm filters traditionally used for sterilization of media and medium ingredients. For the same reason they are difficult to observe in cell cultures. Moreover, despite the richness of standard culture media, mycoplasmas grow slowly, a fact that contributes to the problems with detecting them, visually or by change in turbidity. Although we tend to consider mycoplasmas mainly as a problem in cell culture work, it should be remembered that the mycoplasma family also includes a number of human pathogens.

How will mycoplasma infection affect the cultivated cells? Different mycoplasma species will introduce different problems, but in general, mycoplasma infections tend to influence a number of cellular processes leading to decreased growth rate due to inhibition of protein-, DNA- and RNA synthesis. Additionally, they introduce changes in gene expression that can be measured as down-regulation of cytokine and growth factor secretion, expression of receptors, intracellular signaling molecules, ion channels etc.

How to avoid contamination by mycoplasmas? A few major guidelines should be followed: (1) Strict aseptic techniques are an absolute requirement. Minimize talking, and do not practice mouth pipetting when working with cell cultures. Never pour medium between bottles and flasks, and avoid crowding in the laminar flow hood. Unnecessary equipment in the hood will contribute to turbulence in the laminar airflow thus increasing the risk of contaminating flasks and culture plates. Avoid unnecessary traffic in the vicinity of the sterile hood. (2) All surfaces in laminar flow hoods, incubators and water baths should be regularly cleaned and disinfected (70 % ethanol or isopropanol). (3) Try to avoid unnecessary and excessive use of antibiotics. Routine work may preferably be carried out without addition of antibiotics to the medium. Use of antibiotics will camouflage poor aseptic techniques, and will contribute to worsen the problem. (4) When new cell lines arrive in the laboratory, they should be cultivated separated from other cell cultures, rigorously tested and found to be mycoplasma free before being admitted to the cell culture laboratory. (5) Establish a routine for frequent testing of the cultures. This is an absolute requirement for a responsible scientist. If a culture or cell line is found to be infected, cells, culture trays, medium bottles and any other equipment that has been in contact with the contaminated culture should be destroyed by autoclaving.

There are several ways to detect mycoplasma infection in cell cultures, but today, most people rely on PCR technique which is fast and sensitive. Results can be obtained in hours. Several vendors provide kits with specially designed primers that allow selective amplification of mycoplasma DNA for PCR-detection with a sensitivity that is close to the direct culture approach. Thus, for routine purposes PCR represents the method of choice.

Mycoplasma infection of cell lines and particularly clones of valuable primary cells can be a disaster. It may be possible to "cure" the cells with antibiotics. Typically, antibiotics like ciprofloxacin, minocyclin or a combination of tiamulin and minocyclin, have been demonstrated to be effective. However, toxic effects causing cell death is not uncommon. In general, cleaning up a contaminated culture should only be considered as an alternative for absolutely irreplaceable cultures provided that the source of the contamination has been identified and removed from the laboratory.

References

Armstrong SE, Mariano JA, Lundin DJ (2010) The scope of mycoplasma contamination within the biopharmaceutical industry. Biologicals 38:211–213

Barile MF, Hopps HE et al (1973) The identification and sources of mycoplasmas isolated from contaminated cell cultures. Ann N Y Acad Sci 225:251–264

Drexler HG, Uphoff CC (2002) Mycoplama contamination of cell cultures: Incidence, sources, effects, detection, elimination, prevention. Cytotechnology 39:75–90

Heeneman S, Deutz NE, Buurman WA (1993) The concentrations of glutamine and ammonia in commercially available cell culture media. J Immunol Methods 166:85–91

Roth E, Ollenschlager G, Hamilton G, Simmel A, Langer K, Fekl W, Jakesz R (1988) Influence of two glutamine-containing dipeptides on growth of mammalian cells. In Vitro Cell Dev Biol 24:696–698

Chapter 9
Epithelial Cell Models; General Introduction

Tor Lea

Abstract An extremely important feature of the intestinal epithelium is its function as a physical barrier between the environment and our bodies' internal milieu. At the same time it has to allow for uptake of important nutrients. At least four different transport mechanisms exist that allow selective uptake and transport of macromolecules across the epithelial cell layer, i.e. paracellular transport, passive diffusion of molecules from the apical to the basolateral side, vesicle-mediated transcytosis and carrier-mediated uptake and diffusion through the epithelial cell layer. Each of these transport mechanisms depends on the physicochemical properties of the compound, its ability to interact with the plasma membrane, its molecular weight and size, stability and charge distribution. In vivo, parameters not directly associated with the molecule in question will influence uptake and transepithelial transport. Intestinal motility, interactions with other molecules from the diet and the digestive process like bile salts and enzymes, and solubility in the mucus layer will affect the absorption process. Thus, in vitro models for studying absorption through the intestinal epithelium have several limitations. Still, they are considered useful model systems for such purposes. Similarly, effects of bioactive molecules on the epithelium can be studied by measuring barrier function and effects on transport processes.

Keywords Polarized epithelium • barrier function • tight junctions • transepithelial transport • paracellular transport • transepithelial electrical resistance (TEER) • permeability coefficient

The one cell-layer thick intestinal epithelium has two challenging and opposing missions. The first is to represent a physical barrier between the contents of the gut lumen and the rest of our body. The second is to ascertain efficient absorption of essential nutrients from the gut lumen and produce mucus with protective properties, anti-microbial peptides that affect microbiota composition, and cytokines with

T. Lea (✉)
Department of Chemistry, Biotechnology and Food Science,
Norwegian University of Life Sciences, Ås, Norway
e-mail: tor.lea@nmbu.no

K. Verhoeckx et al. (eds.), *The Impact of Food Bio-Actives on Gut Health*,
DOI 10.1007/978-3-319-16104-4_9

both protective and immune-regulatory properties. The small and large intestine are differently organized, have different functional properties and exhibit differences regarding resistance to injury and disease susceptibility.

The small intestinal epithelium is organized into numerous units of crypts and villi. The villi project into the lumen to maximize nutrient breakdown and absorption. It has been estimated that each crypt domain comprises approximately 250 cells, and an equivalent number of cells are generated every day. The epithelial cell layer includes four major cell types that all derive from adult stem cells located at the crypt bottoms: (1) absorptive enterocytes, (2) goblet cells responsible for mucus secretion, (3) enteroendocrine cells that secrete important hormones, and (4) Paneth cells that secrete antimicrobial peptides. In addition, rare and characteristic M-cells (microfold cells) are found in the epithelium overlaying the organized lymphoid tissue, and Peyers patches, in the small intestine. For various purposes, the intestinal epithelium has a very high turnover rate. In the small intestine, apoptotic absorptive enterocytes are continuously sloughed off into the lumen completely renewing the epithelium every 5–7 days.

An extremely important feature of the intestinal epithelium is its function as an efficient barrier between the environment and our bodies' internal milieu. Firstly, the single layer of columnar epithelial cells acts as a mechanical barrier, and, in addition, throughout the whole digestive tract, the epithelium is covered with a highly viscous mucus layer that both trap antimicrobial peptides and neutralizing secretory IgA antibodies. In addition, the mucus layer reduces the direct contact between the luminal microbiota and the epithelial cell layer. The mucus is made up of mucins, i.e. proteoglycans, among which MUC2 dominates in the intestine. There are two types of mucus organization in the intestine. The small intestine has a single mucus layer, while the colon has a two-layered system with an outer and an inner layer (reviewed by Johansson et al. 2013). Despite the high load of microbiota in the colon, the inner mucus layer is more or less impermeable to the lumenal bacteria.

The epithelial cell layer is selectively permeable to bacterial metabolites and digested nutrients allowing regulated transport of soluble molecules through the paracellular space between the epithelial cells. Paracellular transport is controlled by intercellular tight junction (TJ) structures. Disruption of the intestinal TJ barrier, followed by permeation of lumenal noxious molecules, induces a perturbation of the mucosal immune system and inflammation, and can act as a trigger for the development of intestinal and systemic diseases. Over the past decade, there has been increasing recognition of an association between disrupted intestinal barrier function and the development of autoimmune and inflammatory diseases. Recently, although experiments were carried out in *Drosophila*, barrier function turned out to be a better predictor for life expectancy than nominal age and any other clinical parameter tested for Rera et al. (2012).

The protein networks connecting epithelial cells form three adhesive complexes: desmosomes, adherens junctions, and tight junctions (see Fig. 9.1). These complexes consist of transmembrane proteins that interact extracellularly with adjacent cells and intracellularly with adaptor proteins that link to the cytoskeleton. TJ barrier disruption and increased paracellular permeability, followed by permeation of lumenal proinflammatory molecules, can induce activation of the mucosal immune system, resulting

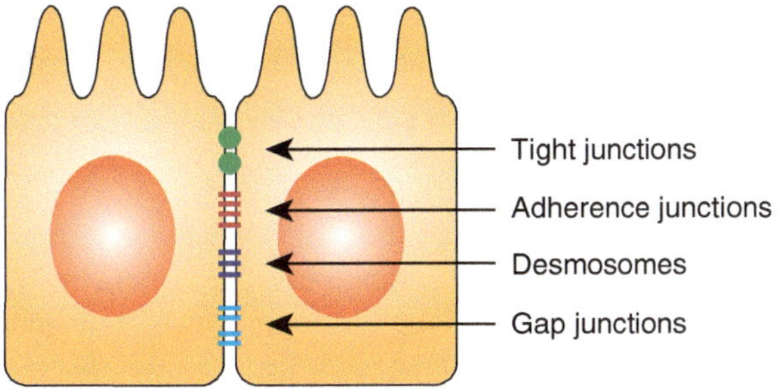

Fig. 9.1 The figure shows a schematic drawing of polarized epithelial cells with different types of intercellular contacts being essential for maintenance of barrier function and communication between neighboring cells

Tight junctions

Adherence junctions

Desmosomes

Gap junctions

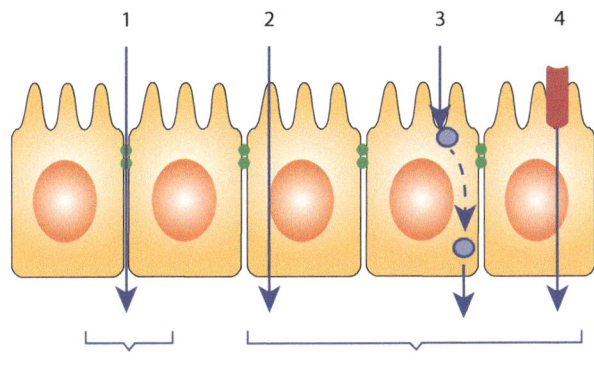

Fig. 9.2 The figure illustrates the different modes of absorption and transport through the intestinal epithelium: (1) paracellular transport, (2) passive diffusion of molecules from the apical to the basolateral side, (3) vesicle-mediated transcytosis and (4) carrier-mediated uptake and diffusion through the epithelial cell layer

Paracellular transport   Transcellular transport

in inflammation and tissue damage. Thus reduced barrier function may have far reaching consequences, not only for gut health but also systemic health.

As mentioned above, the intestinal epithelium is not only a mechanical barrier that prevents entry of molecules into circulation. At least four different transport mechanisms exist that allow selective uptake and transport of macromolecules across the epithelial cell layer (Fig. 9.2). Transport of molecules through the epithelium by paracellular transport is regulated by complex intracellular processes finely tuning the permeability of the tight junction complex (1 in Fig. 9.2). Some nutrients or microbial metabolites have properties allowing them to diffuse passively through the enterocyte plasma membrane and through the cell to be released at the basolateral side (2 in Fig. 9.2). Endocytosis (eventually via membrane receptors) allows vesicle-mediated uptake, transport and release at the basolateral side in a process called transcytosis (3 and 4 in Fig. 9.2). Various membrane associated carrier systems may contribute in internalizing luminal molecules allowing further diffusion through the cell cytoplasm and release at the basolateral side.

Each of these transport mechanisms depends on the physicochemical properties of the compound, its ability to interact with and pass the plasma membrane, its molecular weight and size, stability and charge distribution. In vivo parameters not directly associated with the molecule in question will influence uptake and transepithelial transport. Typically, intestinal motility, interactions with other molecules from the diet and the digestive process like bile salts and enzymes, solubility in the mucus layer which increases in thickness from the proximal to the distal part of the small intestine, will affect the absorption process. Thus, in vitro models for studying absorption through the intestinal epithelium have many limitations. Still, they are considered as useful model systems for such purposes. Similarly, effects of bioactive molecules on the epithelium can be studied by measuring barrier function and influences on transport processes.

Epithelial cells from the intestine have been very difficult to cultivate in vitro as primary cells. Recent technological developments have allowed the growth of intestinal organoids in 3D culture. This approach is described in Chap. 22 of this book. However, 3D organoids are difficult to use for screening purposes. During the last four decades the preferred model of the intestinal epithelium has been transformed, continuously growing epithelial cell lines like Caco-2, and HT29. Both are described in the following chapters. Caco-2 demonstrates many of the properties associated with the enterocytes of the small intestine. When grown on filter supports, after reaching confluence the Caco-2 cells will spontaneously start to differentiate into a polarized cell layer with apical microvilli and intercellular tight junction complexes. Caco-2 cells also express many of the enzyme markers and transport systems of primary epithelial cells. However, Caco-2 cells do not produce mucins, a property present in the HT29 cell line. When HT29 cells were treated with the antimetabolite methotrexate, they differentiated into mature goblet cells (Lesuffleur et al. 1990). Mucus-secreting HT29-MTX cells have been characterized with regard to tight junction formation, development of confluent monolayers and production of the mucus layer. Confluent HT29-MTX cells develop functional tight junction complexes, but not to the same degree as Caco-2 cells. Thus, Caco-2 cells have been more often used to study epithelial barrier function and transepithelial transport than HT29. To compensate for lacking mucus production a co-culture system has been developed based on Caco-2 and the HT29-MTX cell line. This co-culture system is described in detail in Chap. 13. Irrespective of the cell model chosen, e.g. Caco-2, HT29 or Caco-2/HT29 co-cultures, the main readout parameters are effects on the epithelial barrier function, absorption or transepithelial transport of the test compound. The reliability of such studies depends on the uniformity and integrity of the confluent and polarized cell monolayer. Monolayer integrity can be verified in different ways, e.g. by measuring transepithelial electrical resistance or by measuring the passage of the fluorescent dye Lucifer Yellow. Both methods are described below.

9.1 Measurement of Transepithelial Electrical Resistance (TEER)

Determination of transepithelial electrical resistance is a simple and convenient technique that provides information about the uniformity of the Caco-2 cell layer on the filter support, and the integrity of the tight junctions formed between the polarized cells. Thus, TEER measurements are often used to study epithelial barrier function. The electrode is placed in the medium in the upper chamber, and resistance is directly measured by a portable voltmeter like the Millicell-ERS Voltmeter

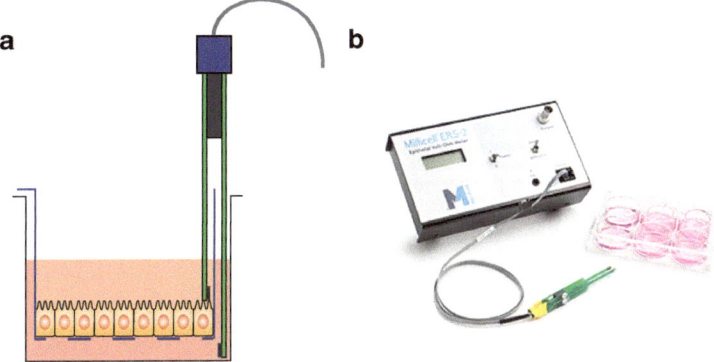

Fig. 9.3 The figure shows The Millicell ERS-2 unit with an STX chopstick electrode. To the left, a drawing of the electrode placed in a tissue culture insert. Note that the shorter tip of the electrode should not be in contact with the cell layer, while the long tip should just touch the bottom of the outer chamber

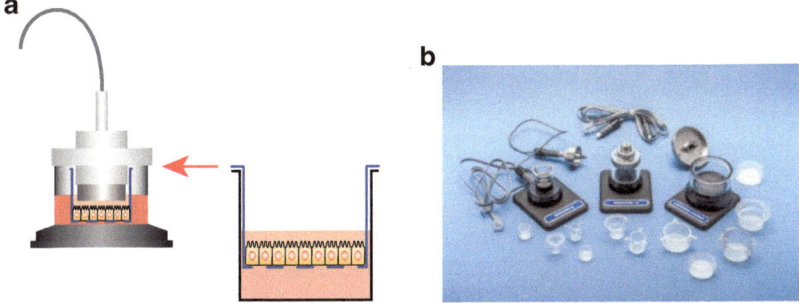

Fig. 9.4 The figure shows different variants of the Endohm chambers with culture inserts for 6, 12 and 24 well plate inserts. To the left is a schematic drawing of the chamber with culture insert. The gap between the upper electrode and the cell layer should be adjusted to 1–2 mm

(Millipore) or the EVOM2, Epithelial Voltohmmeter for TEER (World Precision Instruments Inc., Sarasota, FL) (see Figs. 9.3 and 9.4, respectively).

If the culture is to be continued, the electrode should be sterilized with 70 % ethanol. Before use rinse the electrode in electrolyte solution similar to the culture medium.

9.1.1 Basic Protocol

Measure the resistance of a blank culture cup with no cells that has been thoroughly soaked with an electrolyte solution similar to the culture medium. Move the culture tray from the incubator into the laminar air hood and place it on the heating plate maintained at 37 °C. Change medium in the culture inserts. Depending on the voltmeter and electrode in use, different protocols have to be followed:

A. With a chopstick probe like the Millicell STX probe, the shorter electrode is placed into the upper compartment (apical side) containing the complete culture medium, and the other electrode in the lower compartment (basolateral side, see Fig. 9.3) containing the same medium. The shorter electrode should not be in contact with the cell layer. Ascertain that the electrode is held steady and at a 90° angle to the plate insert.

   This is perhaps the most common way of measuring TEER. However, the readings with the STX electrodes may vary across the cell layer. An alternative is to use the Endohm electrode system developed for the EVOM2 Epithelial Voltohmmeter (Fig. 9.4) which measures resistance or voltage over the whole area covered by the cells. However, this alternative necessitates transfer of the inserts from their culture wells to the Endohm chamber for measurement rather than using the hand-held STX electrodes. The Endohm chamber and the cap each contain a pair of concentric electrodes, a voltage-sensing silver/silver chloride pellet in the center plus an annular current electrode. The height of the top electrode can be adjusted to fit cell culture inserts from different manufacturers. The circular disc electrodes, situated above and beneath the membrane, allow a more uniform current density to flow across the membrane than with STX electrodes.

B. Sterilize the Endohm chamber and electrode with 70 % ethanol for 15 min and let air dry. Measure the resistance of a blank culture cup with no cells that has been thoroughly soaked with an electrolyte solution similar to the culture medium. Add culture medium to the chamber and transfer the insert to the Endohm chamber and mount the upper electrode. The height of the culture medium outside and inside the insert should be at the same level. Measure resistance in ohms (Ω).

9.1.2 Calculating Transepithelial Resistance

If using a chopstick electrode like the Millicell STX electrode, resistance should be measured several times at different places of a monolayer due to variability in the uniformity of the cell layer, and an average value calculated. The average value of the blank, or preferably blanks, should then be subtracted from the resistance reading of the sample ($R_{sample} - R_{blank} = R_{cell\ layer}$). In a typical example the values can be like: $450\,\Omega - 160\,\Omega = 290\,\Omega$, meaning that the resistance of the monolayer is $290\,\Omega$. Then a correction has to be made for the area covered by the monolayer. Different culture inserts may have different shapes, and the filter covers different areas. Also, data from experiments carried out in different plate formats, i.e. 6-, 12- or 24-well plates must be normalized for comparison. Typically, the area covered by a filter insert for a 24-well plate will be $0.4\ cm^2$. Multiply the area covered by the cell monolayer with the resistance found in the experiments: $290\,\Omega \times 0.4\ cm^2 = 116\,\Omega\ cm^2$. This value is independent of the area of the membrane used.

9.2   Verification of Monolayer Integrity by Lucifer Yellow Flux

The permeability of the cell monolayer can also be studied by measuring the passive transport of small hydrophilic molecules across the monolayer. Such molecules mainly pass the monolayer via the paracellular route, i.e. through the tight junctions (TJ), and can be used to get information about the leakiness of the TJs. Lucifer Yellow (LY) is an easily detectable marker. It is particularly useful in studies of transepithelial transport of a test compound where it can be used to determine the highest concentration of the sample that does not disturb the integrity of the cell monolayer. The protocol is based on the exposure of the apical side of monolayer to a 4.5 μM solution of LY, eventually together with the test substance, prior to collecting samples from the basolateral side at different time intervals.

9.2.1   Basic Protocol

Prepare the LY (and test compound) solutions immediately prior to each experiment by diluting stocks in buffered Hanks Balanced Salt Solution (HBSS) assay buffer, for example HEPES- or MES-buffered HBSS. LY should be used in a final concentration of 4.5 μM. Prepare a working solution of the test compound containing 4.5 μM LY. Change the medium on the culture inserts 24 h before the experiment. Wash the inserts with buffered HBSS solution to remove medium components. Add 0.5 ml buffered HBSS to the insert compartment (apical) and 1.2 ml to the outside compartment (this is for inserts in a 24-well plate). Keep inserts at 37 °C on a heating plate (temperature is a critical parameter in this type of experiments. Carefully add an extra 0.5 ml LY working solution with or without the test substance to the apical side of the monolayer and 1.2 ml buffered HBSS in the outside compartment (basolateral side). Place the culture trays with inserts on a rotary plate shaker (200–300 rpm at 37 °C). At different time points, for example 30, 60, 90, 120 min, remove a 0.5 ml sample from the basolateral side and replace with 0.5 ml prewarmed buffered HBSS solution. Based on fluorimetric readings (excitation wavelength 428 nm and emission wavelength 536 nm), measure the concentration of LY in the test samples and make a standard curve. Calculate the apparent permeability coefficient (P_{app}) values of the LY flux through the monolayer. P_{app}, (unit: cm s^{-1}) is determined from the amount of compound transported per time by the equation:

$$P_{app} = (dQ/dt)(1/(AC_0))$$

where dQ/dt is the steady-state flux (μmol s^{-1}), A is the surface area of the filter (cm^2) and C_0 is the initial concentration in the donor chamber (μM). If the value deviates from the normal values obtained, the cell monolayer integrity is disrupted.

In the same way, the effects of a given compound on epithelial LY permeability can be tested by including the test compound with the LY solution in the donor compartment. Then the calculated permeability coefficient to the test compound (P_{appt})

can be compared with the apparent permeability value of the control cells, e.g. without added compound (P_{appc}). Values of $P_{appt} \le P_{appc}$ suggest that the test compound can be regarded as non-toxic and that it can be used to perform studies of effects on epithelial barrier function or absorption experiments.

9.3   Summary

The single layer of epithelial cells in the intestine represents the rate-limiting barrier for transport and uptake of compounds between the lumen and the body interior. Thus, in vitro-differentiated human epithelial cell monolayers allows for testing of absorption and active and passive transport processes through the intestinal epithelium. Among available epithelial cell lines Caco-2 has been found particularly suitable. In food science, problems of interest are not only related to absorption and transport issues, but also to biological effects of dietary compounds that may influence other aspects of epithelial function, typically tight junction structure, pattern recognition receptor activation and intracellular signaling processes that may affect cytokine production and subsequently both innate and adaptive immune responses. In such situations the lack of mucus production by Caco-2 cells may be considered a disadvantage. The HT29-MTX cell line may then represent an alternative, or a co-culture of Caco-2 and HT29-MTX cells, that also may provide information about effects on mucus production, an essential part of the intestinal barrier function. All these culture systems are described in the following chapters.

References

Johansson ME, Sjövall H, Hansson GC (2013) The gastrointestinal mucus system in health and disease. Nat Rev Gastroenterol Hepatol 10(6):352–361
Lesuffleur T, Barbat A, Dussaulx E, Zweibaum A (1990) Growth adaptation to methotrexate of HT-29 human colon carcinoma cells is associated with their ability to differentiate into columnar absorptive and mucus-secreting cells. Cancer Res 50:6334–6343
Rera M, Clark RI, Walker DW (2012) Intestinal barrier dysfunction links metabolic and inflammatory markers of aging to death in Drosophila. Proc Natl Acad Sci USA 109:21528–21533

Chapter 10
Caco-2 Cell Line

Tor Lea

Abstract The human epithelial cell line Caco-2 has been widely used as a model of the intestinal epithelial barrier. The Caco-2 cell line is originally derived from a colon carcinoma. However, one of its most advantageous properties is its ability to spontaneously differentiate into a monolayer of cells with many properties typical of absorptive enterocytes with brush border layer as found in the small intestine. The Caco-2 cell line is heterogeneous and contains cells with slightly different properties. Thus, cultivation conditions can be expected to select for the growth of subpopulations of cells resulting in a cellular model system with properties that may differ from the original cell line. Accordingly, results obtained under similar experimental conditions in different laboratories may not be directly comparable. Due to this, a variety of cloned Caco-2 cell lines has been established, and described in the literature. This chapter will however, focus on describing how to handle and cultivate the original Caco-2 cell line as obtained from cell culture collections like American Type Culture Collection and the European Collection of Cell Cultures. Detailed protocols for handling the Caco-2 cells in the laboratory are provided. Furthermore, in Chap. 9 general protocols for measuring barrier function by transepithelial resistance (TEER), and monolayer integrity by Lucifer Yellow flux are described. Proper testing of the cell monolayer is absolutely critical in exploiting Caco-2 cells to measure interaction, uptake and cellular transport of drugs and food components.

Keywords Caco-2 • Adsorptive enterocytes • Epithelial transport • Tight junctions • Barrier function

T. Lea (✉)
Department of Chemistry, Biotechnology and Food Science, Norwegian University of Life Sciences, PO Box 5003, 1432 Ås, Norway
e-mail: tor.lea@nmbu.no

© The Author(s) 2015
K. Verhoeckx et al. (eds.), *The Impact of Food Bio-Actives on Gut Health*,
DOI 10.1007/978-3-319-16104-4_10

103

10.1 Origin

Back in the 1970s several epithelial cell lines were established from gastrointestinal tumors. The purpose was to study mechanisms in cancer development and effects of cytotherapy. Partly due to the heterogeneity of primary intestinal epithelial cells both in morphology and function, i.e. small bowel enterocytes, goblet cells, entero-endocrine cells, Paneth cells and M-cells, there was a need for differentiating the tumor cells into more specialized cell types. Several of the cell lines could be partly differentiated by addition of synthetic or biological factors to the medium. However, one of them, Caco-2, was rather unique as it was able to differentiate spontaneously when reaching confluence. Caco-2 (*Ca*ncer *co*li-2) was established from a human colorectal adenocarcinoma by Jorgen Fogh at the Sloan-Kettering Cancer Research Institute (Fogh et al. 1977).

10.2 Features and Mechanisms

Early studies revealed that differentiated Caco-2 cells expressed several morphological and functional properties characteristic of small bowel enterocytes. Towards confluence they start to polarize acquiring a characteristic apical brush border with microvilli. Tight junctions form between adjacent cells, and they express enzyme activities typical of enterocytes, i.e. lactase, aminopeptidase N, sucrase-isomaltase and dipeptidylpeptidase IV. However, markers of colonocytes are also present in Caco-2 cells (Engle et al. 1998) (Table 10.1).

Table 10.1 Properties of Caco-2 cells

Growth	Grows in culture as an adherent monolayer of epithelial cells
Differentiation	Takes 14–21 days after confluence under standard culture conditions
Cell morphology	Polarized cells with tight junctions and brush border at the apical side
Electrical parameters	High electrical resistance
Digestive enzymes	Expresses typical digestive enzymes, membrane peptidases and disaccharidases of the small intestine (lactase, aminopeptidase N, sucrase-isomaltase and dipeptidylpeptidase IV)
Active transport	Amino acids, sugars, vitamins, hormones
Membrane ionic transport	Na^+/K^+ ATPase, H^+/K^+ ATPase, Na^+/H^+ exchange, $Na^+/K^+/Cl^-$ cotransport, apical Cl^- channels
Membrane non-ionic transporters	Permeability glycoprotein (P-gp, multidrug resistance protein), multidrug resistance-associated protein, lung cancer-associated resistance protein
Receptors	Vitamin B12, vitamin D3, EGFR (epidermal growth factor receptor), sugar transporters (GLUT1, GLUT2, GLUT3, GLUT5, SGLT1)
Cytokine production	IL-6, IL-8, TNFα, TGF-β1, thymic stromal lymphopoietin (TSLP), IL-15

10.3   Stability, Consistency and Reproducibility

During the 35 years that have passed since its establishment, Caco-2 cells have been propagated in a number of laboratories worldwide. Due to different culture conditions and different numbers of passages, Caco-2 cells have often acquired different properties. Thus, the expression of differentiation markers typical of enterocytes, change with increasing numbers of passages (Artursson et al. 2001). Also, parameters like transepithelial electric resistance (TEER) and proliferation rate have been reported to increase with passage number (Briske Andersson et al. 1997). It has also been documented that late passage cells may start growing in multilayers, a phenomenon that will affect TEER measurements, and make comparisons with results based on early passage cells difficult.

10.4   Relevance to Human In Vivo Situation

To closer mimic the steric conditions in the intestine in vivo, Caco-2 cells are cultured on permeable filter inserts that can be obtained from a number of vendors like Becton Dickenson, Corning, Costar, etc. Cultivation of Caco-2 cells on filter supports improves their morphological and functional differentiation. It has been well documented that polarized Caco-2 monolayers represent a reliable correlate for studies on the absorption of drugs and other compounds after oral intake in humans. Several studies have compared Caco-2 permeability coefficients with absorption data in humans and found high correlation, particularly if the compounds are transported by passive paracellular transport mechanisms (Artursson and Karlsson 1991; Cheng et al. 2008; Sun et al. 2008).

Although Caco-2 cells have been found to express a large number of enzymes and transporter proteins present in normal human intestinal epithelium, recent studies suggest that there are variations between gene expression profiles of transformed epithelial cell lines like Caco-2, HT29 and normal human intestinal epithelium (Bourgine et al. 2011). Furthermore, differences are not only found between the cell lines and normal epithelium, principle component analysis of gene expression profiles reveals clear differences between the Caco-2 and HT29 cell lines, also (Bourgine et al. 2011). This is not unexpected. Normal intestinal epithelium is made up of several different cell types, and differences in gene expression profiles are not only observed in the mucosal epithelium along the whole gastrointestinal tract, but also along the crypt-villus axis (Anderle et al. 2005). Obviously, data analysis of experiments in the Caco-2 cell model cannot be directly compared with the in vivo situation. Still, intestinal epithelial cell models like the Caco-2 model, holds many advantages due to its simplicity and reproducibility allowing inter-laboratory comparison of results. Furthermore, pursuing an effect of a food bioactive in a cell line model opens for studies of molecular mechanisms which may be more difficult to address in vivo.

10.5 General Protocols for Caco-2 Cells

10.5.1 General Maintenance

Caco-2 cells are routinely maintained in RPMI 1640 (alternatively Dulbecco's Modified Eagle Medium, DMEM) with 10 % fetal bovine serum and the following additions: 1 % non-essential amino acids (NEAA), 50 µM thioglycerol, 25 mg/ml gentamycin (complete medium). The cells should be kept at 37 °C in a humidified atmosphere containing 5 % CO_2. For propagation in culture flasks, Caco-2 cells are seeded in a concentration of 10^5 cells/cm^2. Medium should be changed every 3 days. At 80 % confluence, typically after 4–5 days, the cells are split 1:10 before further cultivation. Trypsinize Caco-2 cells by first rinsing with PBS (15 ml pr. 75 cm^2 flask). Add 5 ml trypsin/EDTA solution, rinse the cells and pour off most of the trypsin/EDTA. Only 1 ml trypsin/EDTA solution is necessary to wet the whole cell layer in a 75 cm^2 flask. Incubate the flask at 37 °C for 6–15 min. The incubation with trypsin should be as short as possible as this process will affect cell viability. As soon as the cells are detached, stop trypsinization by adding complete medium containing fetal calf serum. Transfer the cells to a test tube and let cell aggregates and debris sediment. Transfer the supernatant to a new test tube, and take an aliquot to count the cells. Check cell viability. The content of dead cells should not exceed 5 %.

10.5.2 Protocol for Polarizing Caco-2 Cells in Tissue Culture Inserts

Caco-2 cells can be cultured on a filter support (Fig. 10.1). The filter support can be made from polycarbonate, polyester or polyethylene terephthalate. Particularly the latter is claimed to be inert with low non-specific protein-binding properties. The filter supports can be transparent or translucent. If you want to follow the differentiation process in the inverted microscope and prepare the differentiated cells for scanning electron microscopy, transparent filters are preferred. Furthermore,

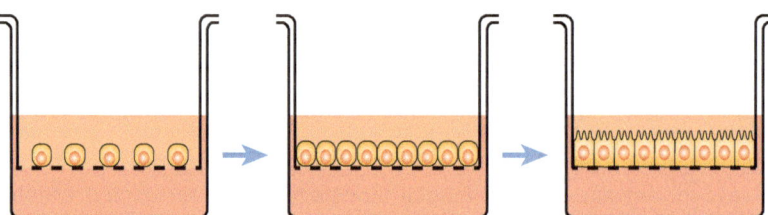

Fig. 10.1 The figure is a drawing of the differentiation of Caco-2 cells on a tissue culture insert. After the Caco-2 cells reach confluence (*middle*) they start to differentiate spontaneously, and after a total culture period of around 21 days they will appear with dense microvilli on the apical side characteristic of small intestinal enterocytes

the filters can come with different pore sizes ranging from 0.4 to 8 μm. To study parameters like transcellular transport and permeability, 0.4 μm filters are recommended. Insert with larger pore sizes can be exploited if direct cell–cell interactions are being scrutinized, i.e. in co-cultures of Caco-2 cells with stromal cells or adherent immune cells.

Depending on inherent properties of the particular Caco-2 cells available, it may be necessary to coat the filters with protein to prevent the cells from detaching in the final stages of the differentiation process. Some vendors offer ready-coated filters. However, efficient coating of the filters can be carried out by covering them with collagen solution Type I from Sigma (1/100 in water). Incubate at room temperature for 3–4 h. Remove the collagen solution and leave the filter inserts to dry overnight.

Place the necessary number of filter inserts in a 12 well plate. Dilute the cell suspension to a concentration of 1×10^6/ml. Seed 0.5 ml of the cell suspension to each insert corresponding to seeding density of 500,000 cells pr. 12 mm filter insert.

This seeding density corresponds to 4×10^5 cells/cm². Add 1.5 ml medium to the basolateral compartment in a 12 well plate, and 0.5 ml to the apical compartment. Depending on experience it is recommended to start with setting up filters in quadruplicate, that is later reduced to triplicate or even duplicate. Medium should be changed on days 4, 8, 12, 16 and 18. Cells will be fully polarized by day 21. See SEM picture in Fig. 10.2.

The differentiation of Caco-2 cells seem to follow a time schedule in the expression of morphological and biochemical properties of the absorptive enterocytes. Due to the cellular heterogeneity of the cell line the differentiation process occurs in a mosaic pattern, with some areas expressing fully differentiated cells with microvilli after 12–14 days while other areas contain less differentiated cells. According to the protocol above, the Caco-2 monolayer will be homogenously differentiated after 18–21 days. When polarized and confluent, the cell layer forms a continuous barrier between the upper and lower compartments (apical/mucosal and basolateral/serosal). The compound of interest, suspended in a physiological fluid, typically PBS,

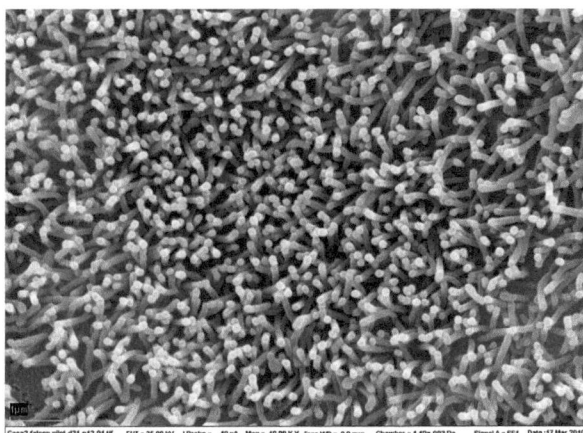

Fig. 10.2 SEM picture of polarized Caco-2 cells after 21 days of culture

is added to the upper compartment, and the increase in concentration in the lower compartment is measured. In this way the permeability, absorption and transepithelial transport of any nutritional compound can be studied. Special emphasis should be paid to compounds or drugs that are poorly soluble in physiological buffers. If stock solutions of the compound of interest has to be made in organic solvents like ethanol or dimethylsulfoxide (DMSO), be aware that organic solvents may affect the cell layer in various ways. Always include a filter/filters for solvent control. In general, avoid using higher solvent concentrations than 1 % (v/v) or else.

10.5.3   Troubleshooting Guide for Transport Experiments Across Caco-2 Monolayers

Problem	Possible explanation	Try the following
Very high P_{app}	Leaky cell monolayers.	Validate integrity of monolayers (see protocol with Lucifer Yellow flux in Chap. 9).
	High expression level of transporter protein for active transport process	Protein expression levels may vary with passage number and culturing conditions. Use the same batch of cells with similar passage number for consistent results.
Very low P_{app}	Test compound degradation	Check stability of test compound under experimental conditions
	Poor solubility	Verify that the test compound is fully dissolved.
	Membrane transporter may have become saturated	Verify the concentration of the test substance. Titrate and use lower concentration.
Poor mass balance	High degradation	Stability of test compound (see above)
	High cellular uptake	Analyze amount of compound in the cell monolayer by lysing the cells and extracting the compound for quantification.
	Adsorption of compound to the filter insert or culture vessel	Monitor compound concentration before and after transfer to the filter insert system.
High standard deviation	Damaged cell monolayers	It is vitally important to avoid damaging the monolayers during sampling

10.6   Applications

Cell culture assays has offered exciting new possibilities in many scientific disciplines. If properly used the Caco-2 cell line can provide information about the biological and biochemical basis of barrier properties of the intestinal mucosa, but may also unravel valuable information about the absorption of drugs and dietary components relevant for both the pharmaceutical and the food industry. Thus, the Caco-2 cell line has been exploited for a range of applications, among them the following:

To study mechanisms and effects of microbiota, microbiota metabolites, food digesta and bioactive food components on the barrier function of the intestinal epithelium (Shimizu 2010).

To elucidate pathways for the transport of drugs or food components (e.g. paracellular versus transcellular or passive versus carrier-mediated mechanisms) across the intestinal epithelium (Knipp et al. 1997).

In studying potential toxic effects of drug candidates or food metabolites in the intestinal mucosa (Chang et al. 1993).

To determine how components of a formulation (e.g. adjuvants or food matrices) may influence intestinal epithelial transport of bioactive molecules (Nerurkar et al. 1996).

In the characterization of the optimal physiochemical properties of a bioactive molecule for passive diffusion via the paracellular or transcellular pathways across the intestinal epithelium (Burton et al. 1996).

To study molecular details and significance of efflux systems (e.g. multi-drug resistance proteins like the P-glycoprotein) in the intestinal epithelium (Burton et al. 1997).

To study and determine interactions between bioactive molecules during transport across the intestinal epithelium (Wacher et al. 1996).

10.7   Advantages and Disadvantages

The Caco-2 cells spontaneously differentiate to express morphological (polarized columnar epithelium) and functional characteristics of mature small intestinal enterocytes. The polarized Caco-2 cell layer shows 4 times higher TEER values compared to HT29 monolayers, i.e. more similar to the in vivo situation. Caco-2 cells express most receptors, transporters and drug metabolizing enzymes like aminopeptidase, esterase and sulfatase found in normal epithelium. However, no P-450 metabolizing enzyme activity has been reported.

In comparison with normal intestinal epithelium the Caco-2 cell model have several limitations. First of all that the normal epithelium contains more than one cell type, not only enterocytes. Secondly, when using the Caco-2 cell model, no mucus and unstirred water layer is present. Furthermore, a number of non-cellular parameters will affect the absorption of a certain compound in cells. Thus, transport of lipophilic molecules is strongly influenced by the presence of bile acids and phospholipids, and also compound solubility in the mucus layer as well as the unstirred water layer close to the epithelium, will strongly influence uptake in vivo. Although Caco-2 cells in general provide a powerful tool for studying properties of the intestinal epithelium, one has to be cautious in extrapolating data from such in vitro models to the in vivo situation.

10.8   Conclusion

For Caco-2 cell monolayer systems to provide reliable information several critical parameters have to be controlled. First of all, it is imperative that the integrity of the monolayers is validated prior to transport studies. The protocol based on TEER measurements of Lucifer Yellow flux (Chap. 9) are crucial for confirming the quality and integrity of the polarized epithelial cell monolayer. Also, if the compound being tested is a candidate substrate for a specific transporter, the monolayer must be checked to verify the expression and activity of that particular transporter. The presence of the transporter at the mRNA level can be verified by PCR, at the protein level by immunoblotting and functionally by employing inhibitors to and known ligands to demonstrate the presence of a functional transporter. It is important to recognize that protein expression may be influenced by a variety of parameters, the most important being culture conditions, passage number or age of the cells in culture. Thus, a positive control should always be included demonstrating the presence of a functional transporter protein or protein complex. It is also important that the effect of contamination of the cell culture is eliminated. Although short time protocols for obtaining a polarized Caco-2 cell monolayer has been published, the standard and most reliable procedure requires 21 days of culture to obtain a polarized monolayer with full expression of tight junctions and other intercellular contacts. During a 3 weeks culture period and with antibiotics in the medium, a subclinical infection not identifiable even by light microscopy, may bias or obscure results completely. Thus, it is of imperative importance.

References

Anderle P, Sengstag T et al (2005) Changes in the transcriptional profile of transporters in the intestine along the anterior-posterior and crypt-villus axes. BMC Genomics 6:69–86

Artursson P, Karlsson J (1991) Correlation between oral drug absorption in humans and apparent drug permeability coefficients in human intestinal epithelial (Caco-2) cells. Biochem Biophys Res Commun 175:880–885

Artursson P, Palm K, Luthman K (2001) Caco-2 monolayers in experimental and theoretical predictions of drug transport. Adv Drug Deliv Rev 46:27–43

Briske Andersson MJ, Finley JW, Newman SM (1997) The influence of culture time and passage number on the morphological and physiological development of Caco-2 cells. Proc Soc Exp Biol Med 214:248–257

Bourgine J, Billaut-Laden I et al (2012) Gene expression profiling of systems involved in the metabolism and the disposition of xenobiotics: comparison between human intestinal biopsy samples and colon cell lines. Drug Metab Dispos 40(4):694–705

Burton PS, Conradi RA et al (1996) How structural features influence the permeability of peptides. J Pharm Sci 85:1336–1340

Burton PS, Goodwin JT et al (1997) In vitro permeability of peptidomimetics: the role of polarized efflux pathways as additional barriers to absorption. Adv Drug Deliv Rev 23:143–156

Chang AS, Chikhale PJ et al (1993) Utilization of a human intestinal epithelial cell culture system (Caco-2) for evaluating cytoprotective agents. Pharm Res 10:1620–1626

Cheng K-C, Li C, Uss AS (2008) Prediction of oral drug absorption in humans from cultured cell lines and experimental animals. Expert Opin Drug Metab Toxicol 4:581–590

Engle MJ, Goetz GS, Alpers DH (1998) Caco-2 cells express a combination of colonocyte and enterocyte phenotypes. J Cell Physiol 174:362–369

Fogh J, Wright JC, Loveless JD (1977) Absence of HeLa cell contamination in 169 cell lines derived from human tumours. J Natl Cancer Inst 21:393–408

Knipp GT, Ho NFH et al (1997) Paracellular diffusion in Caco-2 monolayers: effects of perturbants on the transport of hydrophilic compounds that vary in charge and size. J Pharm Sci 86: 1105–1110

Nerurkar MM, Burton PS, Borchart RT (1996) The use of surfactants to enhance the permeability of peptides through Caco-2 cells by inhibition of an apically polarized efflux system. Pharm Res 13:528–534

Shimizu M (2010) Interaction between food substances and the intestinal epithelium. Review Biosci Biotechnol Biochem 74:232–241

Sun H et al (2008) The Caco-2 cell monolayer: usefulness and limitations. Expert Opin Drug Metab Toxicol 4:395–411

Wacher VJ, Salphati L, Benet LZ (1996) Active secretion and enterocytic drug metabolism barriers to drug absorption. Adv Drug Deliv Rev 20:99–112

Chapter 11
HT29 Cell Line

Daniel Martínez-Maqueda, Beatriz Miralles, and Isidra Recio

Abstract The human colon adenocarcinoma cell line HT29, is not only used to study the biology of human colon cancers, but it is receiving special interest in studies focused on food digestion and bioavailability due to the ability to express characteristics of mature intestinal cells. In the differentiated phenotype, they are able to form a monolayer with tight junctions between cells and a typical apical brush border. In addition, these differentiated cells express brush-border-associated hydrolases typical of the small intestine although the enzymatic activity is lower than that found in vivo. Although they represent a valuable model due to their similarities with enterocytes of the small intestine, their limitations and the relevance to the in vivo situation are also considered in this chapter. The application of this cell line to transport studies of drugs and food compounds is illustrated, especially when the effect of the mucus layer is considered or used as co-culture in combination with Caco-2 cells. They have also been frequently used to study the intestinal immune response to bacterial infection, and microorganism survival, adhesion or invasion. Finally, the use of these cells to evaluate the effect of several food compounds and mucin secretion is summarized.

Keywords HT29 • Cell differentiation • Transport studies • Intestinal cell

11.1 Origin

The human colon adenocarcinoma cell line HT29 was isolated from a primary tumor of a 44 years old Caucasian female in 1964 by Fogh and Trempe (1975). Since then, many cell lines have been derived from human colon cancers. Initially, this cell line was used to study different aspects of the biology of human cancers. However, these cells have attracted attention due to the fact they were able to express characteristics of mature intestinal cells, such as enterocytes or mucus producing cells.

D. Martínez-Maqueda • B. Miralles • I. Recio (✉)
Instituto de Investigación en Ciencias de la Alimentación (CIAL, CSIC-UAM),
Nicolás Cabrera 9, 28049 Madrid, Spain
e-mail: i.recio@csic.es

K. Verhoeckx et al. (eds.), *The Impact of Food Bio-Actives on Gut Health*,
DOI 10.1007/978-3-319-16104-4_11

11.2 Features and Mechanisms

These cells have shown a high rate of glucose consumption and therefore, they require high glucose concentration in the medium. Under these standard growth conditions, i.e., in the presence of 25 mM glucose and 10 % serum, these cells display an undifferentiated phenotype; they grow as a multilayer of unpolarised undifferentiated cells, and functionally they do not express any typical characteristics of intestinal epithelial cells, and they express low amounts of hydrolases. Since the discovery that the replacement of glucose by galactose in the culture medium induced a reversible enterocytic differentiation (Pinto et al. 1982), HT29 cells have become a unique model for studying the molecular mechanisms of intestinal cell differentiation. When these cells are grown under appropriate culture conditions or after treatment with different inducers, like butyrate (Augeron and Laboisse 1984) or acid (Fitzgerald et al. 1997), they can be modulated to express different pathways of enterocyte differentiation. For this reason, HT29 is considered a pluripotent intestinal cell line which can be used for the study of the structural and molecular events involved in cell differentiation.

The polarised phenotype is characterised morphologically, physiologically and by biochemical markers. The differentiation is growth-related, starting after confluence (after 15 days of growth), and the cells form a monolayer with tight junctions between cells and a typical apical brush border. The brush border of these differentiated cells contains proteins which are normally present in the intestinal microvilli, such as villin. In addition, these differentiated cells express brush border-associated hydrolases typical for the small intestine, have brush border microvilli although the enzyme activities is much lower than that of the normal intestine, and do not express lactase. The maximum enzyme activities, such as sucrose-isomaltase, aminopeptidase N, dipeptidyl-peptidase-IV and alkaline phosphatase, are reached after 30 days in culture. Under these conditions, p.e. growth in glucose-free medium, cell differentiation is very similar to that observed in Caco-2 cells but in HT29 the time course of the differentiation process is longer (30 days vs. 15–20 days in Caco-2); the levels of enzymatic activities are lower than in Caco-2; lactase is absent; and only 40–50 % of HT29 cells express sucrose-isomaltase (Zweibaum 1986; Zweibaum et al. 2011). However, one of the main differences between this cell line and Caco-2 is that HT29 cells can produce mucin at a relatively high level (Huet et al. 1987; Maoret et al. 1989). Stepwise adaptation of exponentially growing HT29 cells to increasing concentrations of methotrexate (MTX) (up to 10^{-5} mol) resulted in their transformation into mucus-secreting differentiated cells (Lesuffleur et al. 1990). As occurs under other metabolic stress conditions like glucose-deprivation, after a high rate of mortality, the resistance to MTX is associated with the cells possessing this stable differentiated mucus-secreting phenotype. Interestingly, cells adapted to low-dose MTX consist of a double population of columnar absorptive and mucus-secreting cells and, at a higher dose, cells are almost exclusively of mucus-secreting phenotype. This mucus-secreting phenotype has been used in the transport studies of different compounds, to study the mucus-inducing properties of different food compounds or in studies regarding microorganisms adhesion and survival (see Sect. 11.6).

The factors secreted from cells in culture medium include metabolites, cytokines, growth factors, etc., which are known to promote cell survival. Recently, the soluble factors secreted by HT29 and other tumor cells in basal conditions and after gamma radiation have been reported. HT29 secreted pro-inflammatory cytokines, such as, tumor necrosis factor (TNF) α and interleukins (IL) 1β and IL 6; growth factors such as platelet-derived growth factor AA and transforming growth factors (TGF) α and β; chemokines such as fractalkine, IL-8, monocyte chemoattractant protein-1 and interferon-γ-induced protein 10; pro-angiogenic factors such as IL-15 and vascular endothelial growth factor; and immune-modulatory cytokines such as granulocyte colony-stimulating factor, granulocyte macrophage colony-stimulating factor and IL-3. It was suggested, based on previous reports on biopsy samples, that in vivo, a similar cytokine secretion profile can be observed (Desai et al. 2013).

11.3   Stability, Consistency and Reproducibility

Many tumor-derived cells have been reported to show lack of culture stability and reproducibility based on the existence of irregular growth and non-specific genetic alterations (Lipps et al. 2013). To minimize these handicaps, the establishment of large cell banks is recommended to allow disposing cells at early and similar passages. On the contrary, the consistency of the assumed cell properties should be verified for highly different passages. For instance, Chen et al. (1987) proved that karyotypes were comparable for HT29 cells along more than 100 passages, which represents valuable data in the assessment of the cell line evolution. Likewise, derived cells adapted by specific treatments must be evaluated under different culture conditions as time and presence/absence of responsible agent(s). Lesuffleur et al. (1990) tested the irreversibility of the differentiation of MTX adapted cells by the comparison of the growth curve for several passages. No significant differences were observed for both 10^{-7} and 10^{-6} M MTX-derived cell lines between cells cultured with drug-enriched media and drug-free media. The stability of other differentiation characteristics was also confirmed for MTX-free cells compared with equivalent MTX-treated cells as well as themselves along several passages before.

11.4   Relevance to Human In Vivo Situation

As in other cancer-derived cell line models, significant differences in the gene expression of transporters and metabolic enzymes from the normal human intestinal cells can affect the suitability of the model in reflecting the in vivo permeability (Langerholc et al. 2011). Similar protein expression was found in HT29 cells and the corresponding intestinal scrapings, of which some appear to be characteristic to human intestinal epithelium in vivo (Lenaerts et al. 2007). Nevertheless, several proteins, including transporters, were over- or underexpressed. Bourgine et al. (2012)

compared the expression of 377 genes in HT29 and other intestinal cell lines that are used as in vitro models of the epithelium with the corresponding tissue biopsy. The results showed that differentiated HT29 cells and human colonic tissues do not appear to be significantly different.

The use of HT29 as in vitro model of intestinal cells has some advantages and limitations that have been summarized by Zweibaum et al. (2011). This cell line in its differentiated phenotype is similar to small intestine enterocytes with respect their structure, and the presence of brush border-associated hydrolases and the time course of the differentiation process which is also comparable to that found in the small intestine. In addition, the amount of villin expressed in differentiated HT29 cells is close to the value observed for normal freshly prepared colonocytes. However, these cells also have some limitations, because (1) they are malignant cells with a high rate of glucose consumption and an impairment of the metabolism of glucose; (2) although they mimic some characteristics of small intestine enterocytes, they are colonic cells; (3) they cannot be compared with enterocytes from normal colon since they express brush border-associated hydrolases; (4) but they cannot be compared with absorptive enterocytes because not all hydrolases are present (e.g. lactase and maltase-glucoamylase are absent) and the ion transport properties are different. It has been postulated that these cells are close to human fetal colonic cells because of the type of hydrolases present and the intracellular concentration of glycogen accumulated (Hekmati et al. 1990).

Regarding the expression of cell surface receptors, it has been reported that differentiated HT29 cells express several receptors for peptides, such as vasoactive intestinal peptide, or insulin but also for non-peptide substances. In general, the receptors found in this cell line have their equivalent in normal intestinal cells, except for the receptor to neurotensin, which has been characterised in HT29 cells (Kitabgi et al. 1980), but it is not detectable in normal human colonic epithelium. On the contrary, receptors for peptide YY or neuropeptide Y which are well characterised in normal small intestine epithelial cells have not been reported in HT29. It has to be taken into account that these receptors are located at the small intestine and that HT29 are of colonic origin. It is also remarkable that when using cell cultures, the expression of a particular receptor may depend on the degree of cell differentiation and the growing conditions. Recently, the presence of opioid, serotonin, muscarinic, PPARβ/δ receptors has been reported in this cell line (Zoghbi et al. 2006; Ataee et al. 2010; Belo et al. 2011; Foreman et al. 2011). Information on the studies on receptors carried out in this cell line can be found at Zweibaum et al. (2011).

11.5   General Protocol for HT29-MTX Cells

11.5.1   Cell Maintenance Protocol

HT29-MTX cells are cultured using DMEM medium with high glucose (25 mM), without sodium pyruvate, but with GlutaMAX® (Gibco, Paisley, UK), 10 % of heat-inactivated fetal bovine serum and penicillin/streptomycin (100 units/mL of penicillin and 100 µg/mL of streptomycin) in a humidified incubator with 5 % CO_2 atmosphere at 37 °C. Culture medium is changed every 2 days from the second day after seeding, and cells are harvested in the logarithmic phase of growth after reaching 80–90 % confluency by 0.05 % trypsin/EDTA. Cells are seeded at 2.4×10^4 cells per cm^2 in 75 cm^2 culture flasks. In these conditions, cells would be confluent at day 6–7 of culture.

11.5.2 Experimental Protocol for Test Compounds

11.5.2.1 Study of the Mucin-Stimulating Activity

Cells are seeded in 12-well plates at a density of 5×10^5 cells per well and experiments are performed 21 days after confluency to ensure that the proportion of cells differentiated that express mucus reaches their maximum (Lesuffleur et al. 1993). Culture media is changed every 2 days until 24 h prior to the experiment, when culture media is replaced by serum- and antibiotic-free media (DMEM GlutaMAX®) to minimize any extraneous interference. On the day of the experiment, cells are rinsed twice with PBS and treated with tested compounds dissolved in serum- and antibiotic-free medium for different incubation times at 37 °C in a 5 % CO_2 humidified atmosphere. After treatment, cell supernatants are collected for the determination of mucins, and adhered cells are lysed to obtain the total RNA according to the RNA isolation kit instructions. Mucin secretion is determined in cell supernatants by applying an enzyme-linked lectin assay, based on the strong binding between wheat germ agglutinin and goblet cell mucins (Campo et al. 1988). Standard curves prepared with porcine mucin are employed, and mucin gene expression is evaluated after isolating and reverse-transcribing the RNA from cell lysates. A quantitative PCR assay is performed with specific primers for MUC5AC, the HT29-MTX major secreted mucin gene (Martínez-Maqueda et al. 2013a). Cells are employed between passages 12 and 24 without remarkable basal differences.

11.5.2.2 Evaluation of Transepithelial Absorption by Transwell® Inserts

Transwell® inserts provides a reliable model once the integrity of the cell monolayer is verified. HT29-MTX cells are seeded in 12-well Transwell® inserts at a density of 5×10^5 cells per well and medium is changed every 2 days. Cells are cultured for at least 21 days to ensure a suitable cell differentiation. Prior to the experiment, the integrity of the cell monolayers is verified by measuring the transepithelial electrical resistance (TEER) with a cell electrical resistance system. Only monolayers with initial TEER values above 200 Ω cm^{-2} must be used. On the day of the experiment, cells are rinsed with Hanks' balanced salt solution (HBSS) and test compounds are dissolved in culture medium, which are added to the apical side of the cell layer (HBSS to the basolateral chamber). According to Foltz et al. (2008), cells are maintained at 37 °C in a 5 % CO_2 humidified atmosphere. Transport assay is initiated by adding 0.5 mL of transport medium, containing compounds of interest to the apical compartment. Samples (100 μL) are taken from the basolateral compartment and replaced by transport medium. After the end of the experiment, TEER measurement is repeated and permeability values are only calculated from experiments in which the second TEER value is at least 75 % of the initial TEER value. Fluorescein flux can be measured as a second parameter of monolayer integrity. Monolayers are deemed intact when TEER values were above 200 Ω cm^{-2}, and the permeability coefficients of the paracellular transport marker fluorescein is less than 1×10^{-6} cm s^{-1}. Apical and basolateral solutions can be collected and analyzed by RP-HPLC–MS/MS to evaluate the permeability, including permeability coefficients as indicated by Quirós et al. (2008) and Contreras et al. (2012), and degradation of tested compounds.

11.6   Experimental Read Out

11.6.1   Functionality Studies

HT29 cells have been widely employed to assess the potential anticancer effect of different food compounds, and their gastrointestinal digests, on the basis of the carcinoma origin of this cell line. An illustrative example is found in the study of the antiproliferative activity of the in vitro gastrointestinal digest of sea cucumber wall (Pérez-Vega et al. 2013). Another representative study is found in the assessment of the cytotoxic activity of different extracts from "Racimo" tomato variety and simulated digests against HT29 cell growth, including also the evaluation of the selectivity of the toxic effect against cancer cells (Guil-Guerrero et al. 2011). Other activities have been widely studied on HT29 cells or derived lines, such as immunomodulatory, antioxidant or barrier protective properties. For instance, milk and soy ferments with different strains of lactic acid bacteria (LAB) were evaluated on human intestinal epithelial cells, including HT29 cells, to test their immunomodulatory activity (Wagar et al. 2009). The study was carried out on cells treated with TNF-α and the production of IL-8 was evaluated in the cell supernatant by a commercial enzyme-linked immunosorbent assay. Another investigation focused on the study of the immunomodulatory activity of cereal β-glucan preparations on both HT29 and Caco-2 cells (Rieder et al. 2011). In vitro data of antioxidative activity for food compounds is often presented together with genotoxic tests performed with HT29 cells. Among others, the comet assay has become one of the standard methods for assessing DNA integrity (Collins 2004). Ferguson et al. (2005) demonstrated both in vitro antioxidant activity and antigenotoxic effect in HT29 cells by using a comet assay for two hydroxycinnamic acids with an important percentage in certain plant foods, as spinach or cereals. Interestingly, the protection of resveratrol and quercetin against exogenous pro-oxidative damage was determined in HT29 cells with induced oxidative stress by addition of fatty acid hydroperoxides (Kaindl et al. 2008). HT29-derived cell lines constitute a valuable tool related to the strengthening of the intestinal mucus barrier to study the mucin-stimulating activity of food compounds. The protocol described above has been recently applied to asses the mucin-stimulating effect of certain milk protein hydrolysates and derived peptides (Martínez-Maqueda et al. 2012, 2013a, b). HT29-MTX cells and an analogous experimental design was also implemented to evaluate the regulation of the mucin production by the β-casomorphin-7, a μ-opioid peptide derived from bovine milk (Zoghbi et al. 2006), or by a yoghurt peptide pool (Plaisancié et al. 2013). Other HT29 derived cell lines have been employed due to their capability to form a mucus-layer, e.g. HT29/B6 cells differentiated in a glucose-free culture (Kreusel et al. 1991). Hering et al. (2011) evaluated the expression of tight junction proteins and the related intestinal barrier-protection in HT29/B6 cells, under the effect of TGF-β, a whey protein component. Likewise, monoterpene d-limonene, naturally occurring in the rind of citrus fruit, showed a significant increase on the transepithelial resistance in HT29/B6 cells (D'Alessio et al. 2013).

11.6.2   Transport Studies

The HT29 cell line is not a particularly suitable model for transport studies, but it has been used for comparison with other lines or as co-culture. Monolayers of HT29-MTX cells have been previously used as a permeability model, for example for studying the effect of mucus on the permeation of drugs. However, in most cases, transport studies and permeation enhancement studies are performed with co-cultures of Caco-2 and HT29-MTX as an accurate model to take into account the mucus layer (Yuan et al. 2013).

It has been reported that casein phosphopeptides (CPPs) form aggregated complexes with calcium phosphate and induce Ca^{2+} influx into HT29 cells that have been shown to be differentiated in culture. Taking into account that upon differentiation HT29 cells elicit a transient rise in Ca^{2+}, this model is being used to explore the molecular mechanism by which CPPs elicit their biological activity (Gravaghi et al. 2007). Similarly, the HT29 cell line is a useful model to study key players involved in intestinal iron absorption because it endogenously expresses a number of genes known to be involved in iron transport in the intestine. Using this model, it has been suggested that the hemochromatosis protein HFE can have multiple roles in maintaining iron homeostasis depending on the availability of other proteins (Davies and Enns 2004).

HT29 cells have also been applied to the study of the receptor and internalization transporter of toxins. As for the *Clostridium botulinum* C16S toxin, it was reported that the receptor on the HT29 cell surface targeted is an O-linked sugar chain of mucin-like glycoproteins (Nishikawa et al. 2004).

11.6.3 Microorganisms Survival, Adhesion or Invasion

The HT29 cells represent a well characterised model to study the intestinal epithelial response to bacterial infection. This cell line expresses the features of enterocytes and is useful for attachment and mechanistic studies. Mucin secretion by these cells is of importance because the mucus layer has been suggested to play a role in modulating the adhesion of live organisms to the epithelial surface as well as bacterial components such as LPS (Otte and Podolsky 2004).

The mechanisms underlying enteropathogenesis are one of the most important applications of this cell model. Studies on attachment, adherence and/or internalization of several pathogens such as *Entamoeba histolytica*, *Salmonella enteriditis*, *Escherichia coli*, *Clostridium difficile*, *Campylobacter jejuni*, *Clostridium butyricum* or *Shigella disenteriae*, in some cases with food components intended to limit their access to cells have been reported. For example, in the study of Nickerson and McDonald (2012) it was shown that exposure to maltodextrin induced type I pilus expression, which was required for enhanced biofilm formation by *E. coli* in Crohn's disease patients. Comparison of the results with those found in the HT29 cell model was used to suggest a novel mechanism of epithelial cell adhesion that can contribute to disease susceptibility.

Studies intended to evaluate the beneficial role of probiotics in the human host are another application of the cell model. For the study of probiotic adhesion, the HT29 cell line and some HT29-derived lines have been used to study the molecular mechanisms underpinning host-microbe interactions or the enteric adaptation features of *Bifidobacterium* strains, sometimes in the presence of food components as milk oligosaccharides (Kavanaugh et al. 2013). Probiotic delivery systems by coating biopolymers are also evaluated by using colonic epithelial models that are intended to mimic the colonic mucosa. The intestinal mucus offers numerous ecological advantages for both resident microbiotic bacteria and some pathogenic bacteria present within the lumen and in the intestinal epithelium. It can provide nutrients for bacterial growth, thus promoting intestinal colonization by the adhering bacteria which have the ability to survive and multiply in the outer regions of the mucus layer (Liévin-Le Moal and Servin 2006).

In response to enteropathogens infection, the intestinal epithelium releases the chemokine IL-8 and other pro-inflammatory molecules that provoke an acute inflammatory response. For the protection against enteropathogens infections, the possibility of using food supplements containing probiotic bacteria is increasingly considered. Selected strains have shown to be able to survive under gastrointestinal challenges, while they were shown to adhere to human epithelial intestinal cell monolayers (Caco-2 and HT29), thereby preventing the establishment of enteric pathogens as *E. coli* and *Cronobacter sakazakii* (Serafini et al. 2013). Furthermore, many probiotic strains have been assessed for their immunomodulatory activity on IL-8 production by HT29 cells, protecting them from an acute inflammatory response (Candela et al. 2008).

11.7   Conclusions

Despite several limitations of the human colon carcinoma cell lines, they are valuable tools to study several aspects related with food digestion and bioavailability of food compounds. HT29 cell line and some derived cell lines thereof are of interest to study food component absorption, especially when used in co-culture with other epithelial cell lines. Although the quantitative results can be questioned in comparison with those obtained in vivo, these in vitro models allow to identify modifications that the food compound may suffer during absorption and these results will help to perform further in vivo studies in animal or humans. The use of these human cells also offers a valuable opportunity to perform studies when a variance in response between different species is observed. For instance, the mucus secreting phenotype of HT29 cell line has proved to be a valuable model to screen food compounds or bacteria which may influence mucus secreting properties in the gut. Finally, it has provided results about the mechanisms by which microbes adhere, invade and signal to the host, as well as, to examine the mammalian cell response. Altogether, they are a complementary tool to the in vivo and ex vivo strategies to study food digestion and the effect of food components on the gut.

Acknowledgments   This work was supported by projects AGL2011-24643 from Ministerio de Economía y Competitividad and FP7-SME-2012-315349 (FOFIND). The authors are participants in the FA1005 COST Action INFOGEST on food digestion.

References

Ataee R, Ajdary S, Zarrindast M, Rezayat M, Hayatbakhsh MR (2010) Anti-mitogenic and apoptotic effects of 5-HT1B receptor antagonist on HT29 colorectal cancer cell line. J Cancer Res Clin Oncol 136(10):1461–1469. doi:10.1007/s00432-010-0801-3

Augeron C, Laboisse CL (1984) Emergence of permanently differentiated cell clones in a human colonic cancer cell line in culture after treatment with sodium butyrate. Cancer Res 44(9):3961–3969

Belo A, Cheng KR, Chahdi A, Shant J, Xie GF, Khurana S, Raufman JP (2011) Muscarinic receptor agonists stimulate human colon cancer cell migration and invasion. Am J Physiol Gastrointest Liver Physiol 300(5):G749–G760. doi:10.1152/ajpgi.00306.2010

Bourgine J, Billaut-Laden I, Happillon ML, Lo-Guidice J-M, Maunoury V, Imbenotte M, Broly F (2012) Gene expression profiling of systems involved in the metabolism and the disposition of xenobiotics: comparison between human intestinal biopsy samples and colon cell lines. Drug Metab Dispos 40(4):694–705. doi:10.1124/dmd.111.042465

Campo E, Condom E, Palacín A, Quesada E, Cardesa A (1988) Lectin binding patterns in normal and neoplastic colonic mucosa—a study of dolichos biflorus agglutinin, peanut agglutinin, and wheat germ agglutinin. Dis Colon Rectum 31(11):892–899. doi:10.1007/BF02554856

Candela M, Perna F, Carnevali P, Vitali B, Ciati R, Gionchetti P, Rizzello F, Campieri M, Brigidi P (2008) Interaction of probiotic *Lactobacillus* and *Bifidobacterium* strains with human intestinal epithelial cells: adhesion properties, competition against enteropathogens and modulation of IL-8 production. Int J Food Microbiol 125(3):286–292. doi:10.1016/j.ijfoodmicro.2008.04.012

Chen TR, Drabkowski D, Hay RJ, Macy M, Peterson W Jr (1987) WiDr is a derivative of another colon adenocarcinoma cell line, HT-29. Cancer Genet Cytogenet 27(1):125–134

Collins AR (2004) The comet assay for DNA damage and repair: principles, applications, and limitations. Mol Biotechnol 26(3):249–261. doi:10.1385/MB:26:3:249

Contreras MM, Sancho AI, Recio I, Mills C (2012) Absorption of casein antihypertensive peptides through an *in vitro* model of intestinal epithelium. Food Digestion 3(1–3):16–24. doi:10.1007/s13228-012-0020-2

D'Alessio PA, Ostan R, Bisson J, Schulzke JD, Ursini MV, Béné MC (2013) Oral administration of *d*-limonene controls inflammation in rat colitis and displays anti-inflammatory properties as diet supplementation in humans. Life Sci 92(24–26):1151–1156. doi:10.1016/j.lfs.2013.04.013

Davies PS, Enns CA (2004) Expression of the hereditary hemochromatosis protein HFE increases ferritin levels by inhibiting iron export in HT29 cells. J Biol Chem 279(24):25085–25092. doi:10.1074/jbc.M400537200

Desai S, Kumar A, Laskar S, Pandey BN (2013) Cytokine profile of conditioned medium from human tumor cell lines after acute and fractionated doses of gamma radiation and its effect on survival of bystander tumor cells. Cytokine 61(1):54–62. doi:10.1016/j.cyto.2012.08.022

Ferguson LR, Zhu S, Harris PJ (2005) Antioxidant and antigenotoxic effects of plant cell wall hydroxycinnamic acids in cultured HT-29 cells. Mol Nutr Food Res 49(6):585–593. doi:10.1002/mnfr.200500014

Fitzgerald RC, Omary MB, Triadafilopoulos G (1997) Acid modulation of HT29 cell growth and differentiation. An *in vitro* model for Barrett's oesophagus. J Cell Sci 110(5):663–671

Fogh J, Trempe G (1975) New human tumor cell lines. In: Fogh J (ed) Human tumor cell *in vitro*, 1st edn. Springer, New York, pp 115–159. doi:10.1007/978-1-4757-1647-4_5

Foltz M, Cerstiaens A, van Meensel A, Mols R, van der Pijl PC, Duchateau GSMJE, Augustijns P (2008) The angiotensin converting enzyme inhibitory tripeptides Ile-Pro-Pro and Val-Pro-Pro show increasing permeabilities with increasing physiological relevance of absorption models. Peptides 29(8):1312–1320. doi:10.1016/j.peptides.2008.03.021

Foreman JE, Chang W-CL, Palkar PS, Zhu B, Borland MG, Williams JL, Kramer LR, Clapper ML, Gonzalez FJ, Peters JM (2011) Functional characterization of peroxisome proliferator-activated receptor-β/δ expression in colon cancer. Mol Carcinog 50(11):884–900. doi:10.1002/mc.20757

Gravaghi C, Del Favero E, Cantu L, Donetti E, Bedoni M, Fiorilli A, Tettamanti G, Ferraretto A (2007) Casein phosphopeptide promotion of calcium uptake in HT-29 cells—relationship between biological activity and supramolecular structure. FEBS J 274(19):4999–5011. doi:10.1111/j.1742-4658.2007.06015.x

Guil-Guerrero JL, Ramos-Bueno R, Rodríguez-García I, López-Sánchez C (2011) Cytotoxicity screening of several tomato extracts. J Med Food 14(1–2):40–45. doi:10.1089/jmf.2010.0051

Hekmati M, Polak-Charcon S, Ben Shaul Y (1990) A morphological study of a human adenocarcinoma cell line (HT29) differentiating in culture. Similarities to intestinal embryonic development. Cell Differ Dev 31(3):207–218. doi:10.1016/0922-3371(90)90133-H

Hering NA, Andres S, Fromm A, van Tol EA, Amasheh M, Mankertz J, Fromm M, Schulzke JD (2011) Transforming growth factor-β, a whey protein component, strengthens the intestinal barrier by upregulating claudin-4 in HT-29/B6 cells. J Nutr 141(5):783–789. doi:10.3945/jn.110.137588

Huet C, Sahuquillo-Merino C, Coudrier E, Louvard D (1987) Absorptive and mucus-secreting subclones isolated from a multipotent intestinal cell line (HT-29) provide new models for cell polarity and terminal differentiation. J Cell Biol 105(1):345–358

Kaindl U, Eyberg I, Rohr-Udilova N, Heinzle C, Marian B (2008) The dietary antioxidants resveratrol and quercetin protect cells from exogenous pro-oxidative damage. Food Chem Toxicol 46(4):1320–1326. doi:10.1016/j.fct.2007.09.002

Kavanaugh DW, O'Callaghan J, Buttó LF, Slattery H, Lane J, Clyne M, Kane M, Joshi L, Hickey RM (2013) Exposure of *Bifidobacterium longum subsp. infantis* to milk oligosaccharides increases adhesion to epithelial cells and induces a substantial transcriptional response. PLoS One 8(6):e67224. doi:10.1371/journal.pone.0067224

Kitabgi P, Poustis C, Granier C, Van Rietschoten J, Rivier J, Morgat JL, Freychet P (1980) Neurotensin binding to extraneural and neural receptors: comparison with biological activity and structure-activity relationships. Mol Pharmacol 18(1):11–19

Kreusel K, Fromm M, Schulzke J, Hegel U (1991) Cl- secretion in epithelial monolayers of mucus-forming human colon cells (HT-29/B6). Am J Physiol Cell Physiol 261(4):C574–C582

Langerholc T, Maragkoudakis PA, Wollgast J, Gradisnik L, Cencic A (2011) Novel and established intestinal cell line models—an indispensable tool in food science and nutrition. Trends Food Sci Technol 22(Suppl 1):S11–S20. doi:10.1016/j.tifs.2011.03.010

Lenaerts K, Bouwman F, Lamers W, Renes J, Mariman E (2007) Comparative proteomic analysis of cell lines and scrapings of the human intestinal epithelium. BMC Genomics 8:91. doi:10.1186/1471-2164-8-91

Lesuffleur T, Barbat A, Dussaulx E, Zweibaum A (1990) Growth adaptation to methotrexate of HT-29 human colon carcinoma cells is associated with their ability to differentiate into columnar absorptive and mucus-secreting cells. Cancer Res 50(19):6334–6343

Lesuffleur T, Porchet N, Aubert J, Swallow D, Gum JR, Kim YS, Zweibaum A (1993) Differential expression of the human mucin genes MUC1 to MUC5 in relation to growth and differentiation of different mucus-secreting HT-29 cell subpopulations. J Cell Sci 106(3):771–783

Liévin-Le Moal V, Servin AL (2006) The front line of enteric host defense against unwelcome intrusion of harmful microorganisms: mucins, antimicrobial peptides, and microbiota. Clin Microbiol Rev 19(2):315–337. doi:10.1128/CMR. 19.2.315-337.2006

Lipps C, May T, Hauser H, Wirth D (2013) Eternity and functionality-rational access to physiologically relevant cell lines. Biol Chem 394(12):1637–1648. doi:10.1515/hsz-2013-0158

Maoret JJ, Font J, Augeron C, Codogno P, Bauvy C, Aubery M, Laboisse CL (1989) A mucus-secreting human colonic cancer cell line. Purification and partial secretion of the secreted mucins. Biochem J 258(3):793–799

Martínez-Maqueda D, Miralles B, De Pascual-Teresa S, Reverón I, Muñoz R, Recio I (2012) Food-derived peptides stimulate mucin secretion and gene expression in intestinal cells. J Agric Food Chem 60(35):8600–8605. doi:10.1021/jf301279k

Martínez-Maqueda D, Miralles B, Ramos M, Recio I (2013a) Effect of β-lactoglobulin hydrolysate and β-lactorphin on intestinal mucin secretion and gene expression in human goblet cells. Food Res Int 54(1):1287–1291. doi:10.1016/j.foodres.2012.12.029

Martínez-Maqueda D, Miralles B, Cruz-Huerta E, Recio I (2013b) Casein hydrolysate and derived peptides stimulate mucin secretion and gene expression in human intestinal cells. Int Dairy J 32(1):13–19. doi:10.1016/j.idairyj.2013.03.010

Nickerson KP, McDonald C (2012) Crohn's disease-associated adherent-invasive *Escherichia coli* adhesion is enhanced by exposure to the ubiquitous dietary polysaccharide maltodextrin. PLoS One 7(12):e52132. doi:10.1371/journal.pone.0052132

Nishikawa A, Uotsu N, Arimitsu H, Lee J-C, Miura Y, Fujinaga Y, Nakada H, Watanabe T, Ohyama T, Sakano Y, Oguma K (2004) The receptor and transporter for internalization of *Clostridium botulinum* type C progenitor toxin into HT-29 cells. Biochem Biophys Res Commun 319(2):327–333. doi:10.1016/j.bbrc.2004.04.183

Otte J-M, Podolsky DK (2004) Functional modulation of enterocytes by gram-positive and gram-negative microorganisms. Am J Physiol Gastrointest Liver Physiol 286(4):G613–G626

Pérez-Vega JA, Olivera-Castillo L, Gómez-Ruiz JT, Hernández-Ledesma B (2013) Release of multifunctional peptides by gastrointestinal digestion of sea cucumber (*Isostichopus badionotus*). J Funct Foods 5(2):869–877. doi:10.1016/j.jff.2013.01.036

Pinto M, Appay MD, Simon-Assman P, Chevalier G, Dracopoli N, Fogh J, Zweibaum A (1982) Enterocytic differentiation of cultured human colon cancer cells by replacement of glucose by galactose in the medium. Biol Cell 44(2):193–196

Plaisancié P, Claustre J, Estienne M, Henry G, Boutrou R, Paquet A, Léonil J (2013) A novel bioactive peptide from yoghurts modulates expression of the gel-forming MUC2 mucin as well as population of goblet cells and paneth cells along the small intestine. J Nutr Biochem 24(1):213–221. doi:10.1016/j.jnutbio.2012.05.004

Quirós A, Dávalos A, Lasunción MA, Ramos M, Recio I (2008) Bioavailability of the antihypertensive peptide LHLPLP: transepithelial flux of HLPLP. Int Dairy J 18(3):279–286. doi:10.1016/j.idairyj.2007.09.006

Rieder A, Grimmer S, Kolset SO, Michaelsen TE, Knutsen SH (2011) Cereal β-glucan preparations of different weight average molecular weights induce variable cytokine secretion in human intestinal epithelial cell lines. Food Chem 128(4):1037–1043. doi:10.1016/j.foodchem.2011.04.010

Serafini F, Strati F, Ruas-Madiedo P, Turroni F, Foroni E, Duranti S, Milano F, Perotti A, Viappiani A, Guglielmetti S, Buschini A, Margolles A, van Sinderen D, Ventura M (2013) Evaluation of adhesion properties and antibacterial activities of the infant gut commensal *Bifidobacterium bifidum* PRL2010. Anaerobe 21:9–17. doi:10.1016/j.anaerobe.2013.03.003

Wagar LE, Champagne CP, Buckley ND, Raymond Y, Green-Johnson JM (2009) Immunomodulatory properties of fermented soy and dairy milks prepared with lactic acid bacteria. J Food Sci 74(8):M423–M430. doi:10.1111/j.1750-3841.2009.01308.x

Yuan H, Chen CY, Chai GH, Du YZ, Hu FQ (2013) Improved transport and absorption through gastrointestinal tract by pegylated solid lipid nanoparticles. Mol Pharm 10(5):1865–1873. doi:10.1021/mp300649z

Zoghbi S, Trompette A, Claustre J, El Homsi M, Garzon J, Scoazec JY, Plaisancie P (2006) Beta-Casomorphin-7 regulates the secretion and expression of gastrointestinal mucins through a mu-opioid pathway. Am J Physiol Gastrointest Liver Physiol 290(6):G1105–G1113. doi:10.1152/ajpgi.00455.2005

Zweibaum A (1986) Enterocytic differentiation of cultured human colon cancer cell lines: negative modulation by D-glucose. In: Alvarado F, van Os CH (eds) Ion gradient-coupled transport. Elsevier, Amsterdam, pp 345–353

Zweibaum A, Laburthe M, Grasset E, Louvard D (2011) Use of cultured cell lines in studies of intestinal cell differentiation and function. Compr Physiol (Suppl 19):223–255. doi:10.1002/cphy.cp060407

Chapter 12
The IPEC-J2 Cell Line

Hans Vergauwen

Abstract IPEC-J2 cells are intestinal porcine enterocytes isolated from the jejunum of a neonatal unsuckled piglet. The IPEC-J2 cell line is unique as it is derived from the small intestine and is neither transformed nor tumorigenic in nature. IPEC-J2 cells mimic the human physiology more closely than any other cell line of non-human origin. Therefore, it is an ideal tool to study epithelial transport, interactions with enteric bacteria, effects of probiotic microorganisms and the effect of nutrients and other feedstuffs on a variety of widely used parameters (e.g. transepithelial electrical resistance (TEER), permeability, metabolic activity) reflecting epithelial functionality.

IPEC-J2 cells undergo in culture a process of spontaneous differentiation that leads to the formation of a polarized monolayer with low or high TEER, depending on the type of serum added to the culture medium, within 1–2 weeks. Porcine serum gives rise to low resistance and normal active transport rates, enabling comparison with the in vivo situation. The high resistance caused by fetal bovine serum can be beneficial to use when investigating compounds having a negative effect on the monolayer permeability or tight junction structures.

There are still many opportunities for exploring the use of these cells as the available research is limited. This chapter will cover the origin, characteristics and methods of the use of IPEC-J2 cells as an in vitro model for several research applications, as well as comparisons between IPEC-J2 cells and other epithelial cell lines.

Keywords IPEC-J2 • Cell line • Small intestinal epithelium • Porcine origin • Transepithelial electrical resistance • Non-transformed • Continuous

12.1 Origin

The intestinal porcine enterocyte cell line (IPEC-J2) is a non-transformed, permanent intestinal cell line (Fig. 12.1). These secondary cells were originally isolated from the jejunal epithelium of a neonatal unsuckled piglet in 1989 by Helen Berschneider

H. Vergauwen (✉)
Department of Veterinary Sciences, Laboratory of Applied Veterinary Morphology, University of Antwerp, Universiteitsplein 1, 2610 Wilrijk, Belgium
e-mail: hans.vergauwen@uantwerpen.be; chris.vanginneken@uantwerpen.be

K. Verhoeckx et al. (eds.), *The Impact of Food Bio-Actives on Gut Health*,
DOI 10.1007/978-3-319-16104-4_12

125

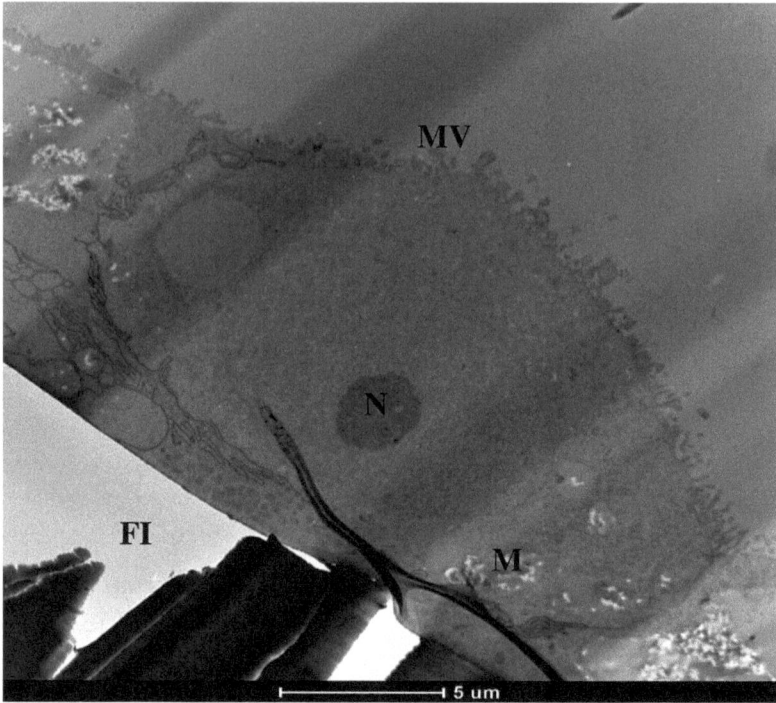

Fig. 12.1 Transmission electron microscopical image of an IPEC-J2 cell grown on a filter insert with fetal bovine serum supplemented to the medium. *M* mucopolysaccharides, *MV* microvilli, *N* nucleus, *FI* filter insert, scale bar: 5 μm

at the University of North Carolina (Berschneider 1989). Given the correct culture conditions, these cells will divide and grow for an infinite number of passages in vitro. To date, they have been cultured continuously for up to 98 passages. IPEC-J2 cells were first used to investigate transepithelial ion transport and enterocyte differentiation (Berschneider 1989). They have already been shown to be a valuable tool in the characterization of epithelial cell interactions with enteric bacteria and viruses providing insight into initial host-pathogen and non-pathogen (e.g. commensal or probiotic) interactions (Brosnahan and Brown 2012).

12.2   Special Features/Morphology/Receptors

The strength of the IPEC-J2 cell line as an in vitro model originates from its morphological and functional similarities with intestinal epithelial cells in vivo. IPEC-J2 cells have microvilli on their apical side and tight junctions sealing neighboring cells together (Schierack et al. 2006). To date, no brush border enzyme activity has been investigated in IPEC-J2 cells. IPEC-J2 cells form polarized monolayers when cultured on 0.4 μm pore-size filters, with or without a collagen basis. High transepithelial electrical resistance (TEER) values and low active transport rates are obtained when culturing the IPEC-J2 cells in fetal bovine serum (FBS) (Geens and Niewold 2011). These atypically high TEER values can be beneficial to use when investigating compounds having a negative effect on the monolayer permeability or tight junction structures (Vergauwen et al. 2015). Porcine serum (PS) resulted in significantly lower TEER values and higher active transport rates comparable to the in vivo situation (Zakrzewski et al. 2013). When IPEC-J2 cells are grown in 10 % PS they will become taller and smaller in diameter, increasing the tight junction ultrastructure (Zakrzewski et al. 2013).

The protein expression of claudin-1, -3, -4, -5, -7, -8, -12, tricellulin, occludin, E-cadherin and zonula occludens-1 (ZO-1) by IPEC-J2 cells has been confirmed by immunoblotting (Zakrzewski et al. 2013). On the other hand, IPEC-J2 cells do not express claudin-2 and -15 resulting in lower cation selectivity, augmenting the ion permeability of tight junctions compared to the porcine jejunum (Zakrzewski et al. 2013; Schierack et al. 2006).

IPEC-J2 cells express and produce cytokines, defensins, toll-like receptors and mucins. The presence of glycocalyx-bound mucins like Muc1 and Muc3 has been confirmed in IPEC-J2 cells, respectively by RT-PCR and ELISA (Schierack et al. 2006). The expression of the Muc2, the major gel-forming mucin in the human small intestine, was not detectable by RT-PCR (Schierack et al. 2006). Thus, the mucus production by IPEC-J2 cells is not comparable to the in vivo situation. The mucus layer is only a very thin layer that is superimposed on the IPEC-J2 cells when cultured with fetal bovine serum. However, PAS staining showed an increase in mucus production when IPEC-J2 cells were cultured with porcine serum (Zakrzewski et al. 2013).

IPEC-J2 cells express proteins such as MHC I and secrete cytokines like GM-CSF and TNF-α that can establish communication between enterocytes and the immune system. Furthermore, mRNA expression of TLR-1, -2, -3, -4, -5, -6, -8, -9, -10 and IL-1α, -1β, -6, -7, -8, -12A, -12B, -18 has been confirmed in IPEC-J2 cells. TLR-2, -3, -5, -9 and IL-6, -8 proteins were also detected in these cells (Brosnahan and Brown 2012; Arce et al. 2010; Burkey et al. 2009).

IPEC-J2 cells express the F4 fimbrial receptor but not the F18 fimbrial receptor as this has been correlated to older pigs (3–23 weeks). These features make IPEC-J2 cells an interesting in vitro model to investigate the pathogenesis of zoonotic enteric infections that also affect humans.

12.3   Stability/Consistency/Reproducibility of the System

In current research IPEC-J2 cells have been used until passage 98. However, most studies do not mention the passage number. Interlaboratory result evaluation presented a consensus that a confluent IPEC-J2 cell monolayer grown in 5 % fetal bovine serum corresponds to a TEER value of $1,000 \ \Omega \times cm^2$. TEER experiments are usually started at $1,000$–$3,000 \ \Omega \times cm^2$. These values are reached after 4–9 days of culturing on an insert with a semi-permeable membrane (0.4 µm pore size) with or without collagen coating. However, when IPEC-J2 cells are incubated with 10 % PS instead of 5 % FBS TEER values of 400–$500 \ \Omega \times cm^2$ are obtained. These TEER values are comparable to those found in Caco-2 cell cultures and in vivo.

The amount of cells seeded on the insert depends on the insert size and brand. It has been demonstrated that seeding density can influence TEER values over time, as well as proliferation capacity and functionality. When a 12-well plate insert is seeded with 1–10×10^5 cells/insert, minimal cell division is needed.

Results obtained from microbial investigations using IPEC-J2 cells show strong correlation with mucosal explants and in vivo. However, various EHEC mutants adhere differently to IPEC-J2 cells and porcine ileal loops. This indicates that different model systems may behave differently (Yin et al. 2009). Nonetheless, the ability of some EHEC mutants to adhere to IPEC-J2 cells and ileal loops was highly correlated.

12.4 Relevance to Human In Vivo Situation

IPEC-J2, IPEC-1 and IPI-2I are three widely used porcine small intestinal cell lines (Arce et al. 2010). IPI-2I cells have been transformed with an SV40 plasmid, whereas IPEC-J2 and IPEC-1 are non-transformed cell lines. IPI-2I cells were isolated from the ileum, IPEC-J2 cells from the jejunum and IPEC-1 cells were isolated from a mixture of ileal and jejunal tissue. IPEC-1 and IPEC-J2 cells were both isolated from one day old piglets, whereas IPI-2I cells were isolated from an adult boar (Berschneider 1989). Furthermore, IPEC-J2 cells, unlike IPI-2I cells, form a polarized monolayer, promoting *S. typhimurium* invasion and replication (Boyen et al. 2009).

Several normal and transformed cell lines are used in food science (Cencic and Langerholc 2010). Caco-2, T84, HT-29, HUTU-80 and SW620 are the most widely used human intestinal cell lines. The majority of human intestinal cell lines is isolated from the colon, and most are tumorigenic. The HUTU-80 cell line is the only widely available small intestinal human cell line. It was derived from the duodenum, but is also cancerous (Brosnahan and Brown 2012).

A drug transport permeability study showed that IPEC-J2 cells form a tighter monolayer compared to Caco-2 cells (Pisal et al. 2008). On the other hand, Caco-2 and T84 cells produce a more pronounced mucus layer compared to IPEC-J2 cells (Navabi et al. 2013). HT29, T84, Caco-2, and SW620 are all negative for IL-2, IL-4, and IFN-γ, while mRNA for IL-1α, IL-8, and TNF-α is variable amongst the human cell lines, but is present in IPEC-J2 cells (Eckmann et al. 1993). Existing cancer-derived cell lines can have limitations such as an altered glycosylation pattern and an aberrant protein expression that define the epithelial character as well as unresponsiveness to hormones or cytokines (Peracaula et al. 2008).

Two major advantages favor IPEC-J2 cells as a model of normal intestinal epithelial cells compared to transformed cell lines: (1) as a permanent cell line they maintain their differentiated characteristics and exhibit strong similarities to primary intestinal epithelial cells, and (2) IPEC-J2 cells have the advantage of being directly comparable to the experimental animal that is used as an in vivo model for humans. Of all non-primates, the gastrointestinal (GI) tract of pigs is the most appropriate in vivo model as it is similar in size, weight, anatomy and physiology as the human GI system (Deglaire and Moughan 2012).

Due to the close similarity between swine and human intestinal function, studies with IPEC-J2 cells provide valuable insights into the pathogenesis of zoonotic enteric infections that also affect humans (Skjolaas et al. 2006). Extrapolating information can be difficult as in vitro, ex vivo and in vivo cells or tissue can respond differently to environmental stimuli (e.g. diet). Indeed, diet-induced gene expression patterns differ between IPEC-J2 cells and intestinal tissue from preterm and newborn piglets, making interpretation rather difficult (Støy et al. 2013). However, an in vitro model is still a valuable tool to investigate a limited number of factors in a standardized, regulated setting. A comparison of normal diploid IPEC-J2 cells with other transformed or tumorigenic cell lines can give more insight when investigating gene expression and to greater extent the effects of bioactives on intestinal health.

12.5 General Protocol

12.5.1 Culture Conditions

The IPEC-J2 cells are cultured in DMEM/F-12 mix (Dulbecco's Modified Eagle Medium, Ham's F-12 mixture) and supplemented with HEPES, fetal bovine serum (FBS) or porcine serum (PS), insulin/transferrin/selenium (ITS), penicillin/streptomycin and cultivated in a humid environment at 37 °C with 5 % CO_2. The IPEC-J2 cells are usually grown for 1–2 weeks before initiating an experiment. When studying TEER and permeability, IPEC-J2 cells are commonly seeded (1×10^5 cells/well) at confluence in a 'Boyden chamber' insert (upper chamber, apical) on a polyethylene terephthalate (PET) membrane (1.12 cm^2, pore size 0.4 μm) in a 12-well plate (lower chamber, basolateral). Cells are seeded (1×10^5/well) in a 12-well plate (flat bottom) to investigate the intracellular oxidative stress and wound healing capacity. Cells are seeded (0.5×10^4 cells/well) in a 96-well plate (flat bottom) to assess viability. The IPEC-J2 cell line is an easy to use and robust cell line, exhibiting structural and functional differentiation pattern characteristics of mature enterocytes (Geens and Niewold 2011).

12.5.2 Experimental Readout

An increasing number of studies use IPEC-J2 cells to investigate interactions of various animal and human pathogens, including *Salmonella enterica* and pathogenic *Escherichia coli* (Boyen et al. 2009; Veldhuizen et al. 2009). Pathogenic permeation (e.g. *E. coli*) is usually presented as colony forming units (CFU). IPEC-J2 cells have also been employed as an initial screening tool for adhesiveness and anti-inflammatory properties of potential probiotic microorganisms. IPEC-J2 cells were also used to investigate the effect of prebiotics on the adhesion of probiotic bacterial strains to these cells. Addition of 200 mM calcium has been shown to increase adhesion (Marcinakova et al. 2010), while magnesium and zinc ions had no influence (Larsen et al. 2007). Innate immune responses (e.g. increase in porcine β-defensin 1 and 2 gene expression) in relation to environmental stimuli (e.g. diet or infection) are investigated with relevance for human and porcine intestinal diseases, specifically in newborns (Schierack et al. 2006; Burkey et al. 2009; Veldhuizen et al. 2009). Numerous studies used IPEC-J2 cells to investigate feedstuffs and antioxidants in relation to inflammation, intestinal permeability and wound healing capacity (Hermes et al. 2011; Ma et al. 2012; Pan et al. 2013; Vergauwen et al. 2015).

Permeability is expressed using TEER, either presented as absolute values or as percentages of control or time point (TEERtx/TEERt0). The net value of the TEER ($\Omega \times cm^2$) needs to be corrected for background resistance by subtracting the contribution of the cell-free filter and the medium (80–150 $\Omega \times cm^2$). Alternatively fluorescein isothiocyanate (FITC)-dextran of 4 kDa (FD-4) permeability can be used to indicate monolayer integrity. FD-4 permeability results are presented as a percentage of control or as absolute quantities or concentrations (e.g. picomoles).

Viability and cytotoxicity are most commonly analyzed using the neutral red method, lactate dehydrogenase release or the MTT reduction assay and presented as percentages of control or absorbance values (Table 12.1).

12.5.3 Sample Preparation

It is important only to incubate sterile filtered samples on the IPEC-J2 cells. Otherwise, interpretation of the results will be ambiguous as the effect cannot be contributed to an impurity, a pathogen or the agent of interest.

Samples that are not readily dissolved in water can be dissolved in ethanol or DMSO. The effect of different concentrations of ethanol or DMSO for either 1 or 18 h on the viability of IPEC-J2 cells was investigated (Fig. 12.2). Results show that short and long term incubation do not favor DMSO or ethanol at concentrations below 1 %. DMSO is favored for long-term incubation at concentrations above 1 %. However, concentrations exceeding 1 % are not recommended. Furthermore, it is always important to minimize the concentration of DMSO or ethanol when solubilizing a compound.

12.6 Conclusion

In summary, the IPEC-J2 cell model provides a perfect tool to investigate intestinal epithelial function. This porcine cell line is closely related to the human in vivo situation and is not cancerous compared to other human (small) intestinal cell lines.

Table 12.1 Experimental read-out of specific assays applicable to the IPEC-J2 cell model

Application	Feedstuff/component/toxin/bacteria/virus	Read-out	Positive/negative controls	References
Barrier/transport	FITC-labelled soybean protein P34	Mean fluorescence intensity (MFI)	Negative control: Incubation at 4 °C → no endocytosis, only binding	Sewekow et al. (2012)
Inflammation	Feedstuffs (wheat bran, casein glycomacropeptide, mannan–oligosaccharides, locust bean and *Aspergillus oryzae* fermentation extract)	Relative abundance (%) (RT-qPCR)	Positive control: sterile PBS, no feedstuffs	Hermes et al. (2011)
Proliferation	Possible toxic potential of Cry1Ab protein, commonly expressed in GM-maize	WST-1 conversion (OD$_{450nm}$)	Negative control: No serum	Bondzio et al. (2013)
Adhesion/translocation	Effect of feedstuffs on the numbers of adhered *E. coli* K88 per well	Detection time (h)	Negative control: PBS	Hermes et al. (2011)
Cross-talk	Gut-mediated changes in gene expression of hepatic CYP enzymes by LPS	Relative CYP gene expression	Negative control: No LPS	Paszti-Gere et al. (2014)
Toxicity/viability	*Fusarium* toxin deoxynivalenol (DON)	Lactate dehydrogenase release (U/L)	Vehicle control: 0.5 % ethanol	Awad et al. (2012)
Wound healing assay	Sodium butyrate (SB)	Wound size (μm)	Negative control: No SB	Ma et al. (2012)
Intracellular oxidative stress	Trolox, ascorbic acid and glutathione monoethyl ester	Mean fluorescence intensity (MFI) using CM-H$_2$DCFDA	Negative control: No antioxidant, no oxidant	Vergauwen et al. (2015)
Intestinal permeability (indirect method)	Soybean agglutinin (SBA)	Alkaline phosphatase (IU/L)	Negative control: No SBA	Pan et al. (2013)

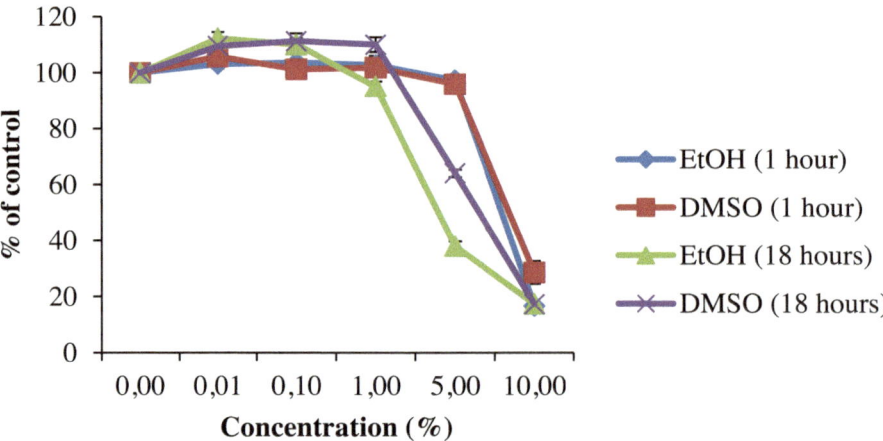

Fig. 12.2 Effect of ethanol and DMSO on the IPEC-J2 cell viability. IPEC-J2 cells were incubated with different concentrations of ethanol (EtOH) and DMSO for 1 or 18 h before assessing the viability using the neutral red method. Results are presented as means ± S.E.M., n = 12

The tumorigenic nature of the human intestinal cell lines can influence gene expression, and transformed cell lines are usually more resistant to stress or cytotoxic insults. This will result in ambiguous information and an underestimation of cytotoxic compounds. Furthermore, IPEC-J2 cell monolayers are ready for experimentation after 1–2 weeks which is a lot faster compared to the 21-day culturing time of Caco-2 cells. In conclusion, IPEC-J2 cells are an ideal small intestinal enterocyte model to study effects of food bioactives in the gut prior to in vivo evaluation.

References

Arce C, Ramirez-Boo M, Lucena C, Garrido JJ (2010) Innate immune activation of swine intestinal epithelial cell lines (IPEC-J2 and IPI-2I) in response to LPS from Salmonella typhimurium. Comp Immunol Microbiol Infect Dis 33(2):161–174. doi:10.1016/j.cimid.2008.08.003

Awad WA, Aschenbach JR, Zentek J (2012) Cytotoxicity and metabolic stress induced by deoxynivalenol in the porcine intestinal IPEC-J2 cell line. J Anim Physiol Anim Nutr 96(4):709–716. doi:10.1111/j.1439-0396.2011.01199.x

Berschneider HM (1989) Development of normal cultured small intestinal epithelial cell lines which transport Na and Cl. Gastroenterology 96:A41

Bondzio A, Lodemann U, Weise C, Einspanier R (2013) Cry1Ab treatment has no effects on viability of cultured porcine intestinal cells, but triggers Hsp70 expression. PLoS One 8(7):e67079. doi:10.1371/journal.pone.0067079

Boyen F, Pasmans F, Van Immerseel F, Donne E, Morgan E, Ducatelle R, Haesebrouck F (2009) Porcine in vitro and in vivo models to assess the virulence of Salmonella enterica serovar Typhimurium for pigs. Lab Anim 43(1):46–52. doi:10.1258/la.2007.007084

Brosnahan AJ, Brown DR (2012) Porcine IPEC-J2 intestinal epithelial cells in microbiological investigations. Vet Microbiol 156(3–4):229–237. doi:10.1016/j.vetmic.2011.10.017

Burkey TE, Skjolaas KA, Dritz SS, Minton JE (2009) Expression of porcine Toll-like receptor 2, 4 and 9 gene transcripts in the presence of lipopolysaccharide and Salmonella enterica serovars Typhimurium and Choleraesuis. Vet Immunol Immunopathol 130(1–2):96–101. doi:10.1016/j.vetimm.2008.12.027

Cai X, Chen X, Wang X, Xu C, Guo Q, Zhu L, Zhu S, Xu J (2013) Pre-protective effect of lipoic acid on injury induced by H(2)O (2) in IPEC-J2 cells. Mol Cell Biochem. doi:10.1007/s11010-013-1595-9

Cencic A, Langerholc T (2010) Functional cell models of the gut and their applications in food microbiology—a review. Int J Food Microbiol 141(Suppl 1):S4–S14. doi:10.1016/j.ijfoodmicro.2010.03.026

Deglaire A, Moughan PJ (2012) Animal models for determining amino acid digestibility in humans – a review. Br J Nutr 108(Suppl 2):S273–S281. doi:10.1017/s0007114512002346

Eckmann L, Jung HC, Schurer-Maly C, Panja A, Morzycka-Wroblewska E, Kagnoff MF (1993) Differential cytokine expression by human intestinal epithelial cell lines: regulated expression of interleukin 8. Gastroenterology 105(6):1689–1697

Geens MM, Niewold TA (2011) Optimizing culture conditions of a porcine epithelial cell line IPEC-J2 through a histological and physiological characterization. Cytotechnology 63(4):415–423. doi:10.1007/s10616-011-9362-9

Hermes RG, Manzanilla EG, Martin-Orue SM, Perez JF, Klasing KC (2011) Influence of dietary ingredients on in vitro inflammatory response of intestinal porcine epithelial cells challenged by an enterotoxigenic Escherichia coli (K88). Comp Immunol Microbiol Infect Dis 34(6):479–488. doi:10.1016/j.cimid.2011.08.006

Larsen N, Nissen P, Willats WG (2007) The effect of calcium ions on adhesion and competitive exclusion of Lactobacillus ssp. and E. coli O138. Int J Food Microbiol 114(1):113–119. doi:10.1016/j.ijfoodmicro.2006.10.033

Ma X, Fan PX, Li LS, Qiao SY, Zhang GL, Li DF (2012) Butyrate promotes the recovering of intestinal wound healing through its positive effect on the tight junctions. J Anim Sci 90(Suppl 4):266–268. doi:10.2527/jas.50965

Marcinakova M, Klingberg TD, Laukova A, Budde BB (2010) The effect of pH, bile and calcium on the adhesion ability of probiotic enterococci of animal origin to the porcine jejunal epithelial cell line IPEC-J2. Anaerobe 16(2):120–124. doi:10.1016/j.anaerobe.2009.05.001

Navabi N, McGuckin MA, Linden SK (2013) Gastrointestinal cell lines form polarized epithelia with an adherent mucus layer when cultured in semi-wet interfaces with mechanical stimulation. PLoS One 8(7):e68761. doi:10.1371/journal.pone.0068761

Pan L, Qin G, Zhao Y, Wang J, Liu F, Che D (2013) Effects of soybean agglutinin on mechanical barrier function and tight junction protein expression in intestinal epithelial cells from piglets. Int J Mol Sci 14(11):21689–21704. doi:10.3390/ijms141121689

Paszti-Gere E, Matis G, Farkas O, Kulcsar A, Palocz O, Csiko G, Neogrady Z, Galfi P (2014) The effects of intestinal LPS exposure on inflammatory responses in a porcine enterohepatic co-culture system. Inflammation 37(1):247–260. doi:10.1007/s10753-013-9735-7

Peracaula R, Barrabes S, Sarrats A, Rudd PM, de Llorens R (2008) Altered glycosylation in tumours focused to cancer diagnosis. Dis Markers 25(4–5):207–218

Pisal DS, Yellepeddi VK, Kumar A, Palakurthi S (2008) Transport of surface engineered polyamidoamine (PAMAM) dendrimers across IPEC-J2 cell monolayers. Drug Deliv 15(8):515–522. doi:10.1080/10717540802321826

Schierack P, Nordhoff M, Pollmann M, Weyrauch KD, Amasheh S, Lodemann U, Jores J, Tachu B, Kleta S, Blikslager A, Tedin K, Wieler LH (2006) Characterization of a porcine intestinal epithelial cell line for in vitro studies of microbial pathogenesis in swine. Histochem Cell Biol 125(3):293–305. doi:10.1007/s00418-005-0067-z

Sewekow E, Bimczok D, Kahne T, Faber-Zuschratter H, Kessler LC, Seidel-Morgenstern A, Rothkotter HJ (2012) The major soyabean allergen P34 resists proteolysis in vitro and is transported through intestinal epithelial cells by a caveolae-mediated mechanism. Br J Nutr 1–9. doi:10.1017/s0007114511007045

Skjolaas KA, Burkey TE, Dritz SS, Minton JE (2006) Effects of Salmonella enterica serovars Typhimurium (ST) and Choleraesuis (SC) on chemokine and cytokine expression in swine ileum and jejunal epithelial cells. Vet Immunol Immunopathol 111(3–4):199–209. doi:10.1016/j.vetimm.2006.01.002

Støy ACF, Heegaard PMH, Sangild PT, Østergaard MV, Skovgaard K (2013) Gene expression analysis of the IPEC-J2 cell line: a simple model for the inflammation-sensitive preterm intestine. ISRN Genomics Volume 2013 (Article ID 980651):7 pages

Veldhuizen EJ, Koomen I, Ultee T, van Dijk A, Haagsman HP (2009) Salmonella serovar specific upregulation of porcine defensins 1 and 2 in a jejunal epithelial cell line. Vet Microbiol 136(1–2):69–75. doi:10.1016/j.vetmic.2008.09.072

Vergauwen H, Tambuyzer B, Jennes K, Degroote J, Wang W, De Smet S, Michiels J, Van Ginneken C (2015) Trolox and ascorbic acid reduce direct and indirect oxidative stress in the IPEC-J2 cells, an in vitro model for the porcine gastrointestinal tract. PLoS One 10(3):e0120485. doi:10.1371/journal.pone.0120485

Yin X, Chambers JR, Wheatcroft R, Johnson RP, Zhu J, Liu B, Gyles CL (2009) Adherence of Escherichia coli O157:H7 mutants in vitro and in ligated pig intestines. Appl Environ Microbiol 75(15):4975–4983. doi:10.1128/aem.00297-09

Zakrzewski SS, Richter JF, Krug SM, Jebautzke B, Lee IF, Rieger J, Sachtleben M, Bondzio A, Schulzke JD, Fromm M, Gunzel D (2013) Improved cell line IPEC-J2, characterized as a model for porcine jejunal epithelium. PLoS One 8(11):e79643. doi:10.1371/journal.pone.0079643

Chapter 13
Co-cultivation of Caco-2 and HT-29MTX

Charlotte R. Kleiveland

Abstract The intestinal epithelium is one of the body's largest mucosal surfaces and it generates a physical barrier against the external environment. The majority of cells lining the epithelium are absorptive enterocytes with mainly metabolic and digestive functions. Hence, the diversity of functions the intestinal epithelium carries out depends on the presence of additional specialized intestinal epithelial cells (IEC). Secretory IEC, goblet cells, enteroendocrine cells and Paneth cells are specialized cells that participate in maintaining the digestive and barrier functions of the epithelium. Goblet cells release mucins, which give rise to a mucus layer on the epithelial surface that functions as physical and biochemical barrier for luminal content. The presence of the different epithelial cell types in an in vitro model will affect how well the model reflects the properties of the intestinal epithelium. We here describe a co-cultivation system of enterocytes and goblet cells, which are the two major epithelial cell types.

Keywords Co-culture • Enterocytes • Goblet cells • Transport • Absorption • Transepithelial electrical resistance (TEER)

13.1 Origin, Features and Mechanisms

The models described in this section include the cell lines Caco-2 and HT-29. Description of the origin, features and mechanisms of these cells are included in previous sections, and further information is found in Chap. 10 for Caco-2, Chap. 11 for HT-29.

C.R. Kleiveland (✉)
Department of Chemistry, Biotechnology and Food Science, Faculty of Veterinary Medicine and Biosciences, Norwegian University of Life Sciences, Ås, Norway

Research Department, Ostfold Hospital Trust, Fredrikstad, Norway
e-mail: charlotte.kleiveland@nmbu.no

© The Author(s) 2015
K. Verhoeckx et al. (eds.), *The Impact of Food Bio-Actives on Gut Health*,
DOI 10.1007/978-3-319-16104-4_13

135

13.2   Stability/Consistency and Reproducibility

The stability and reproducibility of the co-culture model of Caco-2 and HT-29MTX is very similar to the monocultures. The cells are cultivated together during the whole polarization period of 21 days. The co-cultures should be used for experiments between day 21 and 25. See Chap. 10 for Caco-2 cells and for HT29MTX Chap. 11.

13.3   Relevance to the Human In Vivo Situation

The intestinal epithelium is the main barrier preventing molecules from the lumen (e.g. food and toxins) reaching the systemic circulation. The epithelium is composed of several cell types: enterocytes, goblet cells, Paneth cells, enteroendocrine cells and stem cells. However, absorptive and goblet cells constitute the two major cell types of the intestine. The cell lines Caco-2 and HT-29MTX are derived from intestinal absorptive and goblet cell types, respectively. The human intestinal cell line Caco-2 differentiates into enterocytes (Shah et al. 2006), and is extensively used as an in vitro model of the small intestine, particularly to determine the permeability of the intestinal barrier to drug and food components (Sambuy et al. 2005). Caco-2 is of human colonic origin; however, when grown in culture, the cells exhibit many properties of the small intestinal epithelium. They form a well-differentiated polarized monolayer of columnar absorptive cells that express brush border with typical small intestinal enzymes and transporters on their apical surface (Artursson 1991; Hilgendorf et al. 2000). A disadvantages of the Caco-2 model and other monocultures of epithelial cells are that they do not closely simulate the composition of the normal epithelial layer with several types of cells (Hidalgo 1996). The epithelial cell layer is separated from the luminal content by a mucus layer. The mucus layer acts as physical and chemical defence against food particles, chemicals, enzymes and host-secreted products such as bile acids, microbiota and microbial products (Johansson et al. 2008). Only goblet cells are mucus-secreting cells, hence a mucus layer will be lacking in the Caco-2 model. However, the HT-29MTX cells produce both the membrane bound MUC1 and the gel-forming MUC5 (for more information on HT-29 cells see Chap. 11). A co-cultivation of the two cell lines will therefore provide a model constituting the two cell types that are most represented in normal epithelium, enterocytes and goblet cells. In addition the mucin secretion from HT-29 will form a layer on top of the epithelial cells. Presence of mucus in the model system is important for estimation of intestinal permeability as the mucus acts as a barrier against the absorption of certain compounds, particularly those that are lipophilic (Behrens et al. 2001). The lack of mucus allows highly diffusible small molecules easy access to the cells, which often results in an overestimation of permeability of such compounds.

Even though the Caco-2 cells have been widely employed to measure drug and nutrient transport, this model has been criticized because the permeability of the Caco-2 monolayer to hydrophilic compounds, generally transported by paracellular mechanisms, are poor because of the relatively tight junction that are characteristic of these cells (Artursson et al. 2001). A pure Caco-2 cell model also has a overexpression of P-glycoprotein which may lead to higher secretion rates and consequently lower permeability in the absorptive direction (Anderle et al. 1998). The HT-29MTX cell line has less expression of tight junctions. The ability of mannitol to penetrate tight junctions in HT-29 monolayer is 50-fold higher than that of Caco-2 monolayers (Wikman et al. 1993). The permeability of a cell layer resulting from co-cultivation between Caco-2 and HT-29 are more in resemblance with that of the normal intestine. The permeability of the Caco-2/HT-29 co-culture model was correlated with fractions absorbed in humans for selected drugs, and it was found relatively good correlations (Walter et al. 1996).

13.4 General Protocol

Maintain Caco-2 and HT-29MTX cells as described in Chaps. 10 and 11 respectively.

The Caco-2 and HT-29MTX cells are grown separately. The two cell lines are mixed prior to seeding. The most physiological relevant ratios are between 90:10 and 75:25 (Caco-2/HT-29MTX). In this range, the best compromise between model response and the presence of mucus layer will be obtained. Co-cultures of Caco-2 and HT-29MTX cells are seeded onto cell inserts with 0.4 µm pores at a density of 3×10^5 cells per 0.9 cm^2. The co-culture is maintained in Dulbeco's Modified Eagle Medium (DMEM) with 10 % heat inactivated fetal bovine serum, 2 mM L-glutamine, 1 % non-essential amino acids and 100 U/ml penicillin and 100 µg/ml streptomycin, at 37 °C in 5 % CO_2. The culture medium is changed every 2–3 days (both in apical and basolateral compartment) for 21 days.

When investigating intestinal transport and absorption using undigested or digested food compounds it is important to adjust pH (7.4) and osmolality (310 mOsm/kg before addition to the cell culture. Evaluation of transepithelial absorption is described in chapter 2, section 2.2.2 and calculation of the apparent permeability coefficient (P_{app}) is described in Chap. 10.

13.5 Assess Viability

The integrity of the cell layers should always be checked by measurement of TEER values, and filters with a TEER value below 200 Ω cm^2 should not be used for further experiments. A reference compound that is know to be transported over the epithelium (LY, labetalol, propranolol, ranitidine, or colchicine) should always be measured concurrently in each permeability assay to ensure validity of the assay.

Evaluation of transepithelial absorption is described in chapter 2, section 2.2.2 and calculation of the apparent permeability coefficient (P_{app}) is described in Chap. 10.

13.6   Experimental Readout

The Caco-2/HT-29MTX co-culture model is widely used to study intestinal transport and absorption. An in vitro study of intestinal transport of inorganic and methylated arsenic species from in vitro digested rice, garlic and seaweed has compared the Caco-2 monolayer with various proportions of the Caco-2/HT-29MTX co-culture model (Calatayud et al. 2012). They concluded that arsenic absorption increased with increased proportion of HT-29MTX. Another report studying transport of methylmercury and inorganic mercury in various Caco-2 and HT-29MTX models showed that incorporation of HT-29MTX reduced the permeability for mercury (Vazquez et al. 2013). In this case, the layer of mucus secreted by HT-29MTX retained mercury. Iron bioavailability of in vitro digested food (white, red and soy beans, beef and fish) were investigated by the Caco-2/HT-29MTX model (Mahler et al. 2009). They report that addition of HT-29MTX significantly lowered the cell ferritin formation in the presence of high bioavailability iron digests.

The use of Caco-2/HT-29 co-culture to evaluate transport and bioavailability of glutathione-enriched baker's yeast also revealed an increase in transport rates when HT-29MTX was incorporated in the monolayer compared to only using Caco-2 cells (Musatti et al. 2013).

Bacterial adhesion and invasion of *Salmonella enterica* spp. has been studied using the Caco-2/HT-29 co-culture model (Dostal et al. 2014). Bacterial adhesion and invasion was determined by plating of serial dilutions of the disrupted cell suspension after an extensive washing procedure, on agar plates. In the case of invasion, the extracellular bacteria were killed by incubation with gentamicin for 45 min before disrupting the cells. TEER was measured for evaluation of tight junction disruption during bacterial invasion.

13.7   Advantages, Disadvantages and Limitations

The use of cell lines will give good reproducibility of the model system and there are several reports establishing good correlations between this model and human in vivo studies. The two cell lines are easily available and easy to cultivate. The model is quite easy to use and is efficient for screening purposes. The disadvantages of this model is that some of the transporters/carriers found in normal human intestinal epithelium are not expressed by the two cell lines, therefore in the case of transport studies the expression of the necessary molecules should be checked. The ratio between the two cell types will be crucial for the formation of homogeneous mucin layer, which is of great importance for the relevance to the in vivo situation.

13.8 Conclusions

The co-culture of Caco-2 and HT-29MTX is a model that is useful for the investigation of transport over intestinal epithelium and for bacterial adhesion and invasion. For such applications the mucin layer and the permeability of the cell layer is crucial and co-culture will give results that are more in compliance with the in vivo situation than monocultures.

References

Anderle P, Niederer E, Rubas W et al (1998) P-Glycoprotein (P-gp) mediated efflux in Caco-2 cell monolayers: the influence of culturing conditions and drug exposure on P-gp expression levels. J Pharm Sci 87:757–762

Artursson P (1991) Cell cultures as models for drug absorption across the intestinal mucosa. Crit Rev Ther Drug Carrier Syst 8:305–330

Artursson P, Palm K, Luthman K (2001) Caco-2 monolayers in experimental and theoretical predictions of drug transport. Adv Drug Deliv Rev 46:27–43

Behrens I, Stenberg P, Artursson P et al (2001) Transport of lipophilic drug molecules in a new mucus-secreting cell culture model based on HT29-MTX cells. Pharm Res 18:1138–1145

Calatayud M, Vazquez M, Devesa V et al (2012) In vitro study of intestinal transport of inorganic and methylated arsenic species by Caco-2/HT29-MTX cocultures. Chem Res Toxicol 25:2654–2662

Dostal A, Gagnon M, Chassard C et al (2014) Salmonella adhesion, invasion and cellular immune responses are differentially affected by iron concentrations in a combined in vitro gut fermentation-cell model. PLoS One 9:e93549

Hidalgo IJ (1996) Cultured intestinal epithelial cell models. Pharm Biotechnol 8:35–50

Hilgendorf C, Spahn-Langguth H, Regardh CG et al (2000) Caco-2 versus Caco-2/HT29-MTX co-cultured cell lines: permeabilities via diffusion, inside- and outside-directed carrier-mediated transport. J Pharm Sci 89:63–75

Johansson ME, Phillipson M, Petersson J et al (2008) The inner of the two Muc2 mucin-dependent mucus layers in colon is devoid of bacteria. Proc Natl Acad Sci U S A 105:15064–15069

Mahler GJ, Shuler ML, Glahn RP (2009) Characterization of Caco-2 and HT29-MTX cocultures in an in vitro digestion/cell culture model used to predict iron bioavailability. J Nutr Biochem 20:494–502

Musatti A, Devesa V, Calatayud M et al (2013) Glutathione-enriched baker's yeast: production, bioaccessibility and intestinal transport assays. J Appl Microbiol

Sambuy Y, De Angelis I, Ranaldi G et al (2005) The Caco-2 cell line as a model of the intestinal barrier: influence of cell and culture-related factors on Caco-2 cell functional characteristics. Cell Biol Toxicol 21:1–26

Shah P, Jogani V, Bagchi T et al (2006) Role of Caco-2 cell monolayers in prediction of intestinal drug absorption. Biotechnol Prog 22:186–198

Vazquez M, Calatayud M, Velez D et al (2013) Intestinal transport of methylmercury and inorganic mercury in various models of Caco-2 and HT29-MTX cells. Toxicology 311:147–153

Walter E, Janich S, Roessler BJ et al (1996) HT29-MTX/Caco-2 cocultures as an in vitro model for the intestinal epithelium: in vitro–in vivo correlation with permeability data from rats and humans. J Pharm Sci 85:1070–1076

Wikman A, Karlsson J, Carlstedt I et al (1993) A drug absorption model based on the mucus layer producing human intestinal goblet cell line HT29-H. Pharm Res 10:843–852

Part III
Innate and Adaptive Immune Cells: General Introduction

Iván López-Expósito

As an organ specialized in food digestion and nutrient absorption, the intestinal mucosa presents a huge surface area (almost 300 m^2 in comparison with 2 m^2 in skin) to the outside milieu and is continually exposed to foreign antigens derived from dietary constituents and the large number of microbes that reside within the intestinal lumen. In order to maintain the intestinal integrity, it is crucial to possess a fully functional associated immune system able to respond appropriately to such antigens and also to generate protective immunity to potential pathogens that employ the intestine as a primary site of entry and infection. Inappropriate responses to such antigens, apart from infections, are thought to underlie several intestinal pathologies including inflammatory bowel disease as well as food allergies (Bekiaris et al. 2014).

Cells from both the innate and adaptive immune system can be found throughout the intestinal mucosa working together cooperatively with other cells and molecules in order to maintain intestinal functionality. Innate immunity provides effective initial defense mechanisms that take place even before infection and are poised to respond rapidly to microbes. These mechanisms react only to microbes and products of injured cells, and they are specific for structures that are common to a group of related microbes, not being able to distinguish fine differences between foreign substances. On the contrary, adaptive immune responses comprise responses that are stimulated by exposure to antigens of both microbial and non microbial origin and that increase in magnitude and defensive capabilities with each successive exposure. The main characteristics of adaptive immunity are a very high specificity for distinct molecules and the ability to "remember" and respond more vigorously to repeated exposures to the same antigen (Abbas et al. 2007).

When the immune response is triggered, a wide variety of cells from both the immune system and other tissues participate. These include epithelial and

Iván López-Expósito
Departamento de Bioactividad y Análisis de Alimentos,
Instituto de Investigación en Ciencias de la Alimentación (CIAL) (CSIC-UAM),
Madrid, Spain
e-mail: ivan.lopez@csic.es

endothelial cells, neutrophils, monocytes, macrophages, dendritic cells, mast cells, natural killer cells, basophils, eosinophils, B and T cells. All of them have the ability to secrete a wide array of mediators responsible in part for their inflammatory effects (Si-Tahar et al. 2009). Based on the number of publications dealing with the immunomodulant/anti-inflammatory properties of food-derived compounds with bioactive properties, only monocytes, macrophages, dendritic cells, PBMCs and T cells will be covered in this part.

Monocytes and Macrophages

Monocytes and macrophages are essential for the development of inflammation and together with neutrophils are the phagocytic cells involved in the clearance of inert particles and microbial agents. Monocytes are bone marrow-derived cells that are continuously released into the blood. When these cells are recruited by chemotactic molecules, and leave circulation, they become activated and differentiate into macrophages under the stimulus of mediators such as TGF-β or M-CSF, between others (Gordon and Martínez 2009). Macrophages have a number of important functions in body defenses such as (1) capture by phagocytosis and intracellular killing of microorganisms; (2) scavenging of debris potentially harmful to tissues; (3) processing and presentation of antigens for recognition by T cells, and (4) releasing cytokines and chemokines with a major role in immune responses (Davies et al. 2013). The major cytokines produced by macrophages are TNF-α, IL-1β, IL-6, IL-8 and IL-33, all of which are involved in local and systemic responses. Macrophages demonstrate great developmental plasticity and may differentiate into cells with different phenotypes depending on the stimuli received as well as the tissue location. IFN-γ induces classically activated M1 macrophages (CAM), whereas IL-4, IL-13 and IL-10 induce alternatively activated M2-macrophages (AAM) (Mosser and Edwards 2008). In the intestinal mucosa M2-macrophages are more abundant, participating in the resolution of parasite infections, tissue remodeling, immune regulation, allergy development and tumor progression by stimulating a Th2-driven immune response (Takeuchi and Akira 2011). Chapter 14 describes in detail the main characteristics and differentiation protocols for THP-1 and U937, two human monocytic cells lines commonly employed in the study of the anti-inflammatory properties of food bioactives. The main food components tested, together with the methods employed to evaluate their potential anti-inflammatory activity are also explained.

Dendritic Cells

Dendritic cells (DC) are a widely distributed group of cells specialized in antigen sampling. In fact, they also constitute the most efficient antigen presenting cells for T cell activation, hence being the linking bridge between innate and adaptive immune responses. DCs are strategically positioned at body barriers and also organ entry ports, such as the splenic marginal zone, where they remain in an immature form until they encounter an antigen (Mildner and Jung 2014). Once the antigen has been sampled, DCs become activated and travel towards T-cell zones, either within their respective lymphoid organ of residence or towards draining lymph nodes to ensure the activation, proliferation and differentiation of naïve T cells into their corresponding effector cells. This migration depends on the expression of the chemokine receptor CCR7 (Foster et al. 1999). During migration, the mature dendritic cells express high surface levels of class II major histocompatibility complex (MHC) molecules with bound antigenic peptides as well as costimulatory molecules. By the time DCs reach secondary lymphoid organs, they are able to present antigens to populations of naïve and memory T cells. Chapter 17 provides a complete overview of the different DC subsets focusing on the DCs present in the intestinal mucosa as they are among the first immune cells to come into contact with food compounds in the gastrointestinal tract and thus are instrumental in shaping the immune system's response to such exposures. In the same chapter the main DC isolation techniques as well as in vitro/ex vivo culture settings that can be applied for in vitro testing of food compounds with bioactive properties are discussed. Special attention will be paid to the potential of food-derived bioactives in inhibiting DC activation due to the relevance of DCs in initiating the inflammatory processes.

Human Peripheral Blood Mononuclear Cells

Human peripheral blood mononuclear cells (PBMCs) include a mixture of cells composed of lymphocytes (T cells, B cells, and NK cells), monocytes, and dendritic cells obtained from human blood or buffy coats. In humans, the frequencies of these populations vary across individuals, but lymphocytes are the most abundant, constituting in the range of 70–90 %. PBMCs are typically employed in studies where immune-regulatory effects of food bioactives are to be scrutinized. Main read-out systems include proliferation measurements, evaluation of surface activation markers by flow cytometry and quantification of the cytokine profile produced after adding the food bioactive to the cell culture. Chapter 15 fully explains the principal features and isolation procedures of PBMCs from human blood. Moreover, protocols to perform proliferation assays and evaluation of the anti-inflammatory properties with compounds from food origin are also described.

T Lymphocytes or T-Cells

As mentioned earlier, PBMCs are an important source of lymphocytes. Of special interest are T lymphocytes or T-cells (45–70 % of PBMCs in human peripheral blood), which are produced by stem cells in the bone marrow as progenitors and then migrate to the thymus where they mature into T cells. After completing their maturation, T-cells enter the bloodstream and recirculate between blood and secondary lymphoid organs until they encounter their cognate antigen. After antigen presentation by DCs, along with other appropriate stimuli, the cells may proliferate and differentiate into different subsets of effector cells (Santana and Esquivel-Guadarrama 2006). Originally, two main types of effector T cells, called T-helper 1 (*Th1*) and 2 (*Th2*) cells, were distinguished by their cytokine secretion patterns. Th1 cells secrete mainly IL-2, IFNγ and TNFα, and Th2 cells secrete IL-4, IL-13 and IL-5 (Romagnani 2000). Recently, a new lineage of T cells characterized by their ability to secrete a proinflammatory cytokine, IL-17, and thus designated Th17 cells has been discovered. This new T cell type has been related to autoimmune diseases (Jing and Dong 2013). Another subset, named regulatory T cells (*Treg*), acts by inhibiting, between others T cell responses by the production of cytokines, such as IL-10 and TGF-β and/or via cell–cell interactions (Jutel and Akdis 2011). T cell cultures are a valuable tool in food research, especially to perform studies within the food allergy field. To study effects of food bioactives on T cells, it is necessary to activate the T cells by a polyclonal activator, either a mitogen like phytohaemagglutinin or monoclonal antibodies against CD3 and CD28. In food allergy, their main applications include analysis of immunological responses towards food protein antigens to gain further insights into the mechanisms responsible for the development of oral tolerance or for the triggering of food allergies. Chapter 16 describes the main applications in food allergy research, isolation techniques, and culture conditions for PBMC-derived T cells. Furthermore, critical parameters of the model, together with the experimental read outs are discussed.

References

Abbas AK, Lichtman AH, Pillai S (eds) (2007) Cellular and molecular immunology. Saunders, Philadelphia

Bekiaris V, Persson EK, Agace WW (2014) Intestinal dendritic cells in the regulation of mucosal immunity. Immunol Rev 260:86–101

Davies LC, Jenkins SJ, Allen PR, Taylor PR (2013) Tissue resident macrophages. Nat Immunol 14:986–995

Foster R, Schubel A, Breitfeld D, Kremmer E, Renner-Muller I, Wolf E, Lipp M (1999) CCR7 coordinates the primary immune response by establishing functional microenvironments in secondary lymphoid organs. Cell 99:23–33

Gordon S, Martínez FO (2009) Alternative activation of macrophages: mechanism and functions. Immunity 32:593–604

Jing W, Dong C (2013) IL-17 cytokines in immunity and inflammation. Emerg Microbes Infect 2:e60

Jutel M, Akdis CA (2011) T-cell subset regulation in atopy. Curr Allergy Asthma Rep 11:139–145

Mildner A, Jung S (2014) Development and function of dendritic cells subsets. Immunity 40:642–656

Mosser DM, Edwards JP (2008) Exploring the full spectrum of macrophage activation. Nat Rev Immunol 8:958–969

Romagnani S (2000) T cells subsets (Th1 vs Th2). Ann Allergy Asthma Immunol 85:9–18

Santana MA, Esquivel-Guadarrama F (2006) Cell biology of T cell activation and differentiation. Int Rev Cytol 250:217–274

Si-Tahar M, Touqui L, Chignard M (2009) Innate immunity and inflammation: two facets of the same anti-infectious reaction. Clin Exp Immunol 156:194–198

Takeuchi O, Akira S (2011) Epigenetic control of macrophage polarization. Eur J Immunol 41:2490–2493

Chapter 14
THP-1 and U937 Cells

Wasaporn Chanput, Vera Peters, and Harry Wichers

Abstract Monocytes are circulatory precursor cells from myeloid origin that can develop into macrophages or dendritic cells upon migration from the blood stream to tissues. Both macrophages and dendritic cells are professional antigen-presenting cells. Monocytes and their macrophage and dendritic-cell progeny serve three main functions in the immune system. These are phagocytosis, antigen presentation, and cytokine production. THP-1 and U937 are (pro-) monocytic cell lines that can, also in vitro, be differentiated into either various types of macrophages or into dendritic cells.

This chapter describes how to grow THP-1, resp. U937 cells, how to differentiate these into more specialised phenotypes such as various macrophage types or dendritic cells, how to read-out their responses to stimuli and it gives examples of how such cell lines have been used into studying the effects of food compounds.

Keywords THP-1 • U937 • Cell line • Monocyte • Macrophage

14.1 Origin and Some Features of THP-1 and U937 Cells

The THP-1 cell line is a human monocytic leukaemia cell line which was established in 1980 by Tsuchiya et al. (1980). It was derived from the blood of a patient with acute monocytic leukaemia. THP-1 cells resemble primary monocytes and macrophages in morphology and differentiation properties. THP-1 cells show a large, round single-cell morphology and express distinct monocytic markers. Nearly all THP-1 cells start to adhere to culture plates and differentiate into macrophages after exposure to phorbol-12-myristate-13-acetate (PMA, also known as TPA,12-O-tetradecanoylphorbol-13-acetate; see below for details).

W. Chanput
Faculty of Agro-Industry, Department of Food Science and Technology, Kasetsart University, 50 Ngamwongwan Road, Lad Yao, Chatuchak, 10900 Bangkok, Thailand

V. Peters • H. Wichers (✉)
Wageningen University and Research Centre, Food & Biobased Research, Bornse Weilanden 9, PO Box 17, 6700 AA Wageningen, The Netherlands
e-mail: harry.wichers@wur.nl

© The Author(s) 2015
K. Verhoeckx et al. (eds.), *The Impact of Food Bio-Actives on Gut Health*,
DOI 10.1007/978-3-319-16104-4_14

THP-1 cells have some technical advantages over human primary monocytes or macrophages. For instance, their genetic background is homogeneous which minimizes the degree of variability in the cell phenotype. Another technical advantage is that genetic modification of THP-1 cells by small interfering RNAs (siRNAs), in order to down regulate the expression of specific proteins, is relatively simple (Chanput et al. 2014). Furthermore, monocyte-derived macrophages can be polarized into M1, M2a, M2b, and M2c cells. Spencer et al. (2010) published a protocol for THP-1 cell differentiation, showing that THP-1 cells represent a simplified model to study monocyte-macrophage polarization (Qin 2012).

U937 is a pro-monocytic, human myeloid leukaemia cell line and was isolated from the histiocytic lymphoma of a 37 year old male (Sundstrom and Nilsson 1976). This cell line exhibits many characteristics of monocytes and is easy to use. A virtually unlimited number of cells can be prepared and they are relatively uniform. This cell line has been an important tool in the investigation of the mechanisms involved in monocyte-endothelium attachment (Liu et al. 2004). These leukaemia cells have been used as the experimental model to elucidate mechanisms of monocyte and macrophage differentiation. A genetic analysis by Strefford et al. (2001) showed that U937 bears the t(10;11)(p13;q14) translocation. This results in a fusion between the MLLT10 (myeloid/lymphoid or mixed-lineage leukaemia) gene and the Ap-3-like clathrin assembly protein PICALM (Clathrin assembly lymphoid myeloid leukaemia), which is likely important for the tumorous nature of the cell line (Strefford et al. 2001).

14.2 Stability, Consistency and Reproducibility of the System

THP-1 is an immortalized cell line that can be cultured in vitro up to passage 25 (approx. 3 months) without changes of cell sensitivity and activity. As far as our information reaches, U937 is used also at higher passage numbers (see e.g. Strefford et al. 2001). THP-1 as well as U937 cells can be stored for a number of years and, provided an appropriate protocol is followed, the cell lines can be recovered without any obvious effects on monocyte-macrophage features and cell viability (Chanput et al. 2014 and references therein).

14.3 Relevance to Human In Vivo Situation

Cell lines always have a malignant background, which presents a significant risk of experimental bias. The cultivation of cells under controlled conditions and outside their natural environment possibly results in different sensitivity and responses compared to normal somatic cells in their natural environment (Schildberger et al. 2013).

Also, possibly relevant interactions between the target cells and surrounding cells, as in natural tissues, cannot be easily mimicked. In vitro co-cultivation of THP-1 or U937 cells with neighbouring cells might be an option to make this drawback less pronounced (Chanput et al. 2014).

14.4 Other Models with the Same Applicability

Next to THP-1 and U937 cells, ML-2, HL-60 and Mono Mac 6 cells are used in biomedical research. U937 cells are the most frequently used. The basic difference between U937 and THP-1 cells is the origin and maturation stage. U937 cells are of tissue origin, thus at more mature stage, whereas THP-1 cells originate from a blood leukaemia origin at less mature stage (Chanput et al. 2014). Because Mono Mac 6 is able to phagocytose antibody-coated erythrocytes (Ziegler-Heitbrock et al. 1988) and mycobacteria (Friedland et al. 1993; Shattock et al. 1994) it is thought more suitable for the study of phenotypic and functional features of in vivo mature monocytes. Also, it expresses mature monocyte markers that cannot be found on the THP-1 and U937 cell lines, such as M42, LeuM3, 63D3, Mo2 and UCHMI. As THP-1 and U937 are very frequently used, we focus here on these two cell lines.

14.5 General Protocol of Culturing THP-1 Cells

Roswell Park Memorial Institute (RPMI) 1640 medium is a commonly used medium for THP-1 as well as for U937 cells. Alternatively, DMEM (Dulbecco's Modified Eagle's Medium; Morton 1970), also supplemented with 10 % FBS, is used to grow U937 cells. RPMI 1640 medium was originally developed by Moore et al. (1967), at Roswell Park Memorial Institute to culture human leukemic cells in suspension and as a monolayer. The formulation is based on RPMI-1630 medium and uses a bicarbonate buffering system and has alterations in the amounts of amino acids and vitamins. When RPMI is properly supplemented, it has a wide applicability for supporting growth of many types of cell cultures including HeLa, Jurkat, MCF-7, PC12, PBMC, astrocytes and carcinomas.

In most of the studies, the RPMI medium is supplemented with a combination of foetal bovine serum (FBS) and antibiotics. FBS is also known as foetal calf serum (FCS) and is obtained from whole blood by removing blood cells, platelets and fibrinogen. Serum includes all proteins not involved in blood clotting and all electrolytes, antibodies, antigens, hormones and exogenous substances. Foetal bovine serum is obtained via collection at a slaughterhouse. In some studies, foetal bovine serum is heat-inactivated in order to destroy heat-labile complement proteins (Biowest).

In many studies, the antibiotics penicillin and streptomycin are used to supplement the medium in order to prevent bacterial contamination of cell cultures due to their effective combined action against gram-positive and gram-negative bacteria. Penicillin, originally purified from the fungus *Penicillium*, acts directly and indirectly by interfering with the turnover of the bacterial cell wall and by triggering the release of enzymes that further alter the cell wall respectively. Streptomycin was originally purified from *Streptomyces griseus* and acts by binding to the 30S subunit of the bacterial ribosome, which leads to inhibition of protein synthesis and death in susceptible bacteria (Waksman 1953).

In some studies, cells were also supplemented with HEPES (4-(2-hydroxyethyl)-1-piperazineethanesulfonic acid). HEPES is a zwitterionic organic chemical buffer and is widely used in many biochemical reactions and as a buffering agent in some cell culture media. These buffers have pK_a values between 6.0 and 8.0, high solubility, limited effect on biochemical reactions, membrane permeability, are chemically and enzymatically stable and easy to prepare.

14.6 Differentiation of THP-1 and U937 Monocytes into Macrophages

As already mentioned before, nearly all THP-1 cells start to adhere to culture plates and differentiate into macrophages after exposure to PMA. Also U937 pro-monocytes differentiate into mature monocytes or into macrophages upon PMA-treatment. Typical exposure to PMA is for 48 h (Zhang et al. 2010; Gillies et al. 2012; Moreno-Navarrete et al. 2009; Cam and de Mejia 2012). The temperature and atmosphere, if described, was the same for each study, namely 37 °C and 5 % CO_2 respectively. Different stimuli were used, depending on the aim of the study.

Subsequently, macrophages can be further differentiated into subsets. Typical markers for M1-type macrophages are transcription or production of TNF-α, IL-1β, IL12-p40, IL-6, IL-8 and LOX-1, and for M2-type macrophages MRC-1, dectin-1, and DC-SIGN (Chanput et al. 2010).

Based on literature and on our experience, 0.5×10^6 THP-1 monocytes fully differentiate into macrophages after 48 h incubation at a minimal concentration of 100 ng/ml PMA (162 nM), followed by washing twice with culture media without PMA and a resting period of 24 h (Chanput et al. 2014), resulting in macrophages with a high phagocytic capacity for latex beads and expressing cytokine profiles that resembled PBMC monocyte-derived macrophages after exposure to TLR ligands (Chanput et al. 2014).

Human promonocytic leukaemia U937 cells differentiate into monocytes and macrophages by use of various agents such as retinoic acids, 1,25-dihydroxyvitamin D3 (VD3; at 100 nM, i.e. 42 ng/ml, Rots et al. 1999), and 12-O-tetradecanoylphorbol-13-acetate (TPA; at 20 ng/ml, ca. 32 nM) (Chun et al. 2001).

14.7   Differentiation of THP-1 and U937 Monocytes into Dendritic Cells

THP-1 monocytes are described to differentiate into mature dendritic cells by transferring them to serum-free medium, and subsequently treating them with a mixture of IL-4, GM-CSF, TNF-α and ionomycin. These hematopoietic cell line-derived DCs are highly pure and monotypic, and display the morphologic, phenotypic, molecular, and functional properties of DCs generated from human donor-derived monocytes or CD34+ hematopoietic progenitor cells (see also Chap. 17). During differentiation into mature DCs, the cells exhibit de novo cell-surface expression of CD83, CD80, CD86, CD40, CD206, CD209, CD120a, CD120b, and intracellular synthesis of IL-10, increase their endocytotic capacity, and acquire the characteristic stellate morphology (Berges et al. 2005). In the THP-1 monocytes, mRNAs of tight junction molecules, occludin, tricellulin, JAM-A, ZO-1, ZO-2 and claudin-4, -7, -8, and -9 were detected by RT-PCR. In mature DCs that had elongated dendrites, mRNA and protein of JAM-A were significantly increased compared to the monocytes (Ogasawara et al. 2009).

Exposure of U937 to a self-peptide from apolipoprotein E, Ep1.B, induces DC-like morphology and surface marker expression in U937 (Stephens et al. 2008).

14.8   Controls to Test Viability and Performance of the Model

The by far most commonly used and imperative control test for analyses that are based on any kind of THP-1 or U937 derived cell is the possibility of cytotoxic effects of the test samples. For such checks, various methods and kits are in common use in various laboratories. Examples are assessment of functional mitochondrial reductase, with tetrazolium salts such as MTT, XTT and lamar blue. Alternatively, cell membrane integrity, degree of cell lysis via measurement of lactate dehydrogenase and apoptosis kits are used.

14.9   Critical Notes

Regularly, solvents such as ethanol or DMSO are used to facilitate dissolution of test compounds. Also, sometimes a mimic of digestion is applied to the samples prior to analysis. It is highly recommended to determine the toxicity of such solvents or digestion reagents, as well as their effect on the read-out parameters to be used for the cell assay, in a titration that at the least covers the eventual concentration to be used in the assays, and also for a relevant range of higher and lower concentrations (at least a factor 10 lower and higher than eventually applied conditions). Similar recommendations are valid for the use of buffers, salt concentrations and pH-ranges etc.

14.10   Read-Out of the System

Most commonly used read-out systems for responses of THP-1 or U937 cells are transcription or production of relevant cytokines. Which cytokines are relevant depends on specific experimental design. Gene expression of IL-1β, IL-6, IL-8, IL-10, TNF-α, iNOS, COX-2 and NF-κB is measured in many studies. This is done by RT-PCR, Western blot and ELISA. Examples that have been used to read out the effects of various food compounds on resp. THP-1 and U937 cells are presented in Tables 14.1 and 14.2.

Transcription can be measured by various PCR-methods, mostly and preferably quantitative PCR such as qPCR or RT-qPCR. Advantages of using PCR are its ease, speed and sensitivity, and possibly its low cost. A clear disadvantage is that not mRNAs are physiologically relevant, but the eventual gene product i.e. the cytokine.

Cytokines are most commonly measured by ELISA-methods or by FACS, for which a variety of commercial kits are available. An important advantage is relevance as protein products are quantified, and the standardized kits that are on the market that facilitate comparison of read-outs between various laboratories. An important drawback associated to such commercial kits is their cost.

THP-1 cells can also be used to study expression and activation of nuclear receptors. Gillies et al. (2012) studied the effect of omega-3-PUFAs, eicosapentaenoic acid (EPA) and docosahexaenoic acid (DHA) and found that EPA-rich oil activated human peroxisome proliferator-activated receptor α (PPAR α) and PPAR β/α with minimal effects on PPARγ, liver X receptor, retinoid X receptor, farnesoid X receptor, and retinoid acid receptor γ (RARγ).

Notably, several studies looked at the intracellular production of reactive oxygen species (ROS) by THP-1 cells. During inflammation, ROS production is induced by inflammatory cells in order to kill pathogens. However, ROS also act on inflammatory cells themselves, thereby altering the intracellular redox balance and functioning as signalling molecules involved in the regulation of inflammatory and immunomodulatory genes. ROS plays a key role in the control of transcription factors, like NF-κB and activator protein-1 (AP-1), which are involved in the gene expression of both inflammatory and immune mediators. ROS can either activate or inactivate these transcription factors by chemically modifying critical amino acid residues within these proteins or on residues of accessory proteins of the respective signalling pathways (Jabaut and Ckless 2012).

ROS production can be measured by the intracellular ROS assay. In the study of Wu et al. (2011), 2′,7′-dichlorofluorescein (DCF) is used to measure ROS production. The cell-permeant 2′,7′-dichlorodihydrofluorescein diacetate (H_2DCFDA) (also known as dichlorofluorescein diacetate) is a chemically reduced form of fluorescein used as an indicator for ROS in cells. Upon cleavage of the acetate groups by intracellular esterases and oxidation, the non-fluorescent H_2DCFDA is converted to the highly fluorescent 2′,7′-dichlorofluorescein (DCF). The chemically reduced and acetylated form of 2′,7′-dichlorofluorescein (DCF) is non-fluorescent until the acetate groups are removed. The fluorescence can be measured by a flow cytometry, fluorimeter, microplate reader, or fluorescence microscope.

Table 14.1 Food components tested on THP-1 cells

Compound	Analysed response	Method	Result	References
Lunasin	Levels of nitrite PGE2 TNF-α iNOS, COX-2, p-p65, and p-Akt	Griess reaction PGE2 EIA monoclonal assay ELISA Western blot	Reduced COX-2, iNOS, NO, PGE2, TNF-α levels Inhibited activation of p-AKT and NF-κB p65 Reduced αVβ3 intensity	Cam and de Mejia (2012)
Melittin	Serum lipids, pro-inflammatory cytokines TNF-α, NF-κB and IL-1β expression	Serum lipid analysis, ELISA (TNFα, IL8-β), protein analysis, Western blot	Decreased total cholesterol and triglyceride levels, high HDL-C Decreased TNF-α, IL-1β and NF-κB	Kim et al. (2011)
Lactoferrin	NF-κB Lactoferrin IL-6, MCP-1, IL-8, and cyclophilin A	Western blot ELISA RT-PCR	Lactoferrin negatively related to inflammatory markers Decreased IL-6, MCP-1, IL-8 and NF-κB	Moreno-Navarrete et al. (2009)
Casein α s1 (CSN1S1)	CSN1S1	Western blot, Real-time PCR, ELISA, immunocytochemistry, immunofluorescence	Human CSN1S1 may possess an immunomodulatory role beyond its nutritional function in milk	Vordenbaumen et al. (2011)
β-glucans	NO production Hydrogen peroxide (H₂O₂) IL-1β, IL-8, NF-κB and IL-10	Griess reagent Horseradish peroxidise mediated oxidation of phenol red RT-qPCR	No H₂O₂ or NO production Decreased induction of IL-1β, IL-8, NF-κB and IL-10	Chanput et al. (2012)
Barley glucan, quercetin and citrus pectin	Expression and secretion of IL-1β, IL-6, IL-8, IL-10, TNF-α, iNOS, COX-2, NF-κB, AP-1 and SP-1	RT-PCR (gene expression) Cytometric bead array (CBA) analysis (cytokine secretion)	Induced inflammation-related cytokines, COX-2 and NF-κB genes	Chanput et al. (2010)
Saturated fatty acids (SFAs)	NF-κB, TNF-α, IL-8, ROS production	Luciferase assay, Western blot, ELISA flow cytometer and microscopy, LAL endotoxin assay	Increased ROS production, activation of PRR signaling pathways	Huang et al. (2012)

(continued)

Table 14.1 (continued)

Compound	Analysed response	Method	Result	References
Omega-3-PUFAs, eicosapentaenoic acid (EPA), and docosahexaenoic acid (DHA)	Expression and activation of nuclear receptors	RNA extraction, reverse transcription, and real-time PCR, microarray, Human PPARα, PPAR β/δ, PPARγ, LXRβ, RXRα, and FXR reporter assay system	EPA-rich oil activated human peroxisome proliferator-activated receptor α(PPAR α) and PPAR β/α with minimal effects on PPARγ, liver X receptor, retinoid X receptor, farnesoid X receptor, and retinoid acid receptor γ(RARγ)	Gillies et al. (2012)
Proanthocyanidin	IL-6 and TNF-α	ELISA, cytotoxic assay	Decreased IL-6 and TNF-α	Tatsuno et al. (2012)
Curcumin	Expression of lipid genes	Quantitative RT-PCR	Increased expression of the fatty acids transporter CD36/FAT and the fatty acids binding protein 4 FABP4 Increased FOXO3a activity	Zingg et al. (2012)
Caffeic acid (CA), ferulic acid, m-coumaric acid, and chlorogenic acid	ROS TNF-α secretion	DCF-D ELISA RT-PCR, comet assay, cell migration assay	CA induces ROS, IL-1β and TNF-α production CA exhibits pro-oxidative and pro-glycative effects during the glycation process	Wu et al. (2011)
Cyanidin 3-O-beta-glucoside (C3G)	mRNA expression, Iκβα, NF-κB P65, TNF-α, IL-6 secretion	Real-time PCR ELISA, western blot	Increased mRNA expression and secretion of TNF-α and IL-6 Inhibition of Iκβα phosphorylation, thereby inhibiting NF-κB activity	Zhang et al. (2010)
Flavonoids (catechin, EGCG, luteolin, quercetin, rutin)	TNF-α, IL-1β intracellular ROS production NF-κB activity	ELISA Fluorescence RT-PCR, western blotting	All flavonoids inhibited TNF-α, IL-1β, and COX-2 Inhibits intracellular ROS production Inhibition of NF-κB activity	Wu et al. (2009)
Quercetin and catechin (flavonoids), S100B (RAGE ligand)	Gene expression of TNF-α, IL-1β, MCP-1, IP-10, PECAM-1, β2-integrin and COX-2 ROS production	RT-PCR, Western blot, cytokine ELISA assay, intracellular ROS assay	S100B (RAGE ligand) increased gene expression of TNF-α, IL-1β, MCP-1, IP-10, PECAM-1, beta2-integrin and COX-2 Quercetin and catechin inhibited these gene expressions, eliminate ROS	Huang et al. (2006)

Table 14.2 Tested food components on U937 cells

Compound	Analysed response	Method	Result	References
Salmon calcitonin (sCT) Hyaluronic acid (HA)	mRNA expression of NR4A1, NR4A3, and MMP1, -3 and -13	RT-PCR	Both sCT and HA attenuated activated mRNA expression of NR4A1, NR4A2, NR4A3, and MMP1, -3 and -13	Ryan et al. (2013)
Soluble toll-like receptor 2 (sTLR2)	Immunodepletion of sTLR2 IL-8 levels Cell viability	Western blot ELISA Duoset MTT assay	Increased IL-8 production	Henrick et al. (2012)
Quercetin	Expression of inflammatory genes, phosphorylation of JNK and c-Jun, IκBα degradation	PCR, immunoblotting	Attenuated expression of TNF-α, IL-6, IL-8, IL-1β, IFN-γ IP-10, COX-2 and PTP-1B Attenuated phosphorylation of c-Jun N-terminal kinase (JNK), c-Jun, decreased serine residue 307 phosphorylation IRS-1 Attenuated IκBα degradation in MΦs Suppression of insulin-stimulated glucose uptake	Overman et al. (2011)
Human casein α s1 (CSN1S1)	CSN1S1 mRNA and protein amount GM-CSF mRNA	ELISA	Detection of CSN1S1 mRNA and protein Upregulation of GM-CSF mRNA	Vordenbaumen et al. (2011)
Xanthones	Expression of TNFα, IL-6, IP-10, PPARγ, MAPK, AP-1, NF-κB	Quantitative PCR, immunoblotting	α- and γ-MG attenuated expression of TNFα, IL-6, IP-10	Bumrungpert et al. (2010)
Billberry juice (quercetin, epicatechin, and resveratrol)	Lipid peroxidation Levels of CRP, IL-6, IL-15, TNF-α and MIG NF-κB activation	D-ROMs test Sandwich immunoassay, fluorescence HPLC	Decreased CRP, IL-6, IL-15 and MIG (all target genes of NF-κB, also TNF-α) Increased TNF-α Inhibition of NF-κB activation	Karlsen et al. (2010)
Grape powder extract (GPE)	MAPKs, NF-κB AP-1, TNF-α, IL-6, IL-1β, IL-8, IP-10, COX-2, insulin resistance c-Jun, JNK, p38, Elk-1 Iκβα	PCR, immunoblotting, BioPlex suspension array system from BioRad	Induction of TNF-α, IL-6, IL-1β, IL-8, IP-10, COX-2 Attenuated activation of MAPKs, NF-κB, AP-1 Decreased phosphorylation of c-Jun, JNK, p38, Elk-1 Degradation of Iκβα	Overman et al. (2010)

(continued)

Table 14.2 (continued)

Compound	Analysed response	Method	Result	References
Grifola frondosa	NFκB (HT-29 cells) TNF-α, MCP-1 ROS (HT-29 cells)	NF-κB reporter gene dual-luciferase assay ELISA Fluorescence	Suppressed TNF-α expression	Lee et al. (2010)
Carvatrol	COX-2 expression	Transcription assays	Suppressed COX-2 expression	Hotta et al. (2010)
Cysteine	IL-1β levels Precursor IL-1β	ELISA Western blot	Increase in secreted pro-IL-1β and IL-1β levels	Iyer et al. (2009)
Nutrient mixture (NM) Quercetin, naringenin Hespzeretin Tea catechins Lysine, proline, arginine and N-acetylcysteine	COX enzymatic activity COX protein expression Specific mRNA levels NF-κB activation	Western blot analysis RT-PCR Phosphorylated p65 immunoassay	Inhibition of COX-2 enzyme activity Downregulation of COX-2 and pro-inflammatory cytokine protein expression levels by reduced NF-κB activation	Ivanov et al. (2008)
N-acetylcysteine, d-alpha-tocopherol acetate and ascorbic acid (antioxidants) Caffeic acid (CA)	NF-κB binding activity Intracellular peroxides content	Gel shift assay Formation of a fluorescent derivative of DCF	CA inhibits NF-κB binding activity and protein tyrosine kinase activity N-acetylcysteine, d-α-tocopherol acetate and ascorbic acid did not inhibit apoptosis, but affect NF-κB activity	Nardini et al. (2001)

Many studies using U937 have been used to analyse properties of poly-phenolic compounds. Responses like inflammatory gene expression and NF-κB activity are commonly measured. The effect of quercetin on the phosphorylation of JNK and c-Jun and the degradation of IκBα was measured by Overman et al. (2010). The c-Jun N-terminal kinases (JNKs) are members of a group of serine/threonine (Ser/Thr) protein kinases from the mitogen-activated protein kinase (MAPK) family. The JNKs act within a protein kinase cascade and are activated by dual phosphorylation by MAPK kinases, but can also be activated by pro-inflammatory cytokines including TNF-α and IL-1β. Moreover, the JNK pathway is activated in the innate immune response following the activation of various members of the Toll-like receptor family by invading pathogens. The JNK pathways appears to act as a critical intermediate in signalling in the immune system. There is increasing evidence that JNK is activated following sensing of internal stress events, such as protein misfolding. Studies also showed that bacterial, fungal, prion, parasitic, or viral infections activate JNK which may influence important cellular consequences such as alterations in gene expression, cell death, viral replication, persistent infection or progeny release, or altered cellular proliferation. The exact mechanism of JNK activation under each of these circumstances remains not fully clear, but there may be involvement of Toll-like receptors, direct pathway modulation through interaction with upstream protein regulators or the activation following an ER stress response (Bogoyevitch and Kobe 2006).

Overman et al. (2011) found that quercetin attenuated the phosphorylation of JNK and c-Jun, which means a decrease in JNK activation. They also observed attenuation in degradation of IκBα. Overman et al. (2010) showed the same results for grape powder extract. Degradation of IκB is a seminal step in activation of NF-κB. The IκB kinases (IKKs) lie downstream of the NF-κB-inducing kinase (NIK) and activate NF-κB by phosphorylation of IκBα. This leads to IκBα degradation and release of NF-κB. NF-κB is a key regulator of the pro-inflammatory cytokine release, so an inflammatory response is induced. In U937 monocytic cells, IL-1β and TNF-α induced IκB-dependent transcription equally (Nasuhara et al. 1999). These results are contradicting with the results from studies with quercetin on THP-1 cells, in which quercetin showed anti-inflammatory effects. It is possible that these cell lines respond differently to quercetin, simply because the cell lines are not the same.

References

Berges C, Naujokat C, Tinapp S, Wieczorek H, Höh A, Sadeghi M, Opelz G, Daniel V (2005) A cell line model for the differentiation of human dendritic cells. Biochem Biophys Res Commun 333:896–907

Bogoyevitch MA, Kobe B (2006) Uses for JNK: the many and varied substrates of the c-Jun N-terminal kinases. Microbiol Mol Biol Rev 70(4):1061–1095

Bumrungpert A, Kalpravidh RW et al (2010) Xanthones from mangosteen inhibit inflammation in human macrophages and in human adipocytes exposed to macrophage-conditioned media. J Nutr 140(4):842–847

Cam A, de Mejia EG (2012) RGD-peptide lunasin inhibits Akt-mediated NF-kappaB activation in human macrophages through interaction with the alphaVbeta3 integrin. Mol Nutr Food Res 56(10):1569–1581

Chanput W, Mes J et al (2010) Transcription profiles of LPS-stimulated THP-1 monocytes and macrophages: a tool to study inflammation modulating effects of food-derived compounds. Food Funct 1(3):254–261

Chanput W, Reitsma M et al (2012) β-glucans are involved in immune-modulation of THP-1 macrophages. Mol Nutr Food Res 56(5):822–833

Chanput W, Mes JJ, Wichers HJ (2014) THP-1 cell line: an in vitro cell model for immune-modulation approach. Int Immunopharmacol. doi:10.1016/j.intimp.2014.08.002

Chun EM, Park YJ et al (2001) Expression of the apolipoprotein C-II gene during myelomonocytic differentiation of human leukemic cells. J Leukoc Biol 69(4):645–650

Friedland JS, Shattock RJ, Griffin GE (1993) Phagocytosis of *Mycobacterium tuberculosis* or particulates stimuli by human monocytic cells induces equivalent monocyte chemotactic protein-1 gene expression. Cytokine 5(2):150–156

Gillies PJ, Bhatia SK et al (2012) Regulation of inflammatory and lipid metabolism genes by eicosapentaenoic acid-rich oil. J Lipid Res 53(8):1679–1689

Henrick BM, Nag K et al (2012) Milk matters: soluble toll-like receptor 2 (sTLR2) in breast milk significantly inhibits HIV-1 infection and inflammation. PLoS One 7(7)

Hotta M, Nakata R et al (2010) Carvacrol, a component of thyme oil, activates PPAR alpha and gamma and suppresses COX-2 expression. J Lipid Res 51(1):132–139

Huang SM, Wu CH et al (2006) Effects of flavonoids on the expression of the pro-inflammatory response in human monocytes induced by ligation of the receptor for AGEs. Mol Nutr Food Res 50(12):1129–1139

Huang S, Rutkowsky JM et al (2012) Saturated fatty acids activate TLR-mediated proinflammatory signaling pathways. J Lipid Res 53(9):2002–2013

Ivanov V, Cha J et al (2008) Essential nutrients suppress inflammation by modulating key inflammatory gene expression. Int J Mol Med 22(6):731–741

Iyer SS, Accardi CJ et al (2009) Cysteine redox potential determines pro-inflammatory IL-1 beta Levels. PLoS One 4(3)

Jabaut J, Ckless K (2012) Inflammation, immunity and redox signaling. In: Khatami M (ed) Inflammation, chronic diseases and cancer – cell and molecular biology, immunology and clinical bases, Chap 7. InTech, Rijeka. ISBN 978-953-51-0102-4

Karlsen A, Paur I et al (2010) Bilberry juice modulates plasma concentration of NF-kappa B related inflammatory markers in subjects at increased risk of CVD. Eur J Nutr 49(6):345–355

Kim SJ, Park JH et al (2011) Melittin inhibits atherosclerosis in LPS/high-fat treated mice through atheroprotective actions. J Atheroscler Thromb 18(12):1117–1126

Lee JS, Park SY et al (2010) *Grifola frondosa* water extract alleviates intestinal inflammation by suppressing TNF-alpha production and its signaling. Exp Mol Med 42(2):143–154

Liu L, Zubik L et al (2004) The antiatherogenic potential of oat phenolic compounds. Atherosclerosis 175(1):39–49

Moore GE et al (1967) Culture of normal human leukocytes. JAMA 199:519–524

Moreno-Navarrete JM, Ortega FJ et al (2009) Decreased circulating lactoferrin in insulin resistance and altered glucose tolerance as a possible marker of neutrophil dysfunction in type 2 diabetes. J Clin Endocrinol Metab 94(10):4036–4044

Morton HJ (1970) A survey of commercially available tissue culture media. In Vitro 6(2):89–108

Nardini M, Leonardi F et al (2001) Modulation of ceramide-induced NF-kappaB binding activity and apoptotic response by caffeic acid in U937 cells: comparison with other antioxidants. Free Radic Biol Med 30(7):722–733

Nasuhara Y, Adcock IM et al (1999) Differential IkappaB kinase activation and IkappaBalpha degradation by interleukin-1beta and tumor necrosis factor-alpha in human U937 monocytic cells. Evidence for additional regulatory steps in kappaB-dependent transcription. J Biol Chem 274(28):19965–19972

Ogasawara N1, Kojima T et al (2009) Induction of JAM-A during differentiation of human THP-1 dendritic cells. Biochem Biophys Res Commun 389(3):543–549

Overman A, Bumrungpert A et al (2010) Polyphenol-rich grape powder extract (GPE) attenuates inflammation in human macrophages and in human adipocytes exposed to macrophage-conditioned media. Int J Obes (Lond) 34(5):800–808

Overman A, Chuang CC et al (2011) Quercetin attenuates inflammation in human macrophages and adipocytes exposed to macrophage-conditioned media. Int J Obes (Lond) 35(9):1165–1172

Qin Z (2012) The use of THP-1 cells as a model for mimicking the function and regulation of monocytes and macrophages in the vasculature. Atherosclerosis 221(1):2–11

Rots NY, Iavarone A, Bromleigh V, Freedman LP (1999) Induced differentiation of U937 cells by 1,25-dihydroxyvitamin D3 involves cell cycle arrest in G1 that is preceded by a transient proliferative burst and an increase in cyclin expression. Blood 93(8):2721–2729

Ryan SM, McMorrow J et al (2013) An intra-articular salmon calcitonin-based nanocomplex reduces experimental inflammatory arthritis. J Control Release 167(2):120–129

Schildberger A et al (2013) Monocytes, peripheral blood mononuclear cells, and THP-1 cells exhibit different cytokine expression patterns following stimulation with lipopolysaccharide. Mediators Inflamm 2013:10

Shattock RJ, Friedland JS, Griffin GE (1994) Phagocytosis of *Mycobacterium tuberculosis* modulates human immunodeficiency virus replication in human monocytic cells. J Gen Virol 75(4):849–856

Spencer M, Yao-Borengasser A, Unal R, Rasouli N et al (2010) Adipose tissue macrophages in insulin-resistant subjects are associated with collagen VI and fibrosis and demonstrate alternative activation. Am J Physiol Endocrinol Metab 299:E1016–E1027

Stephens TA, Nikoopour E, Rider BJ, Leon-Ponte M, Chau TA, Mikolajczak S, Chaturvedi P, Lee-Chan E, Flavell RA, Haeryfar SM, Madrenas J, Singh B (2008) Dendritic cell differentiation induced by a self-peptide derived from apolipoprotein E. J Immunol 181(10):6859–6871

Strefford JC, Foot NJ, Chaplin T, Neat MJ, Oliver RTD, Young BD, Jones LK (2001) The characterisation of the lymphoma cell line U937, using comparative genomic hybridisation and multiplex FISH. Cytogenet Cell Genet 94:9–14

Sundstrom C, Nilsson K (1976) Establishment and characterisation of a human histocytic lymphoma cell line (U937). Int J Cancer 17:565–577

Tatsuno T, Jinno M et al (2012) Anti-inflammatory and anti-melanogenic proanthocyanidin oligomers from peanut skin. Biol Pharm Bull 35(6):909–916

Tsuchiya S, Yamabe M, Yamaguchi Y, Kobayashi Y, Konno T, Tada K (1980) Establishment and characterization of a human acute monocytic leukemia cell line (THP-1). Int J Cancer 26(2):171–176

Vordenbaumen S, Braukmann A et al (2011) Casein alpha s1 is expressed by human monocytes and upregulates the production of GM-CSF via p38 MAPK. J Immunol 186(1):592–601

Waksman SA (1953) Streptomycin: background, isolation, properties, and utilization. Science 118(3062):259–266

Wu CH, Wu CF et al (2009) Naturally occurring flavonoids attenuate high glucose-induced expression of proinflammatory cytokines in human monocytic THP-1 cells. Mol Nutr Food Res 53(8):984–995

Wu CH, Huang HW et al (2011) The proglycation effect of caffeic acid leads to the elevation of oxidative stress and inflammation in monocytes, macrophages and vascular endothelial cells. J Nutr Biochem 22(6):585–594

Zhang YH, Lian FZ et al (2010) Cyanidin-3-O-beta-glucoside inhibits LPS-induced expression of inflammatory mediators through decreasing I kappa B alpha phosphorylation in THP-1 cells. Inflamm Res 59(9):723–730

Ziegler-Heitbrock HWL et al (1988) Establishment of a human cell line (Mono Mac 6) with characteristics of mature monocytes. Int J Cancer 41(3):456–461

Zingg JM, Hasan ST et al (2012) Regulatory effects of curcumin on lipid accumulation in monocytes/macrophages. J Cell Biochem 113(3):833–884

Chapter 15
Peripheral Blood Mononuclear Cells

Charlotte R. Kleiveland

Abstract Numerous cell types are involved in maintenance of the intestinal tissue. However, the main players are cells of the epithelial lining and the immune system. Human peripheral blood mononuclear cells (PBMCs) are used to investigate the effect of food bioactives on various immune cells. These cells are easily isolated from blood of healthy donors or buffy coats (leukocyte concentrates, a by-product from hospital Blood Banks in the manufacturing of red blood cell and thrombocyte concentrates from anti-coagulated whole blood). PBMCs have a different composition, phenotype and activation status than cells found in intestinal tissue. However, this is a useful test system for investigation of immune modulatory effects of food bioactive compounds. Methods for the isolation of PBMCs and how they are used to investigate effects of bioactive components are discussed in this chapter.

Keywords Peripheral blood mononuclear cells (PBMCs) • Mitogenic lectins • T cells • Cytokines • Surface markers

15.1   Origin

Human peripheral blood mononuclear cells (PBMCs) are isolated from peripheral blood and identified as any blood cell with a round nucleus (i.e. lymphocytes, monocytes, natural killer cells (NK cells) or dendritic cells). The cell fraction corresponding to red blood cells and granulocytes (neutrophils, basophils and eosinophils) is removed from whole blood by density gradient centrifugation. A gradient medium with a density of 1.077 g/ml separates whole blood into two fractions; PBMCs makes up the population of cells that remain in the low density fraction (upper fraction), whilst red blood cells and PMNs have a higher density and are found in the lower fraction (Fig. 15.1).

C.R. Kleiveland (✉)
Faculty of Veterinary Medicine and Biosciences, Department of Chemistry, Biotechnology and Food Science, Norwegian University of Life Sciences, Ås, Norway

Research Department, Ostfold Hospital Trust, Fredrikstad, Norway
e-mail: charlotte.kleiveland@nmbu.no

© The Author(s) 2015
K. Verhoeckx et al. (eds.), *The Impact of Food Bio-Actives on Gut Health*,
DOI 10.1007/978-3-319-16104-4_15

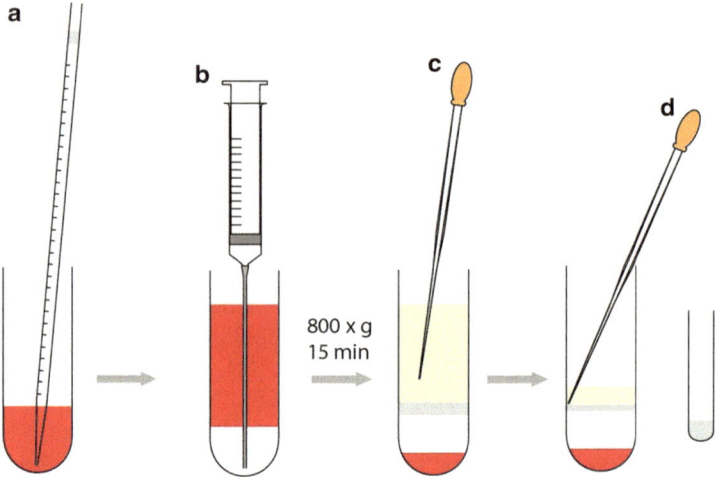

Fig. 15.1 Schematic illustration of how to prepare the density gradient for isolation of PBMCs from blood (**a**, **b**) and where in the gradient the PBMCs (*grey layer*) are found after centrifugation (**c**, **d**)

PBMCs include lymphocytes (T cells, B cells, and NK cells), monocytes, and dendritic cells. In humans, the frequencies of these populations vary across individuals, but typically, lymphocytes are in the range of 70–90 %, monocytes from 10 to 20 %, while dendritic cells are rare, accounting for only 1–2 %. The frequencies of cell types within the lymphocyte population include 70–85 % $CD3^+$ T cells, 5–10 % B cells, and 5–20 % NK cells. The $CD3^+$ lymphocytes are composed of $CD4^+$ and $CD8^+$ T cells, roughly in a 2:1 ratio. After activation the $CD4^+$ T cell subset may develop into diverse effector cell subsets, including Th1, Th2, Th17, Th9, Th22, follicular helper (Tfh) cells and different types of regulatory cells (Akdis et al. 2012; Crotty 2011; Tan and Gery 2012; Sakaguchi et al. 2008). The $CD4^+$ helper T cells are essential mediators of immune homeostasis and inflammation (Hirahara et al. 2013).

15.2 Features and Mechanisms

Most of the PBMCs are naïve or resting cells without effector functions. In the absence of an ongoing immune response T cells, the largest fraction of the isolated PBMCs, are mainly present as naïve or memory T cells. The naïve T-cells have never encountered their cognate antigen before and are commonly characterized by the absence of activation markers like CD25, CD44 or CD69 and the absence of the memory marker CD45RO isoform. Antigen recognition by a naïve T cell may result in activation of the cell (described in detail in Chap. 18), which will then enter a differentiation program and develop effector functions. Activated T cells may further differentiate into memory T cells that are able to mount a faster and stronger immune response the second time the antigen is encountered. B cells are also present in blood as naïve cells with a surface-anchored antigen receptor. Upon encountering antigen, the B cell may become activated and differentiate into antibody-producing plasma cells.

In peripheral blood the frequency of lymphocytes with specificity for a single antigen are low. Polyclonal activators are therefore used for in vitro stimulation as these will activate a large proportion of the lymphocytes, independently of their antigen specificity. The most common activators are mitogenic lectins, carbohydrate-binding proteins that bind to a number of glycoproteins expressed on the plasma membrane of lymphocytes (Ashraf and Khan 2003). Also, polyclonal activation of T cells can be obtained by antibodies that specifically bind to CD3, alone or in combination with CD28. The most commonly used mitogenic lectins are phytohae-magglutinin (PHA) and concanavalin A (con A) that mainly induce T cell prolifera-tion, pokeweed mitogen (PWM) that induce T and B cell proliferation and lipopolysaccharide (LPS) which induce B cell proliferation and activation of mono-cytes. Effects on the immune function of PBMCs are usually investigated by measuring changes in lymphocyte proliferation, characterization of cytokine secretion profiles or changes in gene expression.

15.3 Stability, Consistency and Reproducibility

It is well known that multiple physiological factors such as nutritional status, hormone levels and infections/inflammation will influence the reactivity of immune cells; hence, the composition of PBMCs will depend on donor and the donor's physiological status. Compared to the use of cell lines, this can lead to increased inter-experimental variation when different donors are used. Although the use of different donors leads to increased variance, the strength of reproducing results with cells from several donors will support the generality of the results.

15.4 Relevance to Human In Vivo Situation

PBMCs isolated from whole blood will be different from immune cells isolated from intestinal tissue or lymph nodes draining the intestine. Intraepithelial lympho-cytes are predominantly CD8$^+$ (90 %) cells whilst lymphocytes within the lamina propria are mainly CD4$^+$ cells expressing a CD44^{hi}CD62L$^-$ effector memory phe-notype which indicate that these are antigen-experienced cells (Shale et al. 2013). In addition, several populations of immune cells present in the intestinal mucosa is not present in blood i.e. B1 B cells, natural killer T cells (NKT cells) and innate lymphoid cells (ILCs). Upon inflammatory signals, monocytes from blood will move into the site of infection and differentiate to myeloid antigen-presenting cells which is a heterogeneous population of macrophages and dendritic cells (DCs) (Swiatczak and Rescigno 2012). Because of these differences in the phenotype of cells present in blood and in lamina propria of the gastrointestinal tract it can be expected that mononuclear cells isolated from blood will respond differently from lamina propria mononuclear cells. Care should therefore be taken in drawing conclusions about the mucosal immune response in the intestine from in vitro studies with PBMCs.

15.5 General Protocol

Isolation of PBMCs is done from peripheral blood or buffy coats supplemented with anticoagulants (heparin, EDTA, citrate, ACD-A or citrate phosphate dextrose (CPD)). Blood is diluted with 2–4 volumes phosphate-buffered saline (PBS) pH-7.2. The more diluted blood samples the better purity of the mononuclear cells. Add 35 ml of diluted blood to a 50 ml tube (Fig. 15.1a). Place 10 ml of a density gradient medium with $\rho = 1.077$ g/ml (e.g. Ficoll-Paque PLUS) at the bottom of the tube with a 15 G hypodermic needle (or similar) (Fig. 15.1b) before centrifugation at $400 \times g$ for 25 min at 20 °C in a swinging bucket rotor without brake. Using a brake during centrifugation will disturb the separation of the upper and lower fractions. Aspirate most of the upper layer leaving the mononuclear cells at the interphase (Fig. 15.1c). Carefully transfer the mononuclear cells to a new 50 ml tube, fill the tube with PBS, mix and centrifuge at $300 \times g$ for 10 min at 20 °C (Fig. 15.1d). Remove the supernatant completely and resuspend the cells in a small volume of PBS before diluting the cell suspension with 50 ml PBS and centrifuge at $200 \times g$ for 15 min at 20 °C. This step will remove platelets, and it should be repeated at least once. When most of the platelets are removed, resuspend the cells in complete RPMI 1640 with 10 % heat inactivated fetal calf serum (FCS), and count them. Cells can be used immediately or preserved and stored in liquid nitrogen.

15.5.1 Study of Proliferative/Cytotoxic Activity

Dilute the test compounds in RPMI 1640 with additions as described above. Seed the PBMCs at a density of 1×10^5 cells/well in a 96 well plate. Incubate the PBMCs with serial dilutions of the test compound in triplicates at 37 °C in an atmosphere of 5 % CO_2 and 100 % humidity. Measurement of effects on proliferative/cytotoxic responses must be done in the presence and absence of the mitogenic stimuli. An array of methods and kit-based assay systems are available for this purpose, for example incorporation of ^{3}H-thymidine or measurement of mitochondrial reductase with tetrazolium salts (MTT). The optimal concentrations of the polyclonal activators can be established by cultivating PBMCs in the presence of twofold dilutions of the activators. Effect of the test compound on proliferation/cytotoxicity should be studied in combination with both optimal and suboptimal concentrations of the mitogenic stimuli. The vehicle control (solvent of the test compound stock) should be included too. As background control, RPMI 1640 with the test compound and activators can be used. As negative control, PBMCs in RPMI 1640 without both test compounds and activators should be used.

15.5.2 Study of Inflammatory Responses

Upon stimulation with polyclonal activators PBMCs will produce cytokines and up-regulate activation markers. Characterization of the cytokine profile and changes in activation marker expression, especially on T cells, may provide information whether the response is in the direction of Th1, Th2, Th17 or regulatory T cells. Investigations of the cytokine profile is most easily carried out by ELISA analysis of secreted cytokines in the culture supernatant. Alternatively, the number of cells producing the different cytokines can be studied after intracellular staining and analysis by flow cytometry. Also, changes in gene expression levels by qPCR can be informative. Test compounds should be diluted in RPMI 1640. Plate the PBMCs at a density of $1–2 \times 10^6$ cells/well in a 24 well plate. Incubate the PBMCs with a serial dilution of the test compound in the presence and absence of LPS (100 ng/ml), anti-CD3/CD28 or the above mentioned mitogenic lectins at 37 °C in an atmosphere of 5 % CO_2. Cytokines will be detectable in the culture supernatant after only a few hours. However, cytokines will be secreted with different kinetics, it will therefore be necessary to study the cytokine profile at different time points.

15.6 Assess Viability

In the case of the study of inflammatory responses, the cytotoxicity of the test compound at varying concentrations should be tested for instance by using LDH or MTT.

15.7 Experimental Read Out

PBMCs have been widely employed to assess several aspects of immune regulation. The physiological relevance of in vitro studies with PBMCs is debatable, but several reports have shown that these studies can be predictive for the in vivo situation. de Kivit et al. (2012) reported that increased galectin-9 expression in intestinal epithelial cells and serum induced by dietary symbiotics in mice correlated with reduced acute allergic skin reactions and mast cell degranulation. Parallel studies with stimulation of PBMCs with recombinant galectin-9, showed induction of development of Th1 and Treg cells. CD3/CD28-activated PBMCs were incubated with either medium or increasing concentrations of recombinant galectin-9 for 24 h. The effect on T cell activation was assessed by analysis of surface markers by flow cytometry. Th1 cells were identified as CD4$^+$, CD69$^+$ and CXCR3$^+$ and Treg cells as CD4$^+$, CD25$^+$ and Foxp3$^+$. Further characterization of the cytokine profile was used to support the results. Increased secretion of IFNγ and Il-10 suggest increased levels of Th1 and Treg cells, whilst a decrease in IL-17 supports suppression of Th17 cells.

Cytokine secretion is one of the most used outcomes for evaluating influence on immune responses and has been studied with a variety of different types of compounds. For example, Schroecksnadel et al. reported immune modulatory effects of a vitamin K antagonist (Schroecksnadel et al. 2013). PBMCs were pre-incubated with the vitamin K antagonist for 30 min before the cells were stimulated with PHA for 48 h. The cytokines secreted into the supernatants were analyzed by ELISA. Also, in a recent report cytokine secretion and activation of mitogen-activated protein kinases (MAPK) in response to polysaccharides isolated from *Alchornea cordifolia* (Kouakou et al. 2013) was studied. Evaluation of the effects of polysaccharides on the activation status of an array of different MAPKs was done using a phospho-MAPK kit. PBMCs were incubated with the test compound for 60 min before lysis of the cells. An array of methods and kit based assays are available for evaluation of the activation status of MAPK and other intracellular protein kinases.

Effect of polyphenolic extracts from *Carpobrotus rossii* on cytokine release from PBMCs has been reported (Geraghty et al. 2011). The cytokine release induced by the polyphenolic extract was investigated in the presence and absence of stimulation with PHA and LPS. In this study, a multiplex cytokine profiling kit (Luminex®) was used to determine cytokine concentrations in culture supernatants.

Several reports have used PBMCs to study the immunomodulatory effects of probiotic bacteria. *Lactobacillus plantarum* genes involved in immune modulation were identified in a study with PBMCs (van Hemert et al. 2010). A total of 42 *L. plantarum* strains were evaluated for their capacity to stimulate cytokine production in PBMCs. Comparison of strain-specific cytokine responses by PBMCs resulted in identification of six candidate genetic loci with immune modulatory capacities. Ashraf et al. have recently reported effects of cell-surface extracts and metabolites from a probiotic organism on both cytokine production and induction of CD25 expression on PBMCs (Ashraf et al. 2014).

15.8   Advantages, Disadvantages and Limitations of the System

The advantage of PBMCs is that it is an easy accessible source of human immune cells, as the cells are isolated form full blood or buffy coats.

Disadvantages and limitations of this model system is the phenotypic differences between peripheral mononuclear cells and immune cells of the intestinal mucosa.

15.9 Conclusions

Although PBMCs is an easy accessible source of human immune cells, there are two major points that are important to keep in mind. (1) These cells are blood mononuclear cells that will differ from immune cells found i.e. in intestinal tissues. (2) When using PBMC in in vitro experiments the cells will lack the environmental stimuli they would have been exposed to under normal in vivo conditions. Both of these points are of great importance for how immune cells responds to different stimuli and should be taken in to account when interpreting the results.

References

Akdis M, Palomares O, van de Veen W et al (2012) TH17 and TH22 cells: a confusion of antimicrobial response with tissue inflammation versus protection. J Allergy Clin Immunol 129:1438–1449, quiz 1450–1431

Ashraf MT, Khan RH (2003) Mitogenic lectins. Med Sci Monit 9:RA265–RA269

Ashraf R, Vasiljevic T, Smith SC et al (2014) Effect of cell-surface components and metabolites of lactic acid bacteria and probiotic organisms on cytokine production and induction of CD25 expression in human peripheral mononuclear cells. J Dairy Sci 97:2542–2558

Crotty S (2011) Follicular helper CD4 T cells (TFH). Annu Rev Immunol 29:621–663

de Kivit S, Saeland E, Kraneveld AD et al (2012) Galectin-9 induced by dietary synbiotics is involved in suppression of allergic symptoms in mice and humans. Allergy 67:343–352

Geraghty DP, Ahuja KD, Pittaway J et al (2011) In vitro antioxidant, antiplatelet and anti-inflammatory activity of *Carpobrotus rossii* (pigface) extract. J Ethnopharmacol 134:97–103

Hirahara K, Poholek A, Vahedi G et al (2013) Mechanisms underlying helper T-cell plasticity: implications for immune-mediated disease. J Allergy Clin Immunol 131:1276–1287

Kouakou K, Schepetkin IA, Yapi A et al (2013) Immunomodulatory activity of polysaccharides isolated from *Alchornea cordifolia*. J Ethnopharmacol 146:232–242

Sakaguchi S, Yamaguchi T, Nomura T et al (2008) Regulatory T cells and immune tolerance. Cell 133:775–787

Schroecksnadel S, Gostner J, Jenny M et al (2013) Immunomodulatory effects in vitro of vitamin K antagonist acenocoumarol. Thromb Res 131:e264–e269

Shale M, Schiering C, Powrie F (2013) CD4(+) T-cell subsets in intestinal inflammation. Immunol Rev 252:164–182

Swiatczak B, Rescigno M (2012) How the interplay between antigen presenting cells and microbiota tunes host immune responses in the gut. Semin Immunol 24:43–49

Tan C, Gery I (2012) The unique features of Th9 cells and their products. Crit Rev Immunol 32:1–10

van Hemert S, Meijerink M, Molenaar D et al (2010) Identification of *Lactobacillus plantarum* genes modulating the cytokine response of human peripheral blood mononuclear cells. BMC Microbiol 10:293

Chapter 16
PBMC-Derived T Cells

Daniel Lozano-Ojalvo, Rosina López-Fandiño, and Iván López-Expósito

Abstract T cell cultures are a valuable tool in food research to perform studies within the food allergy field. Their main applications aim to analyze immunological responses towards food protein antigens to gain further insights into the mechanisms responsible for the development of oral tolerance or for the triggering of food allergies. This chapter describes the main applications, isolation techniques, and culture conditions for PBMC-derived T cells. Furthermore, critical parameters of the model, together with the experimental read outs will be discussed.

Keywords T cells cultures • Food allergy • Immunomagnetic cell isolation • PBMCs • CD4+ T cells

16.1 Introduction and Origin

T lymphocytes or T cells are small (<10 μm) resting cells, which are generated in the bone marrow and migrate to the thymus where they become mature. Once matured, they enter the bloodstream and circulate to the secondary lymphoid organs, the sites of lymphocyte activation by the antigens. Actually, T cells are constantly recirculating between these organs until they encounter their specific antigen. Naïve T cells (T cells that have not yet met their specific antigen) bear antigen receptors specific for a single chemical structure, however lymphocytes in the body collectively carry millions of different receptor specificities. Only those that meet an antigen to which their receptors bind will be activated to proliferate and differentiate into effector cells (Janeway et al. 2005).

This chapter deals with T lymphocytes derived from peripheral blood mononuclear cells (PBMCs). T cells comprise, approximately, 45–70 % of PBMCs in human peripheral blood, with a count in healthy subjects in the range of 1 million cells/mL. For many years there was no function ascribed to these cells, until the 1960s, when it gradually became apparent that T lymphocytes were the key mediators

D. Lozano-Ojalvo • R. López-Fandiño • I. López-Expósito (✉)
Departamento de Bioactividad y Análisis de Alimentos, Instituto de Investigación en Ciencias de la Alimentación (CIAL) (CSIC-UAM), Madrid, Spain
e-mail: ivan.lopez@csic.es

© The Author(s) 2015

169

K. Verhoeckx et al. (eds.), *The Impact of Food Bio-Actives on Gut Health*,
DOI 10.1007/978-3-319-16104-4_16

of adaptive immunity. T cell cultures are a valuable tool in food research to perform studies within the food allergy field. Their main applications aim to analyze immunological responses towards food protein antigens to gain further insights into the mechanisms responsible for the development of oral tolerance or for the triggering of food allergies (Martino et al. 2012).

16.2   Features and Mechanisms

T cells express surface antigen receptors called T cell receptors (TCRs), which are inserted in the T cell surface and never secreted. The most common form is made up of α and β chains and it is found on about 95 % of circulating T cells. The TCR is a clonally distributed receptor, meaning that clones of T cells with different specificities express different TCRs. In the functional TCR, the α and β heterodimers are associated with a complex of four other invariant signaling chains, collectively called CD3, which are required for the cell-surface expression of the antigen binding chains and for signaling (Fig. 16.1) (Farber 2011).

There are two major types of T lymphocytes, classified according to the expression of the cell surface co-receptors CD4 and CD8, which are proteins non-covalently associated with the TCR. Their function is to signal to the T cell that the TCR complex has bound the proper antigen. CD4$^+$ T cells are generally referred as helper T cells or Th cells, because they secrete a multitude of cytokines that help or coordinate cellular and humoral immunity. CD8$^+$ T cells are generally referred as cytotoxic T lymphocytes or Tc cells. Cytotoxic T cells recognize as antigens fragments of viral proteins on the surface of virus-infected cells, which they kill by the activation of a cascade of caspases. In humans, circulating CD4$^+$ T cells outnumber CD8$^+$ T cells by approximately 2:1 (Abbas et al. 2007). This chapter will deal with the CD4$^+$ T lymphocytes, as they are the type of T lymphocytes that become activated during the allergic response.

T cell responses are initiated when a mature CD4$^+$ T cell encounters a properly activated antigen presenting cell (APC), such as a dendritic cell, in a secondary

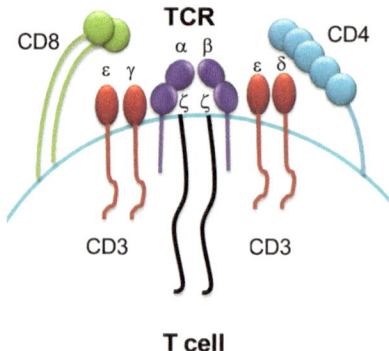

Fig. 16.1 TCR structural components (adapted from Farber 2011)

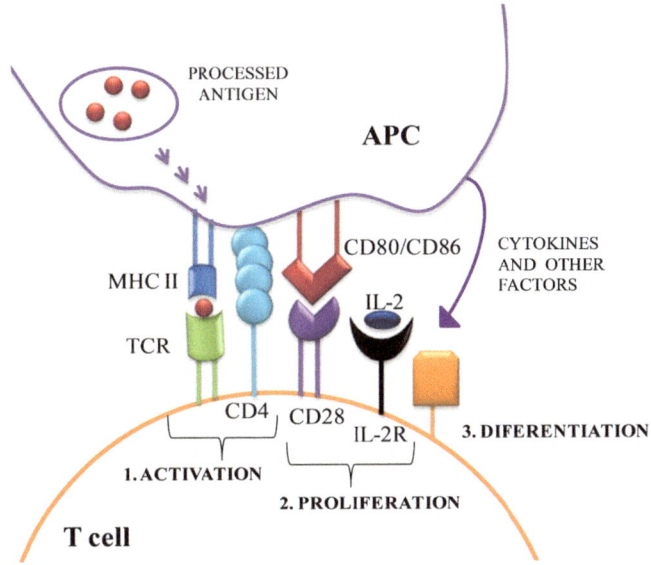

Fig. 16.2 Signals involved in the activation of naïve T cells by APC

lymphoid organ. If the APC displays the appropriate peptide ligand through the major histocompatibility complex (MHC) class II molecule, it is recognized by the TCR. This is essential for activating a naïve T cell, but even if the co-receptor CD4 is also ligated, for the T cell to proliferate and differentiate, two other stimulatory signals delivered by the APC are required. These signals are, on the one hand, the binding of two different ligands on the APC surface, designated as CD80 and CD86, to a surface molecule on the T cell, called CD28, and on the other, those involved in directing T cell differentiation into different subsets of effector T cells, the latter being mainly, but not exclusively, driven by cytokines, such as IL-6, IL-12 and TGF-β (Fig. 16.2). The CD28-dependent co-stimulation of activated T cells leads to production of IL-2 by the activated T cell themselves. Following expression of IL-2, there is also an upregulation of the third component (called α-chain) of the IL-2 receptor, also known as CD25, in addition to other regulatory molecules such as ICOS and CD40L (Farber 2011). Binding of IL-2 to its high affinity receptor promotes cell growth, whilst APCs, mainly dendritic cells generate various cytokines or express surface proteins that induce the differentiation of CD4⁺ T lymphocytes into cytokine producing effector cells, depending on environmental conditions.

Originally, two main types of effector CD4⁺ T cells, called T-helper 1 (*Th1*) and 2 (*Th2*) cells, were distinguished by the pattern of their cytokine secretion. Th1 cells secrete mainly IL-2, IFNγ and TNFα, and Th2 cells secrete IL-4, IL-13 and IL-5 (Romagnani 2000). Recently, this view has been challenged by the discovery of a new lineage of T cells characterized by their ability to secrete a proinflammatory cytokine, IL-17, and thus designated Th17 cells. This new T cell type has been related with autoimmune diseases (Jing and Dong 2013). Another subset of antigen-driven CD4⁺ T cells, named regulatory T cells (*Treg*), acts by inhibiting T cell responses by the production of cytokines, such as IL10 and TGF-β and/or via cell–cell interactions (Jutel and Akdis 2011). Treg are CD4+ cells that also express the α-chain of the IL-2 receptor (CD25+) and the transcription factor FoxP3.

The immune response leading to food allergy is driven by two main groups of signals. The first signal favors the differentiation of naïve T cells into a Th2 phenotype, and the second comprises the Th2 induced activation of B cells (mediated by cytokines such as IL-4 and IL-13 and co-stimulatory signals) to generate allergen-specific IgE antibodies (Mayorga et al. 2013).

16.3 Applications of T Cell Cultures

The main application for T cell cultures is the *characterization of CD4+ T cell epitopes from food allergens*. The mapping of CD4+ T cell epitopes contributes to a better understanding of the pathophysiology of food allergy by enabling the identification of the major peptides that target allergen-specific T cells (Prickett et al. 2011, 2013). This requires the comparison of the epitope recognition patterns of a diverse and representative sample of allergic individuals, since a given T cell epitope elicits a significant response only in subjects expressing MHC-class II molecules able to recognize and bind that particular epitope. In addition, the knowledge of the sequence of T cell epitopes that stimulate CD4+ T lymphocyte responses allows the development of new forms of immunotherapy, which are safer and more effective than those using whole allergens. In this respect, peptide-based vaccines corresponding to T cell epitopes of the allergen of interest retain immunogenicity, but they are of insufficient length to cross-link allergen specific IgE on the surface of effector cells and elicit an allergic response (Worm et al. 2011).

Another important application related with the use of T cell cultures is the *study of the allergen-specific T cell responses during or after immunotherapy*. In this context, evidence has accumulated that immunotherapy treatments shift a patient's immune reaction to a specific allergen from a predominately allergic Th2 response to a Th1 response, while inducing Treg that downregulate Th2 and Th1-cells activity. The characterization of the frequency and phenotype of allergen-specific T cells is useful to understand the immunological changes subjacent to therapeutical interventions in allergic diseases (DeLong et al. 2011; Foster et al. 2011). In this context, the role of Treg producing IL-10 in allergen specific T cell tolerance and immune deviation has been studied by comparing the response of PBMCs and PBMCs depleted of CD25+ in patients undergoing immunotherapy, as well as in tolerant and allergic patients, allowing the identification of potential markers that might be indicators of a favorable prognosis (Shreffler et al. 2009; Bohle et al. 2007). Similarly, allergen-specific signaling can be assessed in activated and purified CD4+ T cells to test whether there is a differential expression in the neonates who subsequently develop allergic diseases (Martino et al. 2012).

In the field of food allergy, T cell gene and protein expression assays are very important for the understanding of the functional characteristics of allergen reactive T cells. The use of PBMCs as a source of circulating T cells, that mimic effects occurring in remote target tissues of antigen exposure, presents the advantage that they can easily obtained by blood extraction and allow the study of parameters that otherwise would require more invasive methods, such a biopsy (de Mello et al. 2012). However, certain applications based on the exclusive employment of PBMC-derived T cells are limited because of the lack of APCs that are keys to the regulation of the complicated set of mechanisms that determine the immune responses, as well as of other environmental influences that play an important role in shaping the APC functions. To overcome this limitation, there are other options, such as the use of co-cultures with APCs (Frischmeyer-Guerrerio et al. 2011; Hofmann et al. 2012) or complex culture systems with other cell types, such as APCs and intestinal epithelial cells (Mileti et al. 2009). Another alternative are cell cultures from tissues of animal models of food allergy (spleen, lymph nodes, Peyer's patches) (López-Expósito et al. 2011). In any case, and because of extension limits, this chapter will only deal with T cells that do not require APCs in the culture (more information available in Chap. 17).

Because of the low frequency of allergen-specific CD4$^+$ T cells, several studies use allergen-specific CD4$^+$ T cell lines or clones (Prickett et al. 2011, 2013), while others use primary CD4$^+$ T cells, without addition of cytokines or repeated stimulation (Frischmeyer-Guerrerio et al. 2011; Hofmann et al. 2012). Although T cell lines overcome the frequency limitation, previous in vitro expansion can alter cell phenotypes or bias the results through the selection of the rapidly proliferating clones (Pascal et al. 2013).

16.4   General Protocol

16.4.1   T Cell Isolation Protocols

Because of the ease of access to peripheral blood, PBMCs, which contain T and B lymphocytes and monocytes, are the major source of human T cells used in most studies. There is a variety of available techniques for isolation and enrichment of T cells, including those based on their unique ability to bind and form rosettes with sheep red blood cells, as well as those based on differential adherence properties T cells to nylon wool. Despite their widespread use, the resulting T cell population is not very pure compared to the level of purity achieved with other procedures, and several reports suggest alterations in T cell functionality when these techniques are used (Wohler and Barnum 2009).

Human T cells can also be purified on the basis of their cell-surface display if specific antigens can be recognized by monoclonal antibodies. This is currently carried out by two main methods: fluorescence-activated cell sorting (FACS) and immunomagnetic cell separation (Martino et al. 2012; Prickett et al. 2013). FACS requires sophisticated technology, highly trained personnel; it is time consuming, expensive and may result in a significant cell loss. Conversely, immunomagnetic separation methods are faster, relatively inexpensive and do not require state of the art technology (Lancioni et al. 2009).

Immunomagnetic separation methods are based on the attachment of small magnetizable particles to cells via antibodies. The physical basis for such separation procedures involve the coupling of antibodies to magnetic beads, which subsequently allows the rapid capture of the cells specifically recognized by these antibodies among a mixed population of cells placed in a magnetic field. There are two types of magnetic cell isolation technologies, column and tube-based. Both work on the same principle, but the strength of the magnetic field required is different because they use beads of different sizes. In particular, the *column-based* technology uses beads consisting of iron oxide and polysaccharide of, approximately 50 nm in diameter, which require a very strong magnetic field. The main disadvantages of this system (exemplified by MACS produced by Milteny Biotech) lay in its high initial and running costs, although it provides cells with high purity and optimal viability and functionality (Li et al. 2012).

The immunomagnetic *tube-based* separation system utilizes micro-sized beads that can be selected using a magnet applied to a tube. The most commonly used beads, produced by Dynal, are 4.5 μm, uniform, spherical beads, which do not have any residual magnetism outside a magnetic field. They consist of an iron-containing core surrounded by a thin polymer shell to which biomolecules may be adsorbed. These beads can be attached to cells via a coating of primary or secondary antibodies. The cells, surrounded by a "rosette" of beads, may then be separated from the unlabeled population in a magnetic field using a relatively small, but powerful magnet (Neurauter et al. 2006).

There are two main strategies for cell isolation using micro-sized beads: *positive* and *negative selection*. Positive selection can be performed using either direct or indirect approaches (Neurauter et al. 2006). Following *direct positive isolation*, the appropriate antibody-coated beads are mixed with the sample and the target cells bound to the coated beads are subsequently collected with the aid of a magnet. For *indirect positive isolation*, a cell population is exposed to saturating amounts of a primary antibody which binds to the target cells and, once the unbound antibodies are washed away, secondary-antibody coated beads bind to the primary antibodies on the surface of the target cells. For further functional applications, and in order to avoid modifications of the phenotype, the cells bound to antibody-coated beads should be detached by exposure to a soluble antiserum against Fab fragments (e.g., DETEACHaBEAD). However, surface bound antibodies may elicit the transmission of signals across the cell membrane. This can be avoided by negative isolation, which does not require attachment of antibodies to the cells of interest at any time (Neurauter et al. 2007). By *negative isolation*, the cells are selected by removing all other cell types from the sample. Generally, a cocktail of monoclonal antibodies is incubated with the sample, followed by depletion of the undesirable cells using secondary antibody-coated beads (Biddison 1998; Mayer et al. 2013). Positive and negative separation strategies can be combined in the sequential sorting of T cell subsets, such as $CD25^+ FoxP3^+$ regulatory T cells, where depletion of unwanted cells is followed by positive selection of $CD25^+$ T cell population (Mayer et al. 2013).

The procedure described in this chapter is based on the indirect positive isolation of a subset of human T cells relevant to food allergy, such as CD4$^+$ T cells, by anti-CD4 and anti-IgG-coated beads (Biddison 1998; Neurauter et al. 2006). This methodology is very efficient in removing target cells due to its fast cell capture kinetics, whereas the use of secondary-antibody coated beads makes it a flexible protocol to isolate any cell of interest (Neurauter et al. 2007).

16.4.2 Indirect Positive Isolation of Human CD4$^+$ T

16.4.2.1 Preparation of Cells and Antibodies

PBMCs could be obtained from whole blood by density gradient of Ficoll-Hypaque (Martino et al. 2012). It is convenient to determine the approximate number of CD4$^+$ T cells in the starting population of PBMCs by flow cytometry using anti-CD4 antibody. It is also helpful to establish the saturating concentration of anti-CD4 mouse-IgG monoclonal antibody to be used by flow cytometry. Generally, 1 µg/mL per 1×10^6 cells works well as a saturating concentration. However, pretitration may allow the use of down to a tenfold lower concentration.

16.4.2.2 Coating of PBMCs with CD4 Antibody

PBMCs ($\leq 200 \times 10^6$ cells) are suspended in PBS/BSA and anti-CD4 mouse IgG antibody is added (10×). After incubation during 45 min at 4 °C with gentle tilting and rotation, cells are centrifuged 10 min at 600×g at room temperature. The supernatant is discarded and the pellet resuspended in pre-cooled PBS/BSA. This step is repeated twice. Finally, washed cells are resuspended in 10 mL of pre-cooled PBS/BSA and kept on ice for 15 min.

16.4.2.3 Magnetic Beads Washing Procedure

On the basis of the number of CD4$^+$ cells estimated by flow cytometry, the number of required goat anti-mouse IgG-coated magnetic beads is calculated. Five to ten magnetic beads are needed for each specific lymphocyte. Taking into account the concentration supplied by the manufacturer, the coated beads are resuspended in PBS and the desired volume of beads are transferred to a tube. PBS is added to the bead suspension and the tube is placed on a magnetic separation device for 2 min. The supernatant is then discarded by leaving the beads clinging to one side of the tube. The tube is removed from the magnet and the beads are washed following the previous step. Finally, the washed beads are resuspended in pre-cooled PBS/BSA and held on ice for 15 min.

16.4.2.4   Separation of T Cells

Anti-CD4-coated PBMCs are mixed with washed anti-IgG-coated beads and incubated for 45 min at 2–8 °C with gentle tilting and rotation. Afterwards, the tubes are placed on a vertical magnet for 2 min and the supernatant is carefully discarded. It is important not to disturb the beads that are clinging to one side of the tube. Once the tube is removed from the magnet, the beads are gently resuspended in 5 mL of Lymphocyte Culture Medium (LCM), which contains RPMI-1640 medium, 2 mM L-glutamine, 100 U/mL penicillin, 100 µg/mL streptomycin and 5 % (v/v) fetal bovine serum (FBS) (heat-inactivated during 1 h at 56 °C) (Prickett et al. 2013). Then, the tube is placed on the magnet for another 2 min and the non-collected cells aspirated, as in the previous step. The same procedure is repeated adding 5 mL of LCM to the tube. Finally, 2.5 mL of supernatant are carefully removed and the tube gently tapped to resuspend the T cells-coated-bead suspension.

16.4.2.5   Detachment of T Cells from Beads

Polyclonal anti-mouse Fab antiserum (200 µL) is added. Approximately 10 µL of Fab antiserum are needed per 10^7 beads, although a smaller amount of Fab antiserum solution may be used when starting with smaller numbers of lymphocytes (see manufacturer's instruction). This is mixed gently by tilting and rotation during an incubation period of 45–60 min at room temperature (incubation at 4 °C reduces the number of detached cells, while 37 °C does not increase detachment efficiency). Due to the relatively small sample volume, care should be taken that the cells remain on the bottom of the tube during agitation. 5 mL of PBS/BSA are added to the tube to resuspend the beads. Tubes are placed on a vertical magnet for 2 min. The supernatant containing the detached cells is then aspirated and left aside, while the beads remain attached to the wall of the tube by the magnet. It is important not to disturb the beads clinging to the side of the tube. This step is repeated four times to improve the isolation yield, saving and combining detached cell-containing supernatants from each separation in the same tube. Then, the tube with the detached T cells is centrifuged 10 min at 600×g at room temperature. The supernatant is discarded and the cells resuspended in LCM. Finally, T cells are counted and the number of viable cells determined by trypan blue exclusion. It is convenient to determine the purity of the CD4+ T cell population by flow cytometry using anti-CD4 antibody.

16.5   Assess Viability

Once PBMC-derived T cells have been isolated and detached from the magnetic beads, they are usually resuspended in LCM (Prickett et al. 2013). Alternatively, Iscove's modified Dulbecco's medium (IMDM), containing 100 U/mL penicillin, 100 µg/mL streptomycin, 5 µg/mL gentamicin and 5 % of heat-inactivated FBS, could also be used (Frischmeyer-Guerrerio et al. 2011). In the above proposed T cell culture media, FBS could be replaced by 5 % of heat-inactivated human serum or 10 % heparinized human plasma. While most protocols use serum to provide optimal conditions for T cell viability, reactivity and expansion, a number of serum-free media have been developed (e.g., AIM-V, Invitrogen Corporation; or X-VIVO, Lonza) (Shreffler et al. 2009). These media are specifically formulated to support the culture of T cells and incorporate defined quantities of purified growth factors, lipoproteins, and other proteins, which are usually provided by the serum.

Commonly, T cells are seeded at a final concentration of $1-2.5 \times 10^6$ cells/mL in 24 or 48-well cell culture plates (Pascal et al. 2013). Although, in T cell proliferation ($[H^3]$-thymidine) and viability (MTT or XTT) assays, 96-well plates are used and only 100 µL of the cell solution are added, resulting in $1-2.5 \times 10^5$ cells per well (Martino et al. 2012). T cell culture assays are usually carried out in sample (three different wells) and biological triplicates (three different plates) (Prickett et al. 2011).

All incubations of T cell cultures must be performed at 37 °C in a humidified 5 % CO_2 incubator. The incubation period for T cell assays varies depending on the initial cell population, percentage of damaged cells and stimuli used. Proliferation following mitogen-induced activation generally peaks after 2–6 days of culture (Bohle et al. 2007). An incubation period of 72 h is also used when proliferative responses to food proteins (such as OVA or peanut extracts) are studied (Prickett et al. 2011; Martino et al. 2012). However, for gene expression and cytokine detection assays, the response may not peak until after 5–10 days. Thus, the effect of Ara h 2 peptides has been analyzed after 4 days of incubation (Pascal et al. 2013), milk allergenic proteins after 5 days (Shreffler et al. 2009) and Ara h 1 peptides after 7 days (Prickett et al. 2013). To ensure that treatments do not affect cell health, a viability test should be performed. An array of methods and kit-based assays are available for cell viability, such as the measurement of mitochondrial reductase with tetrazolium salts (e.g. MTT, XTT and Alamar blue).

16.6 Samples

Hydrophilic samples, such as polar food proteins, are usually added to the cells diluted in PBS or directly in the culture medium at different final concentrations, which range from 10 to 200 µg/mL of culture (Pascal et al. 2013; Wing et al. 2003). When samples contain polyphenols or alcohols, they need to be resuspended in PBS or culture media containing 10 % of the aprotic and highly polar solvent DMSO. Potent T cell activators, such as mitogens (lectins like PHA or Concanavalin A), antigens (tetanus toxoid) or antibodies (anti-CD3/CD28) should be included as positive controls in the assays to test both cell viability and functionality (Bohle et al. 2007; Pascal et al. 2013).

A critical parameter in the evaluation of the response to proteins of T cells is sample quality. This is crucial for recombinant proteins which are produced in bacterial systems. These can carry over contaminants from the host cells to the final recombinant preparation, including irrelevant proteins and lipopolysaccharide (LPS). LPS, synthesized by gram-negative bacteria, has profound effects on T cell responses. LPS drives the development of Th1 subsets and it is also associated with toxicity (McAleer and Vella 2008). Because of these reasons, the purity of the proteins, and particularly the LPS concentration, should be determined by available commercial kits before application to T cell assays (Hofmann et al. 2012).

16.7 Experimental Readouts

In vitro T cell functional measures to stimuli may be monitored by assays that detect proliferation, cytokine secretion and the expression of genes of interest, or characterizing proliferating cells to identify phenotype and frequency. Measurement of the proliferative responses of T cells is fundamental for the assessment of their biological reaction to various stimuli, such as food allergens or hypoallergenic preparations. T cell proliferation is commonly determined by estimating incorporation of $[H^3]$-thymidine into the DNA, a process which is closely related to underlying changes in cell number. $[H^3]$-thymidine methodology has been widely utilized in allergen-specific T cell proliferation studies (Prickett et al. 2011; Shreffler et al. 2009). Alternatively, carboxyfluorescein diacetate succinimidyl ester (CFSE) labeling prior to T cell culture enables direct measurement of cell division and it may be used in combination with cell surface markers to identify the target cells by flow cytometry (Foster et al. 2011; Prickett et al. 2013).

Cytokine secretion in response to a stimulus may be detected by measuring either cytokine production (by ELISA) or enumerating individual cytokine producing T cells (by ELISPOT). ELISA has been used to evaluate T cell behavior towards food allergens in several studies (Glaspole et al. 2011; Hofmann et al. 2012). In contrast to ELISA, ELISPOT allows the visualization of the secretory products of individual activated or responding T cells. Thus, it provides both qualitative (kind of cytokine) and quantitative (number of antigen-specific T cells) information (Prickett et al. 2011; Faresjö 2012). Flow cytometry has been also used for the measurement of T cell cytokine production by fluorescent bead arrays (e.g., Cytometric Bead Array™—CBA, BD Biosciences) (Kücüksezer et al. 2013; Pascal et al. 2013). Flow cytometry can also be used for phenotypic analysis of T cell subsets, frequency determination (Shreffler et al. 2009; Hofmann et al. 2012) and detection of intracellular cytokines (DeLong et al. 2011). Finally, Real-Time quantitative PCR (RT-qPCR) has been largely used for the study the expression of cytokine genes (Kücüksezer et al. 2013). Gene expression has also been assessed in purified CD4+ T cells by DNA microarrays to characterize T cell signaling pathways of allergic and non allergic children (Martino et al. 2012).

Acknowledgements This work was funded by the project AGL2011-24740. D. L.-O. was the recipient of a FPU fellowship from the Spanish Ministry of Education, Culture and Sports. I. L.-E. acknowledges the financial support of CSIC through a JAE Doc grant.

References

Abbas AK, Lichtman AH, Pillai S (eds) (2007) Cellular and molecular immunology. Saunders, Philadelphia

Biddison WE (1998) Preparation and culture of human lymphocytes. Curr Protocols Cell Biol (Suppl):2.2.1–2.2.13

Bohle B, Kinaciyan T, Gerstmayr M, Radakovics A, Jahn-Schmid B, Ebner C (2007) Sublingual immunotherapy induces IL-10-producing T regulatory cells, allergen-specific T-cell tolerance, and immune deviation. J Allergy Clin Immunol 120:707–713

De Mello VDF, Kolehmainen M, Schawb U, Pulkkinen L, Uusitupa M (2012) Gene expression of peripheral blood mononuclear cells as a tool in dietary intervention studies: what do we know so far? Mol Nutr Food Res 56:1160–1172

DeLong JH, Simpson KH, Wambre E, James EA, Robinson D, Kwok WW (2011) Ara h 1-reactive T cells in individuals with peanut allergy. J Allergy Clin Immunol 127:1211–1218

Farber DL (2011) T lymphocytes and cell-mediated immunity. In: Bellanti JA (ed) Immunology IV, clinical applications in health and disease. I Care Press, Bethesda, pp 209–254

Faresjö M (2012) Enzyme linked immuno-spot; a useful tool in the search for elusive immune markers in common pediatric immunological diseases. Cells 29:141–152

Foster B, Foroughi S, Yin Y, Prussin C (2011) Effect of anti-IgE therapy on food allergen specific T cell responses in eosinophil associated gastrointestinal disorders. Clin Mol Allergy 9:7

Frischmeyer-Guerrerio PA, Guerrerio AL, Chichester KL, Bieneman AP, Hamilton RA, Wood RA, Schroeder JT (2011) Dendritic cell and T cell responses in children with food allergy. Clin Exp Allergy 41:61–71

Glaspole IN, de Leon MP, Prickett SR, O'Hehir RE, Rolland JM (2011) Clinical allergy to hazelnut and peanut: identification of T cell cross-reactive allergens. Int Arch Allergy Immunol 155:345–354

Hofmann C, Scheurer S, Rost K, Graulich E, Jamin A, Foetisch K, Saloga J, Vieths S, Steinbrink K, Adler HS (2012) Cor a 1-reactive T cells and IgE are predominantly cross-reactive to Bet v 1 in patients with birch pollen-associated food allergy to hazelnut. J Allergy Clin Immunol 131:1384–1392

Janeway CA, Travers P, Walport M, Shlomchik MJ (eds) (2005) Immunobiology, the immune system in health and disease. Garland Science, New York

Jing W, Dong C (2013) IL-17 cytokines in immunity and inflammation. Emerg Microbes Infect 2:e60

Jutel M, Akdis CA (2011) T-cell subset regulation in atopy. Curr Allergy Asthma Rep 11:139–145

Kücüksezer UC, Palomares O, Rückert B, Jartti T, Puhakka T, Nandy A, Gemicioğlu B, Fahrner HB, Jung A, Deniz G, Akdis CA, Akdis M (2013) Triggering of specific Toll-like receptors and proinflammatory cytokines breaks allergen-specific T-cell tolerance in human tonsils and peripheral blood. J Allergy Clin Immunol 131:875–885

Lancioni CL, Thomas JJ, Rojas RE (2009) Activation requirements and responses to TLR ligands in human CD4+ T cells: comparison of two T cell isolation techniques. J Immunol Methods 344:15–25

Li L, Mak KY, Shi J, Koon HK, Leung CH, Wong CM, Leung CW, Mak CS, Chan NM, Zhong W, Lin KW, Wu EX, Pong PW (2012) Comparative in vitro cytotoxicity study on uncoated magnetic nanoparticles: effects on cell viability, cell morphology, and cellular uptake. J Nanosci Nanotechnol 12:9010–9017

López-Expósito I, Jarvinen K, Castillo A, Seppo AE, Song Y, Li XM (2011) Maternal peanut consumption provides protection in offspring against peanut sensitization that is further enhanced when co-administered with bacterial mucosal adjuvant. Food Res Int 44:1649–1656

Martino DJ, Bosco A, McKenna KL, Hollams E, Mok D, Holt PG, Prescott SL (2012) T-cell activation genes differentially expressed at birth in CD4+ T-cells from children who develop IgE food allergy. Allergy 67:191–200

Mayer CT, Huntenburg J, Nandan A, Schmitt E, Czeloth N, Sparwasser T (2013) CD4 blockade directly inhibits mouse and human CD4(+) T cell functions independent of Foxp3(+) Tregs. J Autoimmun 47:73–82

Mayorga C, Torres MK, Blázquez AB (2013) Peanut allergy in the Mediterranean area. In: López-Expósito I, Blázquez AB (eds) Peanuts, bioactivities and allergies. Nova publishers Inc, New York, pp 153–171

McAleer JP, Vella AT (2008) Understanding how lipopolysaccharide impacts CD4 T cell immunity. Crit Rev Immunol 28:281–299

Mileti E, Matteoli G, Iliev ID, Rescigno M (2009) Comparison of the immunomodulatory properties of three probiotic strains of Lactobacilli using complex culture systems: prediction for in vivo efficacy. PLoS One 4(9):e7056

Neurauter AA, Aarvak T, Norderhaug L, Amellen Ø, Rasmussen AM (2006) Separation and expansion of human T cells. In: Celis JE (ed) Cell biology: a laboratory handbook. Elsevier Science, USA, pp 239–245

Neurauter AA, Bonyhadi M, Lien E, Nøkleby L, Ruud E, Camacho S, Aarvak T (2007) Cell isolation and expansion using Dynabeads. Adv Biochem Eng Biotechnol 106:41–73

Pascal M, Konstantinou GN, Masilamani M, Lieberman J, Sampson HA (2013) In silico prediction of Ara h 2 T cell epitopes in peanut-allergic children. Clin Exp Allergy 43:116–127

Prickett SR, Voskamp AL, Dacumos-Hill A, Symons K, Rolland JM, O'Hehir RE (2011) Ara h 2 peptides containing dominant CD4+ T-cell epitopes: candidates for a peanut allergy therapeutic. J Allergy Clin Immunol 127:608–615

Prickett SR, Voskamp AL, Phan T, Dacumos-Hill A, Mannering SI, Rolland JM, O'Hehir RE (2013) Ara h 1 CD4+ T cell epitope-based peptides: candidates for a peanut allergy therapeutic. Clin Exp Allergy 43:684–697

Romagnani S (2000) T cells subsets (Th1 vs Th2). Ann Allergy Asthma Immunol 85:9–18

Shreffler WG, Wanich N, Moloney M, Nowak-Wegrzyn A, Sampson HA (2009) Association of allergen-specific regulatory T cells with the onset of clinical tolerance to milk protein. J Allergy Clin Immunol 123:43–52

Wing K, Lindgren S, Kollberg G, Lundgren A, Harris RA, Rudin A, Lundin S, Suri-Payer E (2003) CD4 T cell activation by myelin oligodendrocyte glycoprotein is suppressed by adult but not cord blood CD25+ T cells. Eur J Immunol 33:579–587

Wohler JE, Barnum SR (2009) Nylon wool purification alters the activation of T cells. Mol Immunol 46:1007–1010

Worm M, Lee HH, Kleine-Tebbe J, Hafner RP, Laidler P, Healey D, Buhot C, Verhoef A, Maillre B, Kay AB, Larché M (2011) Development and preliminary clinical evaluation of a peptide immunotherapy vaccine for cat allergy. J Allergy Clin Immunol 127:89–97

Chapter 17
Dendritic Cells

Maud Plantinga, Colin de Haar, and Stefan Nierkens

Abstract Dendritic cells (DCs) are the sentinels of the immune system and play a critical role in stimulating immune responses against pathogens and maintaining immune homeostasis to harmless antigens. They can be found in all lymphoid and most non-lymphoid tissues, including mucosal surfaces, like the lung and the gut, where intricate networks of DCs are situated to sense potential harmful exposures. As such, DCs are among the first cells to come into contact with food bioactives in the gastrointestinal tract and thus are instrumental in shaping the immune system's response to such exposures. Here we provide an overview of DC characteristics, with the emphasis on DCs in the mucosal immune system, and discuss in vitro/ex vivo DC culture settings that can be applied for in vitro testing of (food) compounds.

Keywords Dendritic cells • DC subsets • Culture • Mucosa

17.1 Origin

Paul Langerhans was the first to describe DCs in the nineteenth century, when he visualised dendritically shaped cells in the skin (Langerhans 1868). Ralph M. Steinman and Zanvil A. Cohn identified "dendritic cells" as a specific group of white blood cells (Steinman and Cohn 1973), and it was not until 1998 that the central role of DCs as professional antigen presenting cells for the induction of adaptive immunity was defined (Banchereau and Steinman 1998). Over the last few decades the DC-field has developed extensively with enhanced knowledge on DC progenitors, distinctive subsets, and their specific functions depending on the tissues where they reside.

M. Plantinga (✉) • C. de Haar • S. Nierkens
Laboratory of Translational Immunology, UMC Utrecht, Utrecht, The Netherlands
e-mail: M.C.Plantinga-2@umcutrecht.nl; s.nierkens@umcutrecht.nl

© The Author(s) 2015   181
K. Verhoeckx et al. (eds.), *The Impact of Food Bio-Actives on Gut Health*,
DOI 10.1007/978-3-319-16104-4_17

17.2   Features and Mechanisms

DCs continuously sample antigens from their environment by phagocytosis, receptor-mediated endocytosis, and pinocytosis. After uptake of antigens, DCs migrate to the draining lymph node. During migration they may mature and upregulate co-stimulatory molecules like CD40 and CD86, depending on the maturation signals they have encountered. DCs are equipped with unique pattern recognition receptors (PRRs) for which the ligands may activate or inhibit DC maturation and facilitate antigen recognition, uptake and processing. The extensive sets of PRRs (e.g. Toll-like receptors (TLRs), C-type lectins (CLRs), Nod-like receptors (NLRs) and Retinoic acid induced gene-based (RIG)-I like receptors (RLRs)) are expressed on the cell surface or in intracellular compartments, like endosomes or in the cytoplasm to enable recognition of intra- and extracellular exposures.

Dendritic cells are instrumental for the activation of naïve T cells or reactivation of T cells from the memory pool (McLellan et al. 1996). T cell activation requires antigen presentation in MHC class I (for $CD8^+$ T cell) or II (for $CD4^+$ T cells) molecules (signal 1), in combination with expression of co-stimulatory molecules (signal 2). Cytokines provide a third signal and induce T cell differentiation and effector functions. Depending on the cytokine environment provided by DCs, $CD4^+$ T cells differentiate into Thelper1 (Th1) via IL-12, Th2 via IL-4, or Th17 cells via IL-6 (Kapsenberg 2003; Bettelli et al. 2008). In addition, Th cells can also differentiate into regulatory phenotypes (Tregs, Th3 and Tr1) that dampen immune responses (Sakaguchi et al. 2006). On the other hand, DCs instruct $CD8^+$ T cells, which recognize MHC class I (cross)presented antigens, to develop into cytotoxic lymphocytes (CTLs) (Bhardwaj et al. 1994).

The responsiveness of DCs in terms of expression of co-stimulatory/regulatory molecules and cytokine profiles may thus provide insight into the immune modulating properties of food bioactives. Here, we provide an overview of the different DC subsets in blood, and more specifically in the gastrointestinal tract (Fig. 17.1). Then we will provide information on the in vitro and ex vivo possibilities to study effects of food bioactives on DCs.

17.2.1   DC Subsets

To study immune modulatory effect of bioactives on DCs, different models systems may be considered, e.g. DC cell lines, in vitro-generated DCs, or directly isolated primary DCs, each with their own advantages and disadvantages. It is important to realize that different subsets of DCs exist in peripheral blood and (non)lymphoid tissues and that the differentiation and function of each of these subsets is highly dependent on the interacting cells and mediators in the local microenvironment. So, while this chapter provides experimental settings to study the direct effects of food bioactives on DCs, one should always consider the possible indirect effect via stimulation of the bioactives on for instance epithelial cells, and that different DC subsets may respond differently to the same exposure.

		Blood	Mucosa	TLR expression
pDC		CD11c^{int} CD123+ BDCA2+ BDCA4+		TLR 7,9
BDCA3 cDC		CD11c+ CLEC9A+ (Langerin+)	CD103+ Sirpα-	TLR1,2,3,6,8,10
BDCA1 cDCs		CD11c+ CD1c+	CD103+ Sirpα+ CD11b+ RALDH	TLR1,2,3,4,5,6,8,10
Langerin$^+$ DCs			Langerin+	TLR1,2,3,5,6,10
MoDC		CD11c+ FcεR1+	CD103- Sirpα+ CX3CR1	TLR1,2,4,5,8,9

Fig. 17.1 Surface marker phenotype and TLR pattern expression in DC subsets in blood and mucosa

17.2.1.1   Blood DCs

DCs can be divided in conventional DCs (cDCs) and plasmacytoid DCs (pDCs). In human blood, both cDCs and pDCs can be phenotypically characterized by the lack of lineage marker expression (CD3, CD14, CD19, CD56) and a high expression of HLA-DR. The circulating cDCs are CD11c$^+$ and are divided by the expression of CD1c (BDCA1) or CD141 (BDCA3). Human pDCs lack expression of CD11c and are discriminated by the expression of BDCA2, BDCA4 and/or CD123. CD16$^+$ MHC class II$^+$ cells cluster as a distinct population in the blood (Lindstedt et al. 2005) and have previously been associated with DC origins by some. However, CD16$^+$ cells were recently assigned to the monocyte lineage (Robbins et al. 2008; Ziegler-Heitbrock et al. 2010), confirmed by transcript profiling comparing CD16$^+$ cells with BDCA1$^+$ DCs (Frankenberger et al. 2012).

17.2.1.2   Mucosal DCs

The mucosal immune system is highly specialized: it has to be able to avoid invasion of (commensal) bacteria, and tolerate their presence in the intestine. It should also induce tolerance to non-harmful food antigens and protect against potentially harmful pathogens, toxins and xenobiotics. All of this takes place over a single layer of highly specialized epithelial cells (Chapter 2).

The formulation of antigens determines the main site for antigen uptake in the intestinal tract. Considerable evidence suggests that the gut associated lymphoid tissues (GALT), such as Peyers Patches (PP) and isolated lymphoid follicles, are critical for handling of particulate antigens. The microfold cells (M cells), present in the GALT epithelium, are involved in actively transferring the particulate antigens from the gut lumen into the lymphoid areas. The role for GALT in the induction of oral tolerance to soluble antigens is not entirely clear. Normal oral tolerance could be induced in mice lacking PP. It was therefore suggested that the main function of the GALT is to control immune responses to commensal bacteria. Antigen uptake by DCs underlying the villus epithelium of the lamina propria (LP) in the small intestine has been shown to be crucial for induction of oral tolerance to soluble antigens. How the DCs are provided with the antigens over the epithelial barrier will strongly depend on the compound and is beyond the scope of this chapter. Since both the intestinal tract and its draining lymph nodes contain specific DC subsets we will briefly discuss them before providing detailed information of possible DC systems that could be used to mimic these DCs.

Intestinal DCs: Thorough comparative transcriptional and functional profiling in DCs isolated from a human and murine small intestine LP and peripheral blood showed three DC subsets within the CD45$^+$lin$^-$MHCII$^+$ LP population based on the expression of CD103 and Sirpα (Watchmaker et al. 2014). The CD103$^+$Sirpα$^-$ DCs (SP) were related to human blood BDCA3$^+$ DCs, whereas CD103$^+$Sirpα$^+$ (DP) DCs expressed CD11b, CD207 (Langerin), CD209 and high levels of RALDH (coinciding with human blood CD1c$^+$ DCs) and supported the induction of (FoxP3$^+$) regulatory T cells. The CD103$^-$Sirpa$^+$ expressed CX3CR1, but lacked CD64 expression, and clustered with human monocytes indicating that they may have developed from monocytes recruited in response to gut inflammation. Most of these cells are located deeper into the LP when compared to the network of phagocytic cells that is located right beneath the epithelial cells. These phagocytic cells express CD45, HLA-DR, CD14, CD64 and high levels of CX3CR1, and since these cells also do not migrate to the lymph nodes, they have been depicted as intestinal macrophages (Rivollier et al. 2012; Bain et al. 2012; Tamoutounour et al. 2012; Mann et al. 2013).

Mesenteric lymph node DCs: Soluble food bioactives may also be directly available for internalization by DCs in the draining lymph nodes via the conduit system (Gretz et al. 1997; Anderson and Shaw 2005; Sixt et al. 2005; Roozendaal et al. 2009). The mesenteric lymph node DCs are a mixture of cells found in peripheral blood as well as the LP. As such, peripheral blood DCs and their CD34-derived counterparts could represent these lymph node DCs.

17.2.1.3 Monocyte-Derived DCs

Monocyte-derived (mo)DCs are rare under steady-state conditions in both blood and peripheral tissues. They can be found in increased numbers in vivo under inflammatory conditions. The ability to culture moDCs from monocytes boosted their popularity as a model to study human DC biology (Sallusto and Lanzavecchia 1994). The gene profile of in vitro-generated MoDCs is enriched for genes also found in in vivo occurring moDC (Segura et al. 2013) and also emphasizes the monocyte origin (not DC origin) with acquired DC features.

17.3 General Protocols

Primary DCs can be isolated from blood, and lymphoid or peripheral tissues, but the presence of DCs in blood is scarce and they are difficult to obtain in sufficient numbers from tissues. In addition, one may encounter large donor variations. The use of cell lines would overcome these challenges, but may hold insufficient similarity with primary DCs with respect to their response to environmental stimuli, ability to capture antigens, mature, migrate, produce cytokines, present antigens and activate T cells. Here we summarize a series of commonly used functional DC assays below and discuss the performance of cell lines and primary cells (summarized in Table 17.2).

17.3.1 DC Cell Lines

Most DC cell lines were differentiated from leukemia-derived cells, originating from myelogenous or monocytic lineage: the cytokine independent cell lines THP-1, U937, KG-1, HL-60 and the cytokine dependent cell line Mutz-3 are the most commonly used (Table 17.1) (Berges et al. 2005; van Helden et al. 2008; Santegoets et al. 2008). The use of these lines has the advantage of the availability of large numbers of synchronized cells with a long life span, but the association with selective DC subsets in peripheral blood and mucosal surfaces is largely unclear. All of these cells are leukemic, representing DC precursor stages and may often need the addition of specific growth and differentiation factors to induce at least some characteristics of differentiated DCs. MUTZ-3 is suggested to have the highest DC differentiation capacity (Santegoets et al. 2008) and may therefore be most representative for in vivo DCs.

Table 17.1 DC cell lines

Cell line	Derived from	DC differentiation	T cell activation capacity	References
THP-1	Monoblastic leukemia cell line	Low	Yes	Tsuchiya et al. (1980) Chap. 14
U937	Monocytic leukemia cells	Moderate	Yes	Chapter 15
KG-1	CD34 myelomonocytic cell line	Incomplete	Yes	Koeffler et al. (1981)
HL-60	Acute pro-myelocytic leukemia cell line	Moderate	Yes	Koski et al. (1999)
MUTZ-3	Myeloid leukemia cell line	Relatively high	?	Hu et al. (1996)

17.3.2 Isolating Primary DCs from Blood

DCs can be directly isolated from PB using their unique surface markers (Fig. 17.1) and utilizing either magnetic bead separation or flow cytometric cell sorting (Vremec and Shortman 1997; Dzionek et al. 2000; Shortman and Liu 2002). Since the percentage of DCs in PB is generally low (cDC: 0.6 % and pDC 0.2 % of all PBMC), DC-enrichment (by depletion of cells expressing lineage markers CD3, CD14, CD16, CD19, CD20, and CD56) is often used. For magnetic bead sorting, separation/isolation kits are commercially available from companies such as Miltenyi Biotec and StemCell Technologies. Preferably, freshly isolated PBMCs obtained by a density gradient of Ficoll-Hypaque, are suspended in cell isolation buffer as per manufacturer's instructions. To optimize survival of primary DCs after isolation it may be helpful to add survival factors to the culture medium: 800 U/ml GM-CSF for cDCs, like CD1c or CD141 DCs, or 10 ng/ml IL-3 for pDCs.

Culturing peripheral blood-derived CD1c (BDCA1) DCs for 2 days with GM-CSF, vitamin D and Retinoic Acid (RA) induced high expression of RALDH as well as the ability to induce gut-homing receptors resembling the gut-resident $CD103^+Sirp\alpha^+$ (DP) DCs to some extent (Sato et al. 2013).

Although using primary DCs directly isolated from the blood is feasible for most academic research labs and will probably resemble the in vivo DC biology most closely, the limited amount of primary DCs in PB is an obvious obstacle when many conditions and functional effects need to be assessed.

17.3.3 CD34+-Derived DCs

Another alternative is to generate DCs from myeloid precursors. CD34 is a marker expressed on haematopoietic stem cells and early myeloid progenitors that contain high proliferative potential. The $CD34^+$ cells can therefore first be expanded followed by a differentiation towards DCs. $CD34^+$ cells can be isolated from bone marrow (BM), peripheral blood (PB) or from umbilical cord blood (CB) using magnetic labelling (Kato and Radbruch 1993).

Firstly, $CD34^+$ cells can be expanded using different combinations of cytokines and growth factors. Flt-3L, TPO, SCF, IL-3, and IL-6 are early-acting cytokines that support the proliferation of DC precursors from $CD34^+$ HPC. Two different cocktails, Flt3-L, TPO, SCF and Flt3-L, SCF, IL-3, IL-6 (FS36), were compared for their capacity to induce proliferation of $CD34^+$ HPC (Bontkes 2002). FS36 showed the greatest ability to expand the $CD34^+$ HPCs. Expanded $CD34^+$ cells can be differentiated towards different DC subtypes i.e. $BDCA3^+$ DCs up to only 3 % (Poulin et al. 2010), or pDCs up to 5 % (Demoulin et al. 2012). Addition of TGF-beta during differentiation enhances the formation of langerin$^+$ LC (Soulas et al. 2006; Caux et al. 1992; Szabolcs et al. 1995).

Established protocols to culture LCs, dermal DCs, $BDCA3^+$ DCs and pDCs can be found in literature and are summarized below (Fig. 17.2).

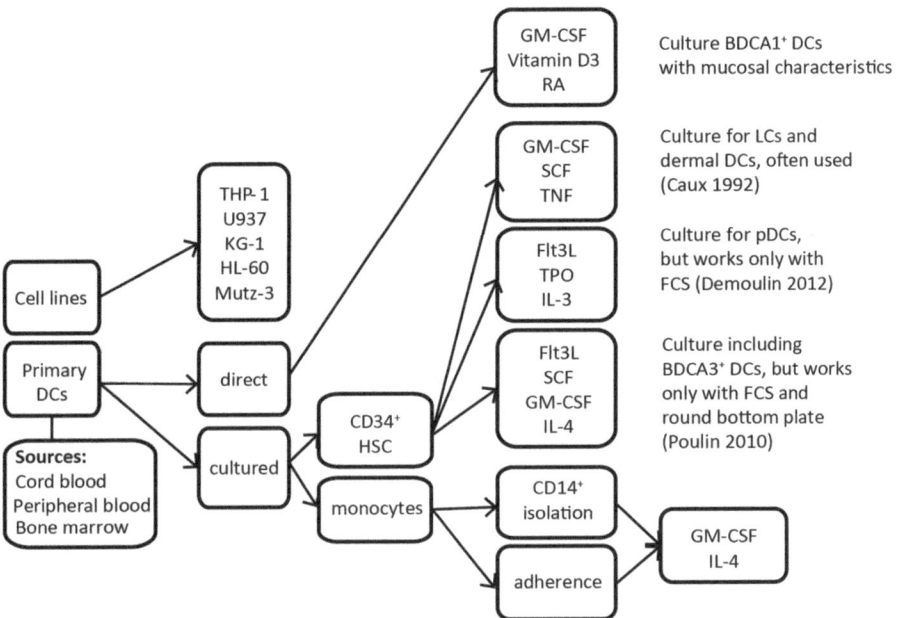

Fig. 17.2 Different DC sources

17.3.4   Monocyte-Derived DCs

Monocytes can be selected via adherence or CD14$^+$ isolation with magnetic beads. For adherence, PBMCs are resuspended in culture medium and placed in a humidified incubator maintained at 37 °C and 5 % CO_2, to allow the monocyte precursors to adhere. After 1 h, non-adherent cells are simply removed by firmly tapping the flask and isolating non-adherent cells. For bead selection, PBMCs are resuspended in cell isolation buffer as per manufacturer's instructions, followed by isolation using a commercially available kit. The isolated monocytes can be frozen for later use.

Once the monocytes are selected, they can be differentiated towards DCs with culture-medium (e.g. X-vivo) supplemented with 800–1,000 U/ml GM-CSF and 300–500 U/ml IL-4 (37 °C and 5 % CO_2) for 7 days. After the 6/7-day culture period, the immature DCs can be used for testing the immune modulatory effects of food bioactives.

To generate mucosal CD103$^+$Sirpα$^+$ (DP) resembling moDCs, RA, the even more potent RARalpha agonist AM580 or TGFbeta, can be added to the culture (Hartog et al. 2013; Martin et al. 2014). RA-induced moDCs express CD103 and can induce increased percentages of FoxP3$^+$ in allogeneic CD4$^+$ cells. RA-induced moDC also express high levels of IL-22BP which is comparable to the CD11b$^+$CD103$^+$ DCs found in mouse LP and MLN (Martin et al. 2014).

17.4 Asses Viability

To validate the generation of DCs in these cultures, the purity of DCs can be assessed by flow cytometric analysis of the surface markers CD11c and HLA-DR. This staining can be expanded with markers, depending on the specific protocol used (for instance CD14 expression on MoDC or CD1c on CD34-derived DC.) Viability can be checked by adding a viability dye (e.g. 7AAD) to the FACS panel, or count the amount of cells with Tryphan blue with a microscope (Table 17.2).

17.5 Experimental Readout

Although bioactive food components have been shown to interact with many different types of cells, it remains a legitimate question whether DCs exposed to food bioactives may in turn affect features of the innate or adaptive immune system, in particular T cells. Some food bioactives may be associated with the development of allergic or autoimmune-like clinical symptoms in susceptible individuals. The clinical manifestation, ranging from mild to severe and life-threatening, vary considerably for different compounds and for the same compound in different subjects and may be idiosyncratic in nature. They comprise pseudo-allergic reactions (interference with immunological effector mechanisms and cells, such as mast cells), compound-allergic reactions (activation of compound-specific T and B cells),

Table 17.2 Summary of (dis)advantages of using differentially generated DCs

Type of DCs	Advantages	Disadvantages
Cell lines (*THP-1, U937, KG-1, HL-60*)	Large number of cells. Relatively long life span.	Association to primary DCs unclear (different stages of differentiation, discrepancies in DC-functions)
Primary DCs		
CD1c	Largely comparable to in vivo counterpart. CD1c can be treated to differentiate into gut-like DCs.	Low cell numbers. Isolation procedures might affect phenotype and function.
BDCA3		
pDCs		
CD34-derived DCs		
Dermal DCs/ LCs	CD34+ cells can be expanded to increase cell number. Resemble primary DCs very well.	Acquisition requires specialized centres. % of pDC and BDCA3 DC are still low in these cultures, thus requiring large numbers of cells for selection.
pDCs		
BDCA3		
Monocyte-derived DCs		
	Large number of cells. Relatively easy to obtain and culture. Very well studied: extensive set of data available.	Represent inflammatory monocytes with DC-features, which may hamper translation for primary DCs.

and compound-induced autoimmune reactions. Although the induction of allergic or autoimmune responses requires DC activation, food bioactives may also inhibit DC activation. This may represent an advantageous effect of food bioactives in reducing (chronic) inflammatory response like asthma, inflammatory bowel disease, etcetera, by reducing the DC activity and possibly inducing regulatory T cells.

Different assays can be performed to test the DC's responsiveness to bioactives for the prediction of the subsequent immune modulatory effect of that specific compound. These assays include the monitoring of phenotypic markers expressed on the surface of DCs indicating their differentiation, activation and their potentially stimulating effect on T cells. Nevertheless, even with a clear outcome of these assays, it remains to be seen whether indeed antigen-specific T cells with the potential to induce adverse effects in vivo will be induced. In this regards it is important to consider whether the food bioactive contains potential antigens, are able to form neo-antigens or merely function as adjuvants for responses against other components in the food or play a role in sensitization to auto-antigens.

17.5.1 Co-stimulation

DCs may undergo a number of phenotypical changes, following maturation by the uptake and processing of an antigen. The process of DC maturation, in general, involves an increase in the surface expression of co-stimulatory molecules, like CD40, CD80, CD83, and CD86, which can be measured by flow cytometry. As a negative control unloaded DCs can be taken along, as well as a DC positively stimulated with a known maturation cocktail, like IL-1beta, IL-6, TNF and PGE-2 or TLR ligands.

17.5.2 Cytokine Production

Besides up-regulation of co-stimulatory molecules, DCs produce a large variety of cytokines that could be either pro/anti-inflammatory or skew the phenotype of T cells. Therefore, a quite simple procedure to test the activity of the DCs towards the food bioactives is the measurement of cytokine levels. A wide variety of cytokines may be expressed by mature DCs including IL-1 alpha, IL-1 beta, IL-4, IL-6, IL-8, IL-10, IL-12, IL-15, IL-16, IL-17, IL-18, IFN-alpha, IFN-beta, IFN-gamma, and TNF. These cytokines can be measured by ELISA, or multiplex assays.

17.5.3 Other DC Readouts

Co-stimulation and cytokine production are effect parameters used in the vast majority of studies on effects of food bioactives on DCs (see Table 17.3 for an overview). Studying the induction of specific T cells is complicated due to the low

Table 17.3 In vitro studies using food-associated antigens in in vitro DC models

DC source	Food bioactive group	Specific food bioactives	Additional DC activator	Expression costimulation	Cytokine production	Other	MLR	References
CD34+ derived DC	Probiotics	L. paracasei CNCM I-4034 B. breve CNCM I-4035	S. typhi CECT 725 or LPS	No	Yes: various	TLR expression	No	Bermudez-Brito et al. (2012, 2013)
pDC (PBMC)	Probiotics	L. lactis JCM5805		Yes: HLA-DR and CD86	Yes: IFN-alpha	IFN and IRF expression	No	Sugimura et al. (2013)
Lamina propria and PBMC	Probiotics	Probiotic preparation VSL#3	LPS	Yes: CD40 and CD80	Yes: IL-10 and IL-12	Effect on T cell responses	No	Hart et al. (2004)
Lamina propria and PBMC	Probiotics	L. plantarum BMCM12		No	Yes: various		Yes	Bernardo et al. (2012)
MoDC	Probiotics	Lactobacillus rhamnosus Lcr35		Yes: CD83, CD86, CD209 and HLA-DR	Yes: various	Microarray gene expression profile	No	Evrard et al. (2011)
MoDC	Polyphenols	Protocatechuic acid	LPS	Yes: CD83, CD86 and HLA-DR	Yes: various	Migration PPARg activation	No	Del Cornò et al. (2014b)
THP-1-derived	Polyphenols	Apple polyphenol extract	OVA (endotoxin?)	Yes: CD86 and HLA-DR	Yes: IL-1b, IL-10, IL-12 and TNF-a	Antigen uptake	No	Katayama et al. (2013)
MoDC	Isoflavones	Genistein and Daidzein	LPS, TNFa, CT	Yes: CD83, CD80, CD86, and HLA-DR	Yes: IL-6, IL-12, TNF-a and IL-10	DC–NK interaction	Yes	Wei et al. (2012)
MoDC	Isoflavones	Genistein, Daidzein and Glycitein	CT	Yes: CD83, CD80 and CD86	Yes: IL-6 and IL-8	Th-cytokine induction (IL-5, IL-9, IL-13)	Yes	Masilamani et al. (2011)

MoDC	Curcuminoids	Curcumin	LPS or Poly I:C	Yes: CD86, CD83, CD54 and HLA-DR	Yes: various	Migration Endocytosis	Yes	Shirley et al. (2008)
MoDC	Curcuminoids	Curcumin	LPS	Yes: CD40, CD83, CD80, CD86 and HLA-DR	Yes: various	FOXP3 induction Negative regulatory molecules (PDL1/2)	Yes	Rogers et al. (2010)
MoDC	Vitamins	1alpha,25-(OH)2 D3	LPS or PGN	Yes: CD40, CD83, CD80, CD86 and HLA-DR	Yes: IL-6, IL-8, IL-12p70 and IL-10		No	Brosbøl-Ravnborg et al. (2013)
MoDC	Vitamins	1,25-(OH)2 D3	LPS or CD40L	Yes: CD40, CD83, CD80, CD86 and HLA-DR	Yes: IL-12p70	Endocytosis	Yes	Piemonti et al. (2000)
MoDC	Proteins	Peptic fragments of gliadin, pRQP and pDAC (decapeptides)		Yes: CD83, CD80, CD86, CCR7 and HLA-DR	Yes: IL-12, TNF-a and IL-10		Yes	Giordani et al. (2014)
MoDC	Proteins	Soluble peanut Ag (PNAg) and purified Ara h 1	LPS and CT (positive controls)	Yes: CD83, CD40, CD86, and HLA-DR	No	Th-cytokine induction (IFN-g, IL4 and IL-13)	Yes	Shreffler et al. (2006)
MoDC	Proteins	Bet v 1, Mal d 1, Api g 1 and Dau c 1		Yes: CD40, CD83, CD80, CD86 and HLA-DR	No	Th-cytokine induction (IFN-g, IL4, IL-10 and IL-13)	No	Smole et al. (2010)
MoDC	Proteins	Peptic fragments of gliadin, soya protein, and OVA		Yes: CD83, CD80, CD86 and HLA-DR	Yes: various	Endocytosis and signal transduction	Yes	Palova-Jelinkova et al. (2005)

frequency of T cell precursors with the corresponding TCR and requires extensive culture conditions and broad read-out tools as the TCR specificity in these conditions is unknown. With regard to the T cell stimulatory **potential** of DCs the mixed lymphocyte reaction (MLR), in which DCs are used to stimulate alloreactive (naïve) T cells, can be considered. Proliferation in combination with cytokine production can be used as a read-out for T cell skewing.

17.6   In Vitro Studies on Food Bioactives Using DCs (Table 17.3)

The immunomodulatory effects of food bioactives can be broadly divided in 'activation', 'suppression' or 'no detectable effect'. Table 17.3 shows that with regard to food proteins immune activation is generally the anticipated endpoint. It should however be considered that other food constituents instead of the protein itself, may stimulate DCs to internalize and present food proteins, and therefore may be involved in the risk for hypersensitivity. Immune suppression in DCs assays is mostly studied in combination with a trigger known to activate the DCs (TLR ligands i.e. LPS, CT or CD40 ligation). Inhibition of DC activation has been studied using probiotics, vitamins, polyphenols (Del Cornò et al. 2014a), isoflavones and curcuminoids. Inhibition of DC activation may be relevant for people suffering from chronic inflammation not only by reducing inflammation directly but also by preventing the risk for developing cancer, which is associated with chronic inflammation. Potential harmful effects of non-specific immune suppression include an increased risk of infection and also here the development of cancer. When there is no direct effect of the food bioactives on the DCs with the assays mentioned above, this does not mean that exposure to the compound is without risk. Indirect effects of the bioactives through their effects on other cells or the effects of their metabolites must be taken into consideration.

17.7   Critical Notes

Extensive comparative studies using different DC assays with proven immune active compounds and food bioactive compound are lacking.

With regard to some of the DC parameters used to study immune modulatory effects of food bioactives, the expression of co-inhibitory molecules (e.g. PDL1 and PDL2) has not been intensively studied. This family of molecules has however a large impact on the immune response and may therefore also be considered in future testing.

As mentioned before, the DCs are the sentinels of the immune system that gather not only by taking up or sensing the food bioactives only, but continuously receive input from cells and factors in their microenvironment. Potential responses should therefore be evaluated in the light of the local environment. For instance, the choice for a particular DC model/subset may be dependent on the availability of certain compounds/entry site, which may relate to the size, solubility and formulation of the compound of interest.

In conclusion, studying the effects of food bioactives on DC (subsets) is an important step in addressing the immune modulating potential of these compounds in vivo, but should be part of a more extended monitoring program involving studies on the compound's availability, factors of the microenvironment, and effects of metabolites/neo-epitopes.

References

Anderson AO, Shaw S (2005) Conduit for privileged communications in the lymph node. Immunity 22:3–5. doi:10.1016/j.immuni.2005.01.003

Bain CC, Scott CL, Uronen-Hansson H et al (2012) Resident and pro-inflammatory macrophages in the colon represent alternative context-dependent fates of the same Ly6C. Mucosal Immunol 6:498–510. doi:10.1038/mi.2012.89

Banchereau J, Steinman RM (1998) Dendritic cells and the control of immunity. Nature 392:245–252. doi:10.1038/32588

Berges C, Naujokat C, Tinapp S et al (2005) A cell line model for the differentiation of human dendritic cells. Biochem Biophys Res Commun 333:896–907. doi:10.1016/j.bbrc.2005.05.171

Bermudez-Brito M, Muñoz-Quezada S, Gomez-Llorente C et al (2012) Human intestinal dendritic cells decrease cytokine release against *Salmonella* infection in the presence of *Lactobacillus paracasei* upon TLR activation. PLoS One 7:e43197. doi:10.1371/journal.pone.0043197

Bermudez-Brito M, Muñoz-Quezada S, Gomez-Llorente C et al (2013) Cell-free culture supernatant of *Bifidobacterium breve* CNCM I-4035 decreases pro-inflammatory cytokines in human dendritic cells challenged with *Salmonella typhi* through TLR activation. PLoS One 8:e59370. doi:10.1371/journal.pone.0059370

Bernardo D, Sánchez B, Al-Hassi HO et al (2012) Microbiota/host crosstalk biomarkers: regulatory response of human intestinal dendritic cells exposed to *Lactobacillus* extracellular encrypted peptide. PLoS One 7:e36262. doi:10.1371/journal.pone.0036262

Bettelli E, Korn T, Oukka M, Kuchroo VK (2008) Induction and effector functions of T(H)17 cells. Nature 453:1051–1057. doi:10.1038/nature07036

Bhardwaj N, Bender A, Gonzalez N et al (1994) Influenza virus-infected dendritic cells stimulate strong proliferative and cytolytic responses from human CD8+ T cells. J Clin Invest 94:797–807. doi:10.1172/JCI117399

Bontkes HJ, De Gruijl TD, Schuurhuis GJ, Scheper RJ, Meijer CJ, Hooijberg E (2002) Expansion of dendritic cell precursors from human CD34(+) progenitor cells isolated from healthy donor blood; growth factor combination determines proliferation rate and functional outcome. J Leukoc Biol 72(2):321–329

Brosbøl-Ravnborg A, Bundgaard B, Höllsberg P (2013) Synergy between vitamin D3 and Toll-like receptor agonists regulates human dendritic cell response during maturation. Clin Dev Immunol 2013:1–8. doi:10.1155/2013/807971

Caux C, Dezutter-Dambuyant C, Schmitt D, Banchereau J (1992) GM-CSF and TNF-alpha cooperate in the generation of dendritic Langerhans cells. Nature 360:258–261. doi:10.1038/360258a0

Del Cornò M, Scazzocchio B, Masella R, Gessani S (2014a) Regulation of dendritic cell function by dietary polyphenols. Crit Rev Food Sci Nutr 140618104333000. doi:10.1080/10408398.2012.713046

Del Cornò M, Varano B, Scazzocchio B et al (2014b) Protocatechuic acid inhibits human dendritic cell functional activation: role of PPARγ up-modulation. Immunobiology 219:416–424. doi:10.1016/j.imbio.2014.01.007

Demoulin S, Roncarati P, Delvenne P, Hubert P (2012) Production of large numbers of plasmacytoid dendritic cells with functional activities from CD34(+) hematopoietic progenitor cells: use of interleukin-3. Exp Hematol 40:268–278. doi:10.1016/j.exphem.2012.01.002

den Hartog G, van Altena C, Savelkoul HFJ, van Neerven RJJ (2013) The mucosal factors retinoic acid and TGF-β1 induce phenotypically and functionally distinct dendritic cell types. Int Arch Allergy Immunol 162:225–236. doi:10.1159/000353243

Dzionek A, Fuchs A, Schmidt P et al (2000) BDCA-2, BDCA-3, and BDCA-4: three markers for distinct subsets of dendritic cells in human peripheral blood. J Immunol 165:6037–6046

Evrard B, Coudeyras S, Dosgilbert A et al (2011) Dose-dependent immunomodulation of human dendritic cells by the probiotic *Lactobacillus rhamnosus* Lcr35. PLoS One 6:e18735. doi:10.1371/journal.pone.0018735

Frankenberger M, Hofer TPJ, Marei A et al (2012) Transcript profiling of CD16-positive monocytes reveals a unique molecular fingerprint. Eur J Immunol 42:957–974. doi:10.1002/eji.201141907

Giordani L, Del Pinto T, Vincentini O et al (2014) Two wheat decapeptides prevent gliadin-dependent maturation of human dendritic cells. Exp Cell Res 321:248–254. doi:10.1016/j.yexcr.2013.11.008

Gretz JE, Anderson AO, Shaw S (1997) Cords, channels, corridors and conduits: critical architectural elements facilitating cell interactions in the lymph node cortex. Immunol Rev 156:11–24

Hart AL, Lammers K, Brigidi P et al (2004) Modulation of human dendritic cell phenotype and function by probiotic bacteria. Gut 53:1602–1609. doi:10.1136/gut.2003.037325

Hu ZB, Ma W, Zaborski M et al (1996) Establishment and characterization of two novel cytokine-responsive acute myeloid and monocytic leukemia cell lines, MUTZ-2 and MUTZ-3. Leukemia 10:1025–1040

Kapsenberg ML (2003) Dendritic-cell control of pathogen-driven T-cell polarization. Nat Rev Immunol 3:984–993. doi:10.1038/nri1246

Katayama S, Kukita T, Ishikawa E et al (2013) Food chemistry. Food Chem 138:757–761. doi:10.1016/j.foodchem.2012.10.076

Kato K, Radbruch A (1993) Isolation and characterization of CD34+ hematopoietic stem cells from human peripheral blood by high-gradient magnetic cell sorting. Cytometry 14:384–392. doi:10.1002/cyto.990140407

Koeffler HP, Bar-Eli M, Territo MC (1981) Phorbol ester effect on differentiation of human myeloid leukemia cell lines blocked at different stages of maturation. Cancer Res 41:919–926

Koski GK, Schwartz GN, Weng DE et al (1999) Calcium ionophore-treated myeloid cells acquire many dendritic cell characteristics independent of prior differentiation state, transformation status, or sensitivity to biologic agents. Blood 94:1359–1371

Langerhans P (1868) Ueber die Nerven der menschlichen Haut. Virch Arch Pathol Anat 44:325–337

Lindstedt M, Lundberg K, Borrebaeck CAK (2005) Gene family clustering identifies functionally associated subsets of human in vivo blood and tonsillar dendritic cells. J Immunol 175:4839–4846

Mann ER, Landy JD, Bernardo D et al (2013) Intestinal dendritic cells: their role in intestinal inflammation, manipulation by the gut microbiota and differences between mice and men. Immunol Lett 150:30–40. doi:10.1016/j.imlet.2013.01.007

Martin JCJ, Bériou G, Heslan M et al (2014) Interleukin-22 binding protein (IL-22BP) is constitutively expressed by a subset of conventional dendritic cells and is strongly induced by retinoic acid. Mucosal Immunol 7:101–113. doi:10.1038/mi.2013.28

Masilamani M, Wei J, Bhatt S, et al (2011) Soybean isoflavones regulate dendritic cell function and suppress allergic sensitization to peanut. J Allergy Clin Immunol 128:1242–1250.e1. doi:10.1016/j.jaci.2011.05.009

McLellan AD, Sorg RV, Williams LA, Hart DN (1996) Human dendritic cells activate T lymphocytes via a CD40: CD40 ligand-dependent pathway. Eur J Immunol 26:1204–1210. doi:10.1002/
eji.1830260603

Palova-Jelinkova L, Rozkova D, Pecharova B et al (2005) Gliadin fragments induce phenotypic
and functional maturation of human dendritic cells. J Immunol 175:7038–7045. doi:10.4049/
jimmunol.175.10.7038

Piemonti L, Monti P, Sironi M et al (2000) Vitamin D3 affects differentiation, maturation, and
function of human monocyte-derived dendritic cells. J Immunol 164:4443–4451

Poulin LF, Salio M, Griessinger E et al (2010) Characterization of human DNGR-1+ BDCA3+
leukocytes as putative equivalents of mouse CD8 + dendritic cells. J Exp Med 207:1261–1271.
doi:10.1084/jem.20092618

Rivollier A, He J, Kole A et al (2012) Inflammation switches the differentiation program of Ly6Chi
monocytes from antiinflammatory macrophages to inflammatory dendritic cells in the colon.
J Exp Med 209:139–155. doi:10.1084/jem.20101387

Robbins SH, Walzer T, Dembélé D et al (2008) Novel insights into the relationships between
dendritic cell subsets in human and mouse revealed by genome-wide expression profiling.
Genome Biol 9:R17. doi:10.1186/gb-2008-9-1-r17

Rogers NM, Kireta S, Coates PTH (2010) Curcumin induces maturation-arrested dendritic cells
that expand regulatory T cells in vitro and in vivo. Clin Exp Immunol 162:460–473.
doi:10.1111/j.1365-2249.2010.04232.x

Roozendaal R, Mempel TR, Pitcher LA et al (2009) Conduits mediate transport of
low-molecular-weight antigen to lymph node follicles. Immunity 30:264–276. doi:10.1016/j.
immuni.2008.12.014

Sakaguchi S, Ono M, Setoguchi R et al (2006) Foxp3+ CD25+ CD4+ natural regulatory T cells
in dominant self-tolerance and autoimmune disease. Immunol Rev 212:8–27.
doi:10.1111/j.0105-2896.2006.00427.x

Sallusto F, Lanzavecchia A (1994) Efficient presentation of soluble antigen by cultured human
dendritic cells is maintained by granulocyte/macrophage colony-stimulating factor plus
interleukin 4 and downregulated by tumor necrosis factor alpha. J Exp Med 179:
1109–1118

Santegoets SJAM, van den Eertwegh AJM, van de Loosdrecht AA et al (2008) Human dendritic
cell line models for DC differentiation and clinical DC vaccination studies. J Leukoc Biol
84:1364–1373. doi:10.1189/jlb.0208092

Sato T, Kitawaki T, Fujita H et al (2013) Human CD1c⁺ myeloid dendritic cells acquire a high level
of retinoic acid-producing capacity in response to vitamin D3. J Immunol 191:3152–3160.
doi:10.4049/jimmunol.1203517

Segura E, Touzot M, Bohineust A et al (2013) Human inflammatory dendritic cells induce Th17
cell differentiation. Immunity 38:336–348. doi:10.1016/j.immuni.2012.10.018

Shirley SA, Montpetit AJ, Lockey RF, Mohapatra SS (2008) Curcumin prevents human dendritic
cell response to immune stimulants. Biochem Biophys Res Commun 374:431–436.
doi:10.1016/j.bbrc.2008.07.051

Shortman K, Liu Y-J (2002) Mouse and human dendritic cell subtypes. Nat Rev Immunol 2:151–161.
doi:10.1038/nri746

Shreffler WG, Castro RR, Kucuk ZY et al (2006) The major glycoprotein allergen from *Arachis
hypogaea*, Ara h 1, is a ligand of dendritic cell-specific ICAM-grabbing nonintegrin and acts as
a Th2 adjuvant in vitro. J Immunol 177:3677–3685

Sixt M, Kanazawa N, Selg M et al (2005) The conduit system transports soluble antigens from the
afferent lymph to resident dendritic cells in the T cell area of the lymph node. Immunity 22:19–29.
doi:10.1016/j.immuni.2004.11.013

Smole U, Wagner S, Balazs N et al (2010) Bet v 1 and its homologous food allergen Api g 1 stimulate
dendritic cells from birch pollen-allergic individuals to induce different Th-cell polarization.
Allergy 65:1388–1396. doi:10.1111/j.1398-9995.2010.02407.x

Soulas C, Arrighi J-F, Saeland S et al (2006) Human CD34+CD11b− cord blood stem cells generate in vitro a CD34−CD11b+ subset that is enriched in langerin+ Langerhans dendritic cell precursors. Exp Hematol 34:1471–1479. doi:10.1016/j.exphem.2006.06.011

Steinman RM, Cohn ZA (1973) Identification of a novel cell type in peripheral lymphoid organs of mice. I. Morphology, quantitation, tissue distribution. J Exp Med 137:1142–1162

Sugimura T, Jounai K, Ohshio K et al (2013) Immunomodulatory effect of *Lactococcus lactis* JCM5805 on human plasmacytoid dendritic cells. Clin Immunol 149:509–518. doi:10.1016/j.clim.2013.10.007

Szabolcs P, Moore MAS, Young JW (1995) Expansion of immunostimulatory dendritic cells among the myeloid progeny of human CD34+ bone marrow precursors cultured with c-kit ligand, granulocyte-macrophage colony-stimulating factor, and TNF-alpha. J Immunol 154:5851–5861

Tamoutounour S, Henri S, Lelouard H et al (2012) CD64 distinguishes macrophages from dendritic cells in the gut and reveals the Th1-inducing role of mesenteric lymph node macrophages during colitis. Eur J Immunol 42:3150–3166. doi:10.1002/eji.201242847

Tsuchiya S, Yamabe M, Yamaguchi Y et al (1980) Establishment and characterization of a human acute monocytic leukemia cell line (THP-1). Int J Cancer 26:171–176

van Helden SFG, van Leeuwen FN, Figdor CG (2008) Human and murine model cell lines for dendritic cell biology evaluated. Immunol Lett 117:191–197. doi:10.1016/j.imlet.2008.02.003

Vremec D, Shortman K (1997) Dendritic cell subtypes in mouse lymphoid organs: cross-correlation of surface markers, changes with incubation, and differences among thymus, spleen, and lymph nodes. J Immunol 159:565–573

Watchmaker PB, Lahl K, Lee M et al (2014) Comparative transcriptional and functional profiling defines conserved programs of intestinal DC differentiation in humans and mice. Nat Immunol 15:98–108. doi:10.1038/ni.2768

Wei J, Bhatt S, Chang LM et al (2012) Isoflavones, genistein and daidzein, regulate mucosal immune response by suppressing dendritic cell function. PLoS One 7:e47979. doi:10.1371/journal.pone.0047979

Ziegler-Heitbrock L, Ancuta P, Crowe S et al (2010) Nomenclature of monocytes and dendritic cells in blood. Blood 116:e74–e80. doi:10.1182/blood-2010-02-258558

Chapter 18
Co-culture Caco-2/Immune Cells

Charlotte R. Kleiveland

Abstract Numerous cell types are involved in maintenance of the intestinal tissue. However, the main players are cells of the epithelial lining and from the immune system. All these cells are communicating with each other and are strongly influenced by interactions with neighboring cells. Mono-cultivation of cells may provide valuable information as to understanding basic biology, but co-cultivation of more than one type of cells gives an opportunity to investigate effects of intercellular communication. Examples of systems for co-cultivation of epithelial cells and immune cells for the investigation of both direct cell–cell contact and communication with soluble factors will be discussed in this chapter. In addition, co-cultivation systems for differentiation to more specialized epithelial cells as microfold cells (M-cells) will be described.

Keywords Epithelial cells • Immune cells • M-cells • Co-culture • Intercellular communication

18.1 Origin, Features and Mechanisms

The models described in this section include the cell lines Caco-2 and THP-1 in addition to peripheral mononuclear cells (PBMCs) and dendritic cells (DCs). Description of the origin, features and mechanisms of these cells are included in previous sections, and further information can be found in Chap. 10 for Caco-2, Chap. 14 for THP-1, Chap. 15 for PBMCs and Chap. 17 for DCs.

C.R. Kleiveland (✉)
Department of Chemistry, Biotechnology and Food Science, Faculty of Veterinary Medicine and Biosciences, Norwegian University of Life Sciences, Ås, Norway

Research Department, Ostfold Hospital Trust, Fredrikstad, Norway
e-mail: charlotte.kleiveland@nmbu.no

© The Author(s) 2015
K. Verhoeckx et al. (eds.), *The Impact of Food Bio-Actives on Gut Health*,
DOI 10.1007/978-3-319-16104-4_18

18.2   Relevance to Human In Vivo Situation

18.2.1   Co-culture Caco-2 and Dendritic Cells

Precise regulation of the intestinal barrier function is important for the maintenance of mucosal homeostasis and prevents the onset of uncontrolled inflammation (Pastorelli et al. 2013). A balanced and fine-tuned cross-talk between intestinal epithelial cells (IECs), immune cells and the intestinal microbiota is necessary to maintain the intestinal homeostasis (Artis 2008). Different immune cells communicate with each other as well as with epithelial cells by an impressive network of regulatory signals. The complexity of the mucosal immune system is difficult to mimic in vitro, but a co-culture system makes it possible to elaborate mechanisms involved in communication between epithelial cells and cells of the immune system. It is possible to study effects of soluble molecules as well as contact dependent mechanisms. DCs are professional antigen-presenting cells central to the regulation of both innate and adaptive immune responses at mucosal surfaces (Coombes and Powrie 2008). The intestinal epithelium is not only a barrier to microorganisms entering via the oral route, but epithelial cells are directly influencing the activating properties of mucosal DCs (Rimoldi et al. 2004). Communication between immune cells and epithelial cells can be studied by co-cultivation of Caco-2 cells with CD14[+] monocyte-derived dendritic cells (mDC), PBMCs or THP-1 cells, a human monocytic cell line (described in Chap. 14). CD14[+] monocytes isolated from blood are differentiated into DCs by stimulation with GM-CSF and IL-4. The phenotype of these in vitro differentiated DC resembles that of inflammatory DCs (Segura et al. 2013). Inflammatory DCs differentiate from monocytes recruited to inflamed tissue from blood during an inflammation. Therefore the CD14[+] monocytes are phenotypically distinct from the bone marrow-derived DC precursors which give rise to tissue resident DCs (Steinman and Idoyaga 2010; Miller et al. 2012) (see Chap. 17 for further information on DCs).

18.2.2   Co-culture Caco-2 and B-cells (Raji)

The follicle-associated epithelium (FAE) overlying the organized lymphoid follicles of the gut-associated lymphoid tissues (GALT) including Peyer's patches, are specialized for sampling luminal content. Approximately 10 % of the cells within the FAE are microfold (M) cells. M cells have unique morphological features that include a reduction in both glycocalyx and microvilli compared to other parts of the epithelial lining. They are highly specialized for phagocytosis and transcytosis of gut luminal content across the epithelium. Beneath the M-cell's basolateral membrane there is an intraepithelial pocket where the luminal content samples are released after transcytosis. This dome structure contains various populations of lymphocytes, macrophages and dendritic cells. M-cells therefore provide an efficient delivery of gut luminal microorganisms and antigens to the underlying mucosal tissue and are therefore essential for mucosal immune responses (Gullberg et al. 2000; Kerneis et al. 1997). Cells with a M cell-similar phenotype can be obtained by co-cultivation of Caco-2 cells with Raji B cells (Gullberg et al. 2000).

18.3 Stability, Consistency and Reproducibility

Co-culture systems of epithelial cells and immune cells have not been much in use for investigation of food bioactive components, despite several methodological reports on such systems. This might indicate that the system gives inconsistent data or has a low level of reproducibility. Some of the co-culture systems described here combine the use of cell lines and primary cells. The cells lines will represent a stable and reproducible system with low variability from experiment to experiment. However, the response of PBMCs and monocyte derived DCs will depend on the donor and can lead to increased variability between experiments.

The Caco-2/Raji B cell model is a simplified model that induces cells with M cell-like morphology. A gene expression study comparing the Caco-2/Raji model with human FAE suggest that the Caco-2 differentiation model is associated with some functional features of M cells. However the genes induced reflect the acquisition of a more general FAE phenotype (Lo et al. 2004). However, this is an easy accessible in vitro model system to study M cell-mediated translocation.

18.4 General Protocol

18.4.1 Co-culture of Caco-2/Human Monocyte Derived DCs (Include Contact Dependent Events)

In this co-cultivation system, the cells will be able to make direct cell–cell contacts between the immune cells and the basolateral side of the epithelial cells. Co-cultivation of the cells in this way makes it possible to investigate effects of both direct interactions between DCs and epithelial cells and communication by soluble factors.

Caco-2 cells are seeded on the backside of filter inserts with 3 μm pores. 3×10^5 Caco-2 cells in 100 μL RPMI 1640 with 10 % FCS, are placed on the bottom side of the inserts placed in an inverted orientation. After overnight incubation, to allow attachment of the Caco-2 cells, the inserts are transferred to their normal orientation in a 24 well plate. The Caco-2 cells now face the lower chamber and are cultured as this for the next 14 days. Every other day the medium has to be exchanged with fresh RPMI 1640 with 10 % FCS. At day 14 add 1×10^6 CD14$^+$ human monocytes in 500 μL RPMI 1640 with 10 % FCS, 25 ng/mL IL-4 and 50 ng/mL GM-CSF. At day 4 of co-cultivation remove the medium in the upper chamber and replace it with 500 μL RPMI 1640 with 25 ng/mL IL-4 and 50 ng/mL GM-CSF. The RPMI 1640 in the lower chamber should be exchanged every other day. Maintain the co-culture for 6 days before the experiments are conducted. Depending on the test compound, it could be added to either compartment. The lower chamber will reflect the apical side and the upper chamber the basolateral side of the intestinal epithelium. Incubation time for the test compound will depend on the type of compound and the read out system to be used and will have to be decided for each compound.

18.4.2   Caco-2/Human Monocyte Derived DCs (Soluble Factors)

In comparison with the above mentioned system co-cultivation of cells in this way will only make it possible to study effect of communication caused by soluble factors. The epithelial cells and the DCs will not be able to be in direct contact with each other and communication can therefore only happen by soluble factors diffusing through the filter.

300 μL of a 1×10^6/mL Caco-2 cells suspended in RPMI 1640 with 10 % FCS, are seeded on filter inserts with 0.4 μm pores. The cells are maintained at 37 °C in 5 % CO_2 atmosphere for 14 days. The medium in both compartments should be exchanged with fresh RPMI 1640 with 10 % FCS, every other day. At day 14, 1×10^6 CD14$^+$ human monocytes in RPMI-1640 with 10 % FCS, 25 ng/mL IL-4 and 50 ng/mL GM-CSF are added to the lower chamber, facing the basolateral side of the epithelial monolayer. The medium in the lower chamber is exchanged with fresh RPMI 1640 with 10 % FCS, 25 ng/mL IL-4 and 50 ng/mL GM-CSF at day 4 of the co-cultivation. Maintain the co-culture for 6 days before experiments are conducted. Continue to exchange the medium in the upper chamber with fresh RPMI 1640 with 10 % FCS every second day.

The upper chamber will reflect the apical side and the lower chamber the basolateral side of the intestinal epithelium. Incubation time for the test compound will depend on type of compound and the read out system to be used and will have to be decided for each compound.

18.4.3   Caco-2/THP-1 (Soluble Factors)

300 μL of a 1×10^6/mL Caco-2 cells suspended in RPMI 1640 with 10 % FCS, are seeded on filter inserts with 0.4 μm pores. The cells are maintained at 37 °C in 5 % CO_2 atmosphere for 21 days, exchanging the medium with fresh RPMI 1640 containing 10 % FSC every other day. At day 18 the THP-1 cells are plated at a density of 2×10^5 cells/mL in RPMI 1640 with 10 % FCS, 200 ng/mL IL-4, 100 ng/mL GM-CSF, 20 ng/mL TNF-α and 200 ng/mL ionomycin in a 24 well plate for 3 days. At day 21 the medium of the THP-1 cells is exchanged with RPMI 1640 with 10 %

FCS and the filter inserts with polarized Caco-2 monolayers are placed into the wells together with the differentiated THP-1 cells.

The upper chamber will reflect the apical side and the lower chamber the basolateral side of the intestinal epithelium. Incubation time for the test compound will depend on type of compound and the read out system to be used and will have to be decided for each compound.

18.4.4   Caco-2/PBMCs (Soluble Factors)

300 µL of a 1×10^6/mL Caco-2 cells suspended in RPMI-1640 with 10 % FCS are seeded on filter inserts with 0.4 µm pores. The cells are maintained at 37 °C in 5 % CO_2 atmosphere for 21 days with exchanges of the medium with fresh RPMI 1640 with 10 % FSC every other day. At day 20, PBMCs are plated in the basolateral compartment at a density of $1–2 \times 10^6$ cells/mL RPMI-1640 with 10 % FCS. PBMCs can be activated by anti-CD3/CD28, mitogenic lectins or LPS (see Chap. 15 PBMCs).

The upper chamber will reflect the apical side and the lower chamber the basolateral side of the intestinal epithelium. Incubation time for the test compound will depend on type of compound and the read out system to be used and will have to be decided for each compound.

18.4.5   Caco-2/B Cells

300 µL of a 1×10^6/mL Caco-2 cells suspended in RPMI-1640 with 10 % FCS are seeded on filter inserts with 3.0 µm pores. The inserts are placed in a 24 well plate with 1 mL of RPMI 1640 with 10 % FCS. The cells are maintained at 37 °C in 5 % CO_2 atmosphere for 14 days with exchanges of the medium with fresh RPMI 1640 with 10 % FSC, every other day. At day 14, 1×10^6 Raji cells in RPMI 1640 with 10 % FCS are added to the lower chamber, facing the basolateral side of the epithelial monolayer. Maintain the co-culture for 6 days. Continue to exchange the medium in the upper chamber with fresh RPMI 1640 with 10 % FCS every second day.

The upper chamber will reflect the apical side and the lower chamber the basolateral side of the intestinal epithelium. To investigate uptake through M-cells the test compound should be added to the apical side. The incubation time for the test compound must be investigated different compounds, but are usually in the range of minutes to a few hours.

In all of the above-mentioned systems, the effects of test compounds on co-cultures should be compared with effects on matched filters with Caco-2 monocultures. This will make it possible to identify whether the effects of the test compounds are related to cell–cell communication in the co-culture systems.

18.5   Assess Viability

The integrity of the Caco-2 monolayers should always be checked by measurement of TEER values, and filters with a TEER value below $300\ \Omega\ cm^2$ should not be used for further experiments.

The paracellular transport marker Lucifer Yellow (LY) is used to measure changes in the integrity of the Caco-2 cell layer. LY is added to the apical compartment at a concentration of 100 μM, in control wells and wells treated with test compound. The fluorescence of the LY transported to the basolateral side is then measured with a fluorescence microplate reader (Calatayud et al. 2012, and Chap. 10).

In comparison with polarized Caco-2 cells M-cells will lack microvilli on their apical surface which can be inspected by transmission electron microscope (TEM) or scanning electron microscope (SEM). Staining of the Caco-2 monolayer with fluoro-chrome-labeled Wheat Germ Agglutinin (WGA) or phalloidin (actin label) will give a continuous staining in a monoculture of Caco-2, however after co-cultivation of Caco-2 cells with Raji, M-cells are identified as a discontinuations in the staining pattern. This is due to the absence of microvilli. Corresponding monocultures of Caco-2 cells on matched filter inserts should be used as controls.

Lactate dehydrogenase (LDH) is also a marker of cytotoxicity and its release can be determined in the supernatant of the cell cultures (Decker and Lohmann-Matthes 1988). There are several commercial kits for determination of LDH concentration in culture supernatants.

18.6   Experimental Readout

Several methodological reports are describing co-culture conditions between Caco-2 and CD14[+] monocyte derived DCs, THP-1 or PBMCs (Araujo and Sarmento 2013; Leonard et al. 2010; Schimpel et al. 2014; Antunes et al. 2013; Kerneis et al. 2000; Parlesak et al. 2004). However, there are still few reports where these co-culture systems have been used to scrutinize communication between epithelial cells and immune cells.

Apical TLR ligation on intestinal epithelial cells and its effect on PBMCs have been studied by co-cultivation of HT-29 and PBMCs (de Kivit et al. 2011). The TLR ligands are added to the apical side of the epithelial monolayer, whilst the PBMCs are stimulated with anti-CD3/CD28 antibodies and LPS. The release of cytokines to the culture supernatants, both in the basolateral and apical compartments, were investigated by ELISA. There is also a report on the effect of the anti-inflammatory drugs ibuprofen and prednisolone in a co-culture system of Caco-2 monolayers in co-culture with whole blood (Schmohl et al. 2012).

There is one report using a co-culture model of Caco-2 monolayers and THP-1 cells to study the function of the immediate-early response gene 1 (IEX-1) (Ishimoto et al. 2011). Overexpression or knock down of the IEX-1 gene in Caco-2 cells were induced by Lentivirus infection. Caco-2 cells were differentiated for 14 days on semi-permeable filter inserts. The THP-1 cells were differentiated for 4 days and then the filter inserts with Caco-2 were placed in the well with differentiated THP-1 cells. After co-cultivation, mRNA was extracted from both Caco-2 cells and THP-1. Changes in expression of genes coding for cytokines were investigated by qPCR.

The Caco-2/B cell model is widely used to study translocation of bacteria over M-cells. Translocation of fluorescence-labelled carboxylated latex or dextran particles was used to confirm in vitro induction of the M-cell like phenotype. Translocation of bacteria was measured by serial dilutions of basolateral culture medium on agar plates 2 h after addition of bacteria to the apical compartment (Finn et al. 2014). How dietary components affect translocation of bacteria over M-cells have also been investigated (Roberts et al. 2013). The test compound was added to the apical compartment for 30 min before infection with bacteria. After 2–4 h, the basolateral medium was harvested and bacteria enumerated following serial dilution on agar plates.

18.7   Advantages, Disadvantages and Limitations

These co-culture systems give the opportunity to study cell–cell communication, and most cell laboratories will have the possibility to establish at least some of these co-cultivation systems. The use of monocyte-derived DCs or PBMCs have the advantage of responding well with the production of cytokines, whilst THP-1 on the other side produce a limited number of cytokines. These protocols are time-consuming and therefore not suitable for high throughput screening. As for other in vitro culturing systems the data should be interpreted with care as they do not represent the complexity found in the in vivo situation.

18.8   Conclusions

The co-cultivation systems described here needs to be carefully validated in order to be able to harvest reliable experimental data related to communication between immune cells and epithelial cells. They can be used for screening purposes, however, due to time consumption they are generally unsuitable for high throughput screening. Co-culture of Caco-2 and Raji cells generates cells with a M-cell like phenotype. Such cells can give useful information regarding transport of test compounds through M-cells, however such studies will eventually have to be confirmed in an in vivo system.

References

Antunes F, Andrade F, Araujo F et al (2013) Establishment of a triple co-culture in vitro cell models to study intestinal absorption of peptide drugs. Eur J Pharm Biopharm 83:427–435

Araujo F, Sarmento B (2013) Towards the characterization of an in vitro triple co-culture intestine cell model for permeability studies. Int J Pharm 458:128–134

Artis D (2008) Epithelial-cell recognition of commensal bacteria and maintenance of immune homeostasis in the gut. Nat Rev Immunol 8:411–420

Calatayud M, Vazquez M, Devesa V et al (2012) In vitro study of intestinal transport of inorganic and methylated arsenic species by Caco-2/HT29-MTX cocultures. Chem Res Toxicol 25: 2654–2662

Coombes JL, Powrie F (2008) Dendritic cells in intestinal immune regulation. Nat Rev Immunol 8:435–446

de Kivit S, van Hoffen E, Korthagen N et al (2011) Apical TLR ligation of intestinal epithelial cells drives a Th1-polarized regulatory or inflammatory type effector response in vitro. Immunobiology 216:518–527

Decker T, Lohmann-Matthes ML (1988) A quick and simple method for the quantitation of lactate dehydrogenase release in measurements of cellular cytotoxicity and tumor necrosis factor (TNF) activity. J Immunol Methods 115:61–69

Finn R, Ahmad T, Coffey ET et al (2014) Translocation of *Vibrio parahaemolyticus* across an in vitro M cell model. FEMS Microbiol Lett 350:65–71

Gullberg E, Leonard M, Karlsson J et al (2000) Expression of specific markers and particle transport in a new human intestinal M-cell model. Biochem Biophys Res Commun 279:808–813

Ishimoto Y, Satsu H, Totsuka M et al (2011) IEX-1 suppresses apoptotic damage in human intestinal epithelial Caco-2 cells induced by co-culturing with macrophage-like THP-1 cells. Biosci Rep 31:345–351

Kerneis S, Bogdanova A, Kraehenbuhl JP et al (1997) Conversion by Peyer's patch lymphocytes of human enterocytes into M cells that transport bacteria. Science 277:949–952

Kerneis S, Caliot E, Stubbe H et al (2000) Molecular studies of the intestinal mucosal barrier physiopathology using cocultures of epithelial and immune cells: a technical update. Microbes Infect 2:1119–1124

Leonard F, Collnot EM, Lehr CM (2010) A three-dimensional coculture of enterocytes, monocytes and dendritic cells to model inflamed intestinal mucosa in vitro. Mol Pharm 7:2103–2119

Lo D, Tynan W, Dickerson J et al (2004) Cell culture modeling of specialized tissue: identification of genes expressed specifically by follicle-associated epithelium of Peyer's patch by expression profiling of Caco-2/Raji co-cultures. Int Immunol 16:91–99

Miller JC, Brown BD, Shay T et al (2012) Deciphering the transcriptional network of the dendritic cell lineage. Nat Immunol 13:888–899

Mora JR, Bono MR, Manjunath N et al (2003) Selective imprinting of gut-homing T cells by Peyer's patch dendritic cells. Nature 424:88–93

Mowat AM (2003) Anatomical basis of tolerance and immunity to intestinal antigens. Nat Rev Immunol 3:331–341

Parlesak A, Haller D, Brinz S et al (2004) Modulation of cytokine release by differentiated CACO-2 cells in a compartmentalized coculture model with mononuclear leucocytes and nonpathogenic bacteria. Scand J Immunol 60:477–485

Pastorelli L, De Salvo C, Mercado JR et al (2013) Central role of the gut epithelial barrier in the pathogenesis of chronic intestinal inflammation: lessons learned from animal models and human genetics. Front Immunol 4:280

Rimoldi M, Chieppa M, Vulcano M et al (2004) Intestinal epithelial cells control dendritic cell function. Ann N Y Acad Sci 1029:66–74

Roberts CL, Keita AV, Parsons BN et al (2013) Soluble plantain fibre blocks adhesion and M-cell translocation of intestinal pathogens. J Nutr Biochem 24:97–103

Schimpel C, Teubl B, Absenger M et al (2014) Development of an advanced intestinal in vitro triple culture permeability model to study transport of nanoparticles. Mol Pharm 11(3): 808–818

Schmohl M, Schneiderhan-Marra N, Baur N et al (2012) Characterization of immunologically active drugs in a novel organotypic co-culture model of the human gut and whole blood. Int Immunopharmacol 14:722–728

Segura E, Touzot M, Bohineust A et al (2013) Human inflammatory dendritic cells induce Th17 cell differentiation. Immunity 38:336–348

Stagg AJ, Kamm MA, Knight SC (2002) Intestinal dendritic cells increase T cell expression of alpha4beta7 integrin. Eur J Immunol 32:1445–1454

Steinman RM (2007) Dendritic cells: understanding immunogenicity. Eur J Immunol 37(Suppl 1): S53–S60

Steinman RM (2012) Decisions about dendritic cells: past, present, and future. Annu Rev Immunol 30:1–22

Steinman RM, Idoyaga J (2010) Features of the dendritic cell lineage. Immunol Rev 234:5–17

Part IV
Enteroendocrine Cell Models: General Introduction

During and after a meal, ingested nutrients modulate the release of a variety of hormones, many of which serve roles to enhance nutrient digestion. A number of these gut hormones, most notably cholecystokinin (CCK), glucagon-like peptide-1 (GLP-1) and peptide YY (PYY), are key players in influencing eating behaviour and food intake, largely through induction of satiety (Begg and Woods 2013). Some of these hormones (glucose-dependent insulinotropic peptide (GIP) and GLP-1)) are also incretins, enhancing glucose-dependent insulin secretion. Identifying food components that alter the gastrointestinal hormonal milieu may aid in our 'battle of the bulge' as well as in the treatment of related metabolic diseases (Bruen et al. 2012).

Mammalian enteroendocrine cell lines provide the starting framework to measure gut hormonal responses to food. In vitro models of enteroendocrine cell lines have proven difficult to develop, primarily because enteroendocrine cells are sparsely dispersed throughout the gastro-intestinal tract and are co-localized with abundant enterocytes. However, several groups have successfully established enteroendocrine cell models that are available for in vitro screening bioassays (Table 1).

The NCI-H716 cell line is the only human enteroendocrine cell line available and it provides a unique L-cell model to screen food components for their ability to modulate secretion of the incretin hormone, GLP-1. This cell line was originally derived from cells present in the ascites fluid of a male 33 year old patient with a poorly differentiated adenocarcinoma of the cecum (Park et al. 1987). When propagated in vitro, these cells have been described to exhibit endocrine differentiation including the expression of secretory granules and chromogranin A (de Bruine et al. 1993). However, they do require attachment matrices (de Bruine et al. 1993; Anini and Brubaker 2003) which could limit their ease of use in high throughput assays. The NCI-H716 cells express both GLP-1 and PYY but not CCK (Reimer et al. 2001; Jang et al. 2007). GLP-1 expression in these cells is mostly homogenous but PYY expression is much less uniform, with only ~30 % of cells containing PYY immunoreactivity (Jang et al. 2007). The NCI-H716 cell line is discussed in more detail in Chap. 20.

Table 1 Enteroendocrine cell models

Model	Species	Cell source	Characteristics	Advantages	Disadvantages
NCI-H716	Human	Spontaneous colorectal tumour (Park et al. 1987)	Undifferentiated tumour cell line with endocrine morphology; secretes GLP-1, PYY and mucin	Representative of distal L-cell; all lab techniques possible; easy to transfect	Possibly heterogeneous as not single-cell cloned; is tumour-derived so may not recapitulate primary human L-cell; requires basement membrane to attach to culture plate
STC-1	Mouse	Duodenal secretin tumour cells from proinsulin-SV40 Large T antigen/proinsulin-polyoma x Small T antigen transgenic mice (Rindi et al. 1990)	Poorly differentiated; secretes CCK, GLP-1, GIP, PYY, pancreatic polypeptide, neurotensin, GLP-2 and oxyntomodulin	Representative of proximal L-cell; all lab techniques possible; secretes several gut hormones	Heterogeneous cell population; murine- and tumour-derived so may not recapitulate primary human L-cells
GLUTag	Mouse	Large bowel tumour in proglucagon-SV40 Large T antigen transgenic mice (Drucker et al. 1994)	Single-cell subcloned population of L-cells; secretes GLP-1 and CCK	Representative of distal L-cell; subcloned homogenous cells; all lab techniques possible; easy to transfect	Murine- and tumour-derived so may not recapitulate human L-cells; requires basement membrane to attach to some surfaces
FRIC	Rat	Small and large intestines combined from fetal rat	Heterogeneous population of intestinal cells (L-cells ≅ 1 %)	Primary cells derived from normal intestine	Heterogeneous cell population; obtained from fetal rats and may therefore not be representative of the adult human L-cell; small number of L-cells precludes cell-specific lab techniques (i.e. RT-PCR, western blot); difficult to transfect
AMIC	Mouse	Duodenum, jejunum, ileum or colon from adult mouse	Heterogeneous population of intestinal cells (L-cells ≅ 1 %)	Primary cells derived from normal intestine; genetically-modified mice can be used as tissue source; can be used to study single cells if derived from L-cell reporter mice	Difficult to culture; small number of L-cells precludes cell-specific lab techniques (i.e. RT-PCR, western blot); may be difficult to transfect

The NCI-H716 cells are available from American Type Culture Collection (ATCC). STC-1 cells are available from Dr. D. Hanahan (University of California, San Francisco, CA), and GLUTag cells can be requested from Dr. Daniel J. Drucker (University of Toronto). Additionally, primary L-cell models can be produced in-house. Fetal rat intestinal culture (FRIC) protocols and adult mouse intestinal culture (AMIC) protocols are widely available (Anini and Brubaker 2003; Reimann et al. 2008)

To study GLP-1 secretion, researchers also routinely use the alternative, albeit non-human, cell lines; STC-1 and GLUTag, which were developed from murine genetically-induced proximal and distal intestinal tumours, respectively. The STC-1 (secretin tumour cell) cell line is a heterogeneous enteroendocrine cell line which expresses several satiety and incretin hormones, including CCK, GIP, PYY, pancreatic polypeptide, neurotensin and the proglucagon-derived peptides: GLP-1, glucagon-like peptide-2 (GLP-2) and oxyntomodulin (Rindi et al. 1990). The plurihormonal nature of these cells appears to be consistent with recent findings on the primary murine duodenal L-cell (Habib et al. 2012, 2013). Molecular mechanisms underlying stimulus-hormone secretion coupling have been well defined in the STC-1 cell line. Primarily because of its ability to secrete several satiety hormones, it remains a popular choice for primary screening platforms, as detailed in the accompanying chapter.

In contrast to STC-1 cells, the GLUTag (proglucagon SV40-large T antigen) cell line appears quite differentiated and demonstrates a stable pattern of *proglucagon* gene expression over a 4–8 week period (Drucker et al. 1994). However, unlike the NCI-H716 and STC-1 cells, the GLUTag cells express GLP-1 and CCK, but not PYY (Drucker et al. 1994). The absence of PYY is somewhat surprising given the colonic origin of the GLUTag cell line and the fact that most colonic L-cells express this hormone in vivo (Habib et al. 2013). Notwithstanding, GLUTag cells recapitulate the responsiveness of both of the other cell lines as well as of primary murine L-cells and fetal rat intestinal cultures, secreting multiple fully-processed glucagon-like peptides in response to known physiological and pharmacological secretagogues (Lim et al. 2009). For use in high throughput assays, it is worth noting that GLUTag cells require a matrix for proper attachment to multi-well plates. Further details regarding the GLUTag cell line can be found in the accompanying chapter.

It is important to note that species-specific differences in nutrient sensitivity in vivo do exist and non-human cell lines may therefore not be appropriate beyond primary screens (Brubaker 1991; Brubaker et al. 1998; Habib et al. 2013). For example, while GIP is an effective stimulator of GLP-1 release in canine and rodent L-cells (Brubaker 1991; Brubaker et al. 1998) it has little effect on GLP-1 secretion in human L-cells in vivo (Kreymann et al. 1987). Moreover, it has been demonstrated that the transcription factors and promoter regions important for human *proglucagon* gene expression are distinct from those utilised by the rat *proglucagon* gene (Nian et al. 1999). Findings made in any cell line may not completely recapitulate those made in vivo, due to the extremely high likelihood that the L-cell constantly integrates multiple signals, from luminal, endocrine and neural inputs. It is thus imperative that all findings made in vitro ultimately be confirmed in the in vivo setting. With these limitations in mind, the enteroendocrine cell models still provide a set of powerful tools that enable a detailed examination of hormonal responses to food bioactives.

References

Anini Y, Brubaker PL (2003) Role of leptin in the regulation of glucagon-like peptide-1 secretion. Diabetes 52(2):252–259

Begg DP, Woods SC (2013) The endocrinology of food intake. Nat Rev Endocrinol 9(10): 584–597

Brubaker PL (1991) Regulation of intestinal proglucagon-derived peptide secretion by intestinal regulatory peptides. Endocrinology 128(6):3175–3182

Brubaker PL, Schloos J, Drucker DJ (1998) Regulation of glucagon-like peptide-1 synthesis and secretion in the GLUTag enteroendocrine cell line. Endocrinology 139(10):4108–4114. doi:10.1210/endo.139.10.6228

Bruen CM, O'Halloran F, Cashman KD, Giblin L (2012) The effects of food components on hormonal signalling in gastrointestinal enteroendocrine cells. Food Funct 3(11):1131–1143

de Bruine AP, Dinjens WN, van der Linden EP, Pijls MM, Moerkerk PT, Bosman FT (1993) Extracellular matrix components induce endocrine differentiation in vitro in NCI-H716 cells. Am J Pathol 142(3):773–782

Drucker DJ, Jin T, Asa SL, Young TA, Brubaker PL (1994) Activation of proglucagon gene transcription by protein kinase-A in a novel mouse enteroendocrine cell line. Mol Endocrinol 8(12):1646–1655

Habib AM, Richards P, Cairns LS, Rogers GJ, Bannon CA, Parker HE, Morley TC, Yeo GS, Reimann F, Gribble FM (2012) Overlap of endocrine hormone expression in the mouse intestine revealed by transcriptional profiling and flow cytometry. Endocrinology 153(7):3054–3065

Habib AM, Richards P, Rogers GJ, Reimann F, Gribble FM (2013) Co-localisation and secretion of glucagon-like peptide 1 and peptide YY from primary cultured human L cells. Diabetologia 56(6):1413–1416

Jang HJ, Kokrashvili Z, Theodorakis MJ, Carlson OD, Kim BJ, Zhou J, Kim HH, Xu X, Chan SL, Juhaszova M, Bernier M, Mosinger B, Margolskee RF, Egan JM (2007) Gut-expressed gustducin and taste receptors regulate secretion of glucagon-like peptide-1. Proc Natl Acad Sci U S A 104(38):15069–15074

Kreymann B, Williams G, Ghatei MA, Bloom SR (1987) Glucagon-like peptide-1 7–36: a physiological incretin in man. Lancet 2(8571):1300–1304

Lim GE, Huang GJ, Flora N, LeRoith D, Rhodes CJ, Brubaker PL (2009) Insulin regulates glucagon-like peptide-1 secretion from the enteroendocrine L cell. Endocrinology 150(2):580–591

Nian M, Drucker DJ, Irwin D (1999) Divergent regulation of human and rat proglucagon gene promoters in vivo. Am J Physiol 277(4 Pt 1):G829–G837

Park JG, Kramer BS, Steinberg SM, Carmichael J, Collins JM, Minna JD, Gazdar AF (1987) Chemosensitivity testing of human colorectal carcinoma cell lines using a tetrazolium-based colorimetric assay. Cancer Res 47(22):5875–5879

Reimann F, Habib AM, Tolhurst G, Parker HE, Rogers GJ, Gribble FM (2008) Glucose sensing in L cells: a primary cell study. Cell Metab 8(6):532–539

Reimer RA, Darimont C, Gremlich S, Nicolas-Metral V, Ruegg UT, Mace K (2001) A human cellular model for studying the regulation of glucagon-like peptide-1 secretion. Endocrinology 142(10):4522–4528

Rindi G, Grant S, Yiangou Y, Ghatei M, Bloom S, Bautch V, Solcia E, Polak J (1990) Development of neuroendocrine tumors in the gastrointestinal tract of transgenic mice. Heterogeneity of hormone expression. Am J Pathol 136(6):1349–1363

Chapter 19
STC-1 Cells

Triona McCarthy, Brian D. Green, Danielle Calderwood, Anna Gillespie, John F. Cryan, and Linda Giblin

Abstract Gastrointestinal hormones such as cholecystokinin (CCK), glucagon like peptide 1 (GLP-1), and peptide YY (PYY) play an important role in suppressing hunger and controlling food intake. These satiety hormones are secreted from enteroendocrine cells present throughout the intestinal tract. The intestinal secretin tumor cell line (STC-1) possesses many features of native intestinal enteroendocrine cells. As such, STC-1 cells are routinely used in screening platforms to identify foods or compounds that modulate secretion of gastrointestinal hormones in vitro. This chapter describes this intestinal cell model focussing on it's applications, advantages and limitations. A general protocol is provided for challenging STC-1 cells with test compounds.

Keywords Enteroendocrine cells • Secretin tumor cell line (STC-1) • Satiety hormones • GLP-1 • CCK • PYY • Food components • G protein coupled receptors • Nutrient sensing

19.1   Origin

STC-1 was originally established from cells present in murine enteroendocrine tumours. These tumours arose in the duodenum of double transgenic mice expressing the rat insulin promoter linked to the simian virus 40 large T antigen and to the

T. McCarthy
Food for Health Ireland, Department of Anatomy and Neuroscience, University College Cork, Cork, Ireland

Food for Health Ireland, Teagasc Food Research Centre, Moorepark Fermoy, Cork, Ireland

B.D. Green • D. Calderwood • A. Gillespie
Institute for Global Food Security, School of Biological Sciences, Queen's University Belfast, Belfast, Northern Ireland

J.F. Cryan
Food for Health Ireland, Department of Anatomy and Neuroscience, University College Cork, Cork, Ireland

L. Giblin (✉)
Food for Health Ireland, Teagasc Food Research Centre, Moorepark Fermoy, Cork, Ireland
e-mail: linda.giblin@teagasc.ie

© The Author(s) 2015
K. Verhoeckx et al. (eds.), *The Impact of Food Bio-Actives on Gut Health*,
DOI 10.1007/978-3-319-16104-4_19

polyomavirus small T antigen (Rindi et al. 1990). STC-1 is a relatively slow growing cell line, with a doubling time of 54 h in standard growth media. They are not easily amenable to transfection with several studies reporting low transfection efficiencies (stable and transient). Under culturing conditions, this cell line is heterogenous. However most subpopulations demonstrate immunoreactivity to the gut hormone, cholecystokinin (CCK) (Rindi et al. 1990). As such, STC-1 cells were originally used as a model of native CCK-producing I-cells. Over the past two decades, the uses of the STC-1 cell line has been expanded to investigate (a) the cellular signaling mechanisms involved in secretion and gene transcription of other gut hormones (Cordier-Bussat et al. 1998; Geraedts et al. 2009), (b) enteroendocrine cell differentiation (Ratineau et al. 1997), (c) tumor cell growth (Bollard et al. 2013) and (d) intestinal immune responses (Palazzo et al. 2007).

19.2   Features and Mechanisms

In addition to CCK, STC-1 cells express and secrete a wide range of gut hormones known for their roles in metabolism, feeding and satiety, including glucose dependent insulinotropic polypeptide (GIP), peptide YY (PYY), pancreatic polypeptide, neurotensin and the proglucagon derived peptides [glucogon-like peptide-1 (GLP-1), glucogon-like peptide-2 (GLP-2) and oxyntomodulin] (Rindi et al. 1990; Hand et al. 2013). STC-1 cells secrete these hormones in response to a range of physiological stimuli, although levels may differ to native enteroendocrine cells (Reimann et al. 2008). This ability to secrete a variety of gut hormones ensures that the STC-1 cell line remains a popular 'look-see' choice with food researchers.

Monosaccharides (Mangel et al. 1994), fatty acids (Hand et al. 2010), aromatic amino acids (Cordier-Bussat 1998; Wang et al. 2011), peptidomimetic compounds (Geraedts et al. 2011) and bitter tastants (Miyata et al. 2014) have all been demonstrated to dose dependently elicit CCK and GLP-1 secretion from STC-1 cells. Several studies have also reported that sucralose and other sweeteners increase CCK and GLP-1 secretion in STC-1 cells (Geraedts et al. 2012). Interestingly, these substances do not appear to be GLP-1 stimulants in primary cultures of murine intestinal epithelium (Reimann et al. 2008). Although PYY secretion has been less widely investigated, STC-1 cells reliably produce and secrete PYY in response to a variety of fatty acids including valeric acid, linoleic acid and conjugated linoleic acid 9,11 (Hand et al. 2013). GIP expression and production by STC-1 cells is poor. Nevertheless as there are few in vitro alternatives, GIP subclones of STC-1 cells with increased *GIP* expression have been generated (Kieffer et al. 1995).

Several neural and hormonal stimuli which play an important role in coordinating early responses to food ingestion in vivo also induce gut peptide release from STC-1 cells. These include the neurotransmitters gamma-aminobutyric acid (GABA), acetylcholine, orexin, bombesin, apelin, gastrin releasing peptide as well as bile acids and leptin.

Hormone secretion from this cell line appears to be dependent on a rise in cytoplasmic cAMP and Ca^{2+} which ultimately leads to alteration in hormone gene transcription levels or hormone peptide release. Increased Ca^{2+} levels can occur as a result of (a) an influx of Ca^{2+} across the plasma membrane through activation of voltage-gated calcium channels or (b) inositol 1,4,5-triphosphate (IP3) triggered Ca^{2+} release from intracellular stores. In native enteroendocrine cells, intracellular levels of Ca^{2+} and cAMP are also altered by the activity of a number of G-protein coupled receptors (GPCR's) and nutrient transporters. STC-1 cells have been found to intrinsically express many of these important GPCRs including GPR40 and GPR120 (Tanaka et al. 2008). Hormone secretion is dependent on GPR40 in native murine I-cells, as linolenic acid induced CCK secretion was absent in cells lacking this receptor (Liou et al. 2011b). In contrast, GPR40 knockdown had little effect on the ability of STC-1 cells to sense fatty acids. Rather, it is GPR120 that appears to play a more important role in fatty acid sensing in STC-1 cells. Fatty acid-induced hormone release is impaired in *GPR120* siRNA silencing experiments (Tanaka et al. 2008). Wang *et al.* demonstrated that phenylalanine induced Ca^{2+} mobilization and hormone secretion occurs via the calcium-sensing receptor which is highly expressed in native murine enteroendocrine cells (Wang et al. 2011). This receptor is however poorly expressed in STC-1 cells, although it is functional. Several taste signalling elements such as alpha-gustducin, phospholipase C-β2, transient receptor potential channel M5 and the taste receptor T1R/T2R subunits are present and functional in STC-1 cells similar to native enteroendocrine cells (Dyer et al. 2005).

Recent evidence suggests that ATP-sensitive K^+ channels and transporters [e.g. sodium-coupled glucose transporters (SGLTs) and peptide transporters (PEPTs)] contribute to hormone secretion in enteroendocrine cells (Tolhurst et al. 2009). However STC-1 cells express low levels of the K_{ATP} channel subunits, Kir6.2 and SUR1. Endogenous expression of PEPTs is also poor in STC-1 cells (Liou et al. 2011a).

In addition to stimulating gut hormone release, common luminal nutrients added to the growth media are also capable of modulating gene expression in STC-1 cells. As such this cell line has been widely used to investigate the regulation of *CCK* gene expression. However, for the *proglucagon* gene, aberrant posttranslational processing has been documented (Blache et al. 1994). In vivo, the synthesis of proglucagon-derived peptides is tissue specific due to the action of various prohormone convertase enzymes (e.g. PC1/3). In pancreatic α cells, posttranslational processing of proglucagon liberates glucagon, glicentin-related pancreatic peptide, and the major proglucagon fragment, which contains the sequences for both GLP-1 and GLP-2. In contrast, processing of proglucagon in the intestinal enteroendocrine L-cell generates GLP-1 and GLP-2, as well as the glucagon-containing peptides, glicentin and oxyntomodulin. Blache *et al.* reported that proglucagon processing in STC-1 cells results in the production of significant quantities of glucagon in addition to glicentin and oxyntomodulin (Blache et al. 1994). Thus STC-1 cells appear to process the proglucagon fragment in a manner which is intermediate between intestinal L-cells and pancreatic α cells.

19.3   Stability, Consistency and Reproducibility

When used appropriately, STC-1 cell line has been shown to be a reliable and reproducible enteroendocrine cell model. However, considerable variances in hormone
secretion levels have been reported by different laboratories for the same substance
(Geraedts et al. 2009; Hand et al. 2013). For example, Cordier-Bussat *et al.* showed
that addition of protein hydrolysates from meat, casein and soybean to STC-1 cells
increased CCK and GLP-1 release compared to undigested proteins (Cordier-Bussat
et al. 1997). In contrast, other studies have demonstrated that intact protein is a
much stronger stimulant for CCK and GLP-1 release in STC-1 cells than hydrolysates or specific peptides (Geraedts et al. 2011). In addition to differences between
different laboratories, significant inter-experimental variability in the amount of
CCK secreted has also been reported within study groups. Differences in culture
protocols such as seeding density, cell feeding routine, washing steps, test buffer
and passage number may account in part for this variability. However, inter-
experimental variability is most likely a direct result of the heterogenous non-stable
nature of the cell line. Cultured STC-1 cells demonstrate 95 % immunoreactivity to
CCK, while only 7 % stain positively for GIP (Rindi et al. 1990). Glassmeier *et al.*
reported that the sensitivity of STC-1 cells to the neurotransmitter GABA varied
considerably from cell to cell, due to the heterogeneous density and expression of
$GABA_A$ receptors in this cell line (Glassmeier et al. 1998). STC-1 cells may switch
to multiple differentiated states during proliferation (Rindi et al. 1990). Consequently
levels of hormone secretion may significantly differ from one test to another.
To improve the homogeneity and the stability of STC-1 cells, several studies have
attempted to isolate and characterize clonal cell lines from parental STC-1 cells.
However even these cells do not express a stable phenotype. For example, while
attempting to clone a pure GIP-expressing cell, Kieffer *et al.* observed that only
30 % of the expanded clone still expressed GIP immunoreactivity due to dediffer-
entiation of the cells (Kieffer et al. 1995). These observations highlight the importance of including adequate controls and replicates for each independent experiment.
As early passage number STC-1 cells have greater heterogeneity, cells that have
been passaged more than 10 times are recommended. It also underlines the importance of verifying results with experiments independent of STC-1 cells.

19.4   Relevance to Human In Vivo Situation

A major requirement of any in vitro model is that it should demonstrate a strong
predictive power, yielding data that support interpretation of results for the in vivo
situation. Although STC-1 cells do respond to a range of physiological stimuli
shown to elicit gut hormone secretion in vivo, some inconsistencies have been
observed. Hall *et al.* demonstrated that while whey consumption in humans resulted
in a greater increase in GLP-1, GIP, and CCK plasma levels compared to other protein sources (Hall et al. 2003), whey protein hydrolysate are ineffective stimulants

of CCK release in STC-1 cells (Foltz et al. 2008). STC-1 and the native state also differ in PYY secretion to fatty acids (Hand et al. 2013).

In vivo gut hormone release occurs through direct sensing of luminal nutrients but also indirectly by vagal and humoral stimulation (Gribble 2012). Evidence shows that the release of bile acids, gastric acid secretions, mechanical stimulation by peristaltic contractions, gastric emptying and hormonal stimulation all result in gut hormone modulation. While STC-1 cells harbour many features of native intestinal hormone secreting cells, they are cultivated as monolayers on plastic surfaces and lack a normal cellular environment. They therefore cannot reproduce the integrative physiological interplay present in vivo. Thus many components which stimulate gut hormone secretion through indirect mechanisms will not be detected by this model. A typical example of this limitation is that potato protease inhibitors fail to elevate CCK expression or release in STC-1 cells (Komarnytsky et al. 2011). However in vivo potato protease inhibitors reduce food intake, and increase plasma and duodenal mucosal mRNA levels of *CCK* in rats indirectly by inhibiting trypsin mediated deactivation of endogenously produced CCK releasing factors (Komarnytsky et al. 2011).

It is also worth noting that compounds which elicit gut hormone secretion from STC-1 cells and in vivo, do not always effect food intake or the perception of the feeling of fullness. For example, Korean pine nut oil effectively increases CCK secretion from STC-1 cells and in plasma levels of humans. This increase in satiety hormone levels however did not correspond to a change in subjective appetite sensations as assessed by visual analogue scales (Pasman et al. 2008). This lack of relationship between increased plasma concentrations of gut hormones and appetite ratings has been observed in other studies (Veldhorst et al. 2009).

19.5 General Protocol

19.5.1 Cell Maintenance Protocol

STC-1 cells should be cultured in supplemented DMEM media (DMEM containing 4.5 g/l D-glucose, without sodium pyruvate) (GlutaMAX, GIBCO, Paisley, UK) with 17.5 % foetal bovine serum (FBS), 100 U/ml penicillin, 100 mg/l streptomycin and incubated in a 5 % CO_2 humidified atmosphere at 37 °C. Cells should be passaged at 80–90 % confluence. Passage number should be between 15 and 40.

19.5.2 Experimental Protocol for Test Compounds

Test compounds should be prepared in pre-warmed buffer. Buffer should be tested to ensure compatibility with the particular immunoassay.

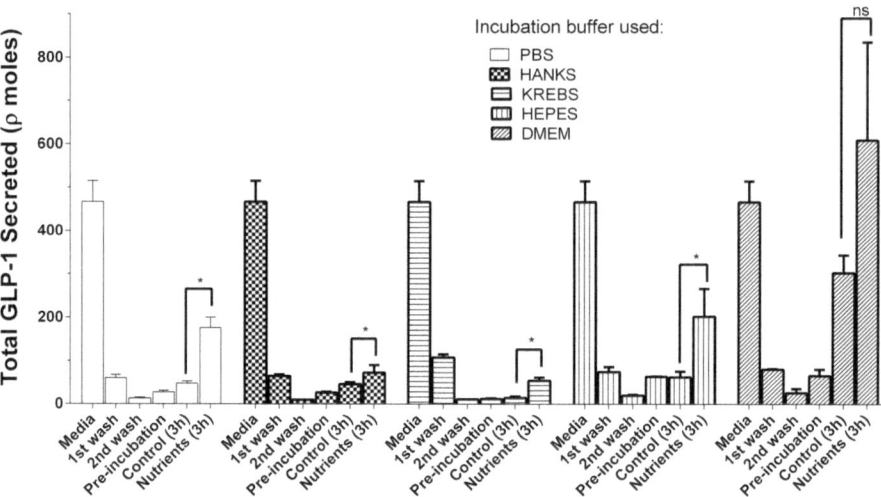

Fig. 19.1 Effect of various incubation buffers on levels of Total GLP-1 secreted from STC-1 cells. STC-1 cells (2×10^6) were challenged with mixed nutrients [glutamine + valine + lysine + glycine + glucose + fructose (all at 40 mM)] for 3 h. Samples were prepared in PBS or HANKS or KREBS or HEPES buffers or DMEM media (GlutaMAX, GIBCO, Paisley, UK). The composition of the buffers are as follows; PBS (136.9 mM NaCl, 2.7 mM KCl, 10.1 mM Na_2HPO_4, 1.8 mM KH_2PO_4), HANKS (136.9 mM NaCl, 5.4 mM KCl, 1.3 mM $CaCl_2$, 1 mM $MgSO_4$, 0.2 mM $NaHPO_4$, 0.4 mM KH_2PO_4, 4.2 mM $NaHCO_3$, 5.6 mM glucose), KREBS (118 mM NaCl, 4.7 mM KCl, 25 mM $NaHCO_3$, 1.25 mM $CaCl_2$, 1.2 mM $MgSO_4$, 1.2 mM KH_2PO_4, 11 mM glucose) and HEPES (140 mM NaCl, 4.5 mM KCl, 1.2 mM $CaCl_2$, 1.2 mM $MgCl_2$, 20 mM HEPES). Total GLP-1 was measured by a specific GLP-1 radioimmunoassay. Statistical comparisons (unpaired Student's t-test) were only performed on 3 h incubations with or without nutrient challenge for each selected buffer. Significant differences are indicated on the graph ($*P < 0.05$)

Figure 19.1 indicates that dramatically different results can be obtained depending on the buffer system used. Use of buffer solutions which are highly stimulatory (e.g. DMEM media which contains amino acids and vitamins), should be avoided because hormone levels for vehicle controls will be greatly elevated and can distort results. In previous experiments, we have favoured KREBS buffer. The test solution pH should be adjusted to 7.0–7.4 and filter sterilised (0.45 µm filter). If solubility of test compounds is an issue (e.g. in the case of lipids or fatty acids) careful monitoring is needed to ensure test compounds remain in solution for the duration of experiments. To measure acute hormone secretion, STC-1 cells should be seeded into 6 well plates at 1.5×10^6 cells in supplemented DMEM media. After 18 h at 37 °C in 5 % CO_2, media should be aspirated and cell monolayers washed 1–2 times with selected buffer. The cells should be acclimatised in buffer for 1 h. Culture media and wash buffers can be kept to ensure that cells have reached basal levels of hormone secretion (Fig. 19.1). After 1 h, the buffer is aspirated and 1 ml test solution added to wells. Cells are then incubated for 1–4 h at 37 °C, 5 % CO_2 (in our experience a period of 3–4 h is optimal for hormone secretion, Fig. 19.2). Positive controls known to stimulate gut hormone secretion and negative controls (buffer alone)

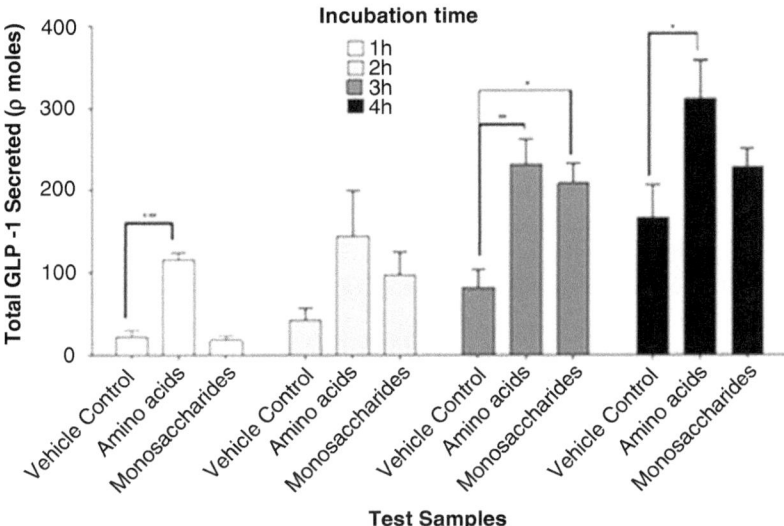

Fig. 19.2 Determining the optimal incubation time to measure levels of Total GLP-1 secreted from STC-1 cells. STC-1 cells (2×10^6) were exposed for various incubation times with stimulatory amino acid solution (40 mM glutamine, 40 mM valine, 40 mM lysine and 40 mM glycine) and monosaccharide solution (40 mM glucose and 40 mM fructose) which were prepared in HEPES buffer. Data were statistically compared using the unpaired Student's t-test and significant differences to vehicle controls are indicated *P < 0.05, **P < 0.01 and ***P < 0.001

should always be included in each experimental unit. Post-incubation, we recommend the addition of 10 μl 10X Halt Protease and Phosphatase Inhibitor (Thermo Fisher Scientific, USA) to protect against gut hormone degradation. Cellular supernatants are collected and store at −80 °C prior to further analysis. It is best practice to immediately transfer solutions to tubes placed on ice and to centrifugate (900×g for 5 min) to remove cellular debris. At least three biological replicates should be performed.

19.6 Assess Viability

Prior to hormone studies, viability or integrity of STC-1 cells challenged with test compounds at various concentrations should be tested. Where test compounds are digested, viability tests should be performed with the inactivated and pH adjusted (7.0–7.4) digestion matrix. An array of methods and kit-based assays are available for cell viability (e.g. measurement of mitochondrial reductase with tetrazolium salts, measurement of cell membrane integrity or signs of cytolysis and/or detection of increased apoptosis). Due the ability of healthy STC-1 cells to change morphology and grow extensions, systems based on electrical impedance may be unsuitable for such tests.

19.7   Experimental Read out

Intracellular levels of the secondary messengers (cAMP, IP3 and Ca^{2+}) are often used as rapid indirect indicators of gut hormone secretion. Several types of assays are currently available to measure cAMP including radioimmunoassay approaches, time-resolved FRET-bioluminescence resonance energy transfer based assays, enzyme fragmentation complementation and cyclic nucleotide gated ion channel coupled assays. Intracellular Ca^{2+} concentrations can be determined directly using membrane permanent calcium-sensitive dyes (e.g. Fluo-3, Fluo-4 or Fura-2) that fluoresce upon exposure to free Ca^{2+}. IP1, a downstream metabolite of IP3, can be used as a surrogate for direct determination of Ca^{2+} as it has a longer half-life. A competitive immunoassay is commercially available that uses terium cryptate-labeled anti-IP1 and D2 labeled IP1 (Cisbso Bioassays, France). IP1 levels only reflect release of Ca^{2+} from intracellular stores and do not account for increased Ca^{2+} levels resulting from influx across the cellular membrane.

Specific assays for individual hormones include qRT-PCR assays and immunoassays (ELISA and RIA). For qRT-PCR, standard methods apply. Result interpretation is important as increases in mRNA levels and intracellular peptide levels may not result in corresponding increases in secretion levels of gut peptides (Hand et al. 2010). A number of sensitive and accurate commercial immunoassays (ELISA and RIA) are available for the detection of CCK, PYY and GLP-1, although multiplexing these analytes has proved difficult. Antibodies may not discriminate between bioactive and inactive forms of gut peptides. Moreover, a large cross-reactivity with gastrin has been observed for many commercially available assays. To overcome these limitations Foltz *et al.* developed an assay based on the ability of dietary components to induce CCK release from STC-1 cells and in turn activate the CCK receptor overexpressed in the chinese hamster ovary cell line (Foltz et al. 2008). Upon activation of the CCK receptor, Ca^{2+} levels increase, which can be visualized with a fluorescent dye. Thus this assay rapidly and directly determines the effect of a nutritional intervention on activation of CCK receptor yielding results which are more physiological relevant.

19.8   Conclusions

The murine STC-1 cell line is regarded as 'best in class' to evaluate secretion of several satiety hormones in response to food components in vitro. The gut hormones CCK, GLP-1, GIP, PYY, pancreatic polypeptide, neurotensin, GLP-2 and oxyntomodulin are all secreted by STC-1 cells. However the heterogeneous nature of STC-1 cells can introduce discrepancies from experiment to experiment. In addition, it's response to stimulants can vary from the native enteroendocrine cell which underlines the importance of substantiating results in vivo.

Acknowledgements TMcC was in receipt of a Teagasc Walsh Fellowship. STC-1 studies in LG laboratory were supported by Enterprise Ireland under Grant Number CC20080001 Food for Health Ireland.

References

Blache P, Le-Nguyen D, Boegner-Lemoine C, Cohen-Solal A, Bataille D, Kervran A (1994) Immunological detection of prohormone convertases in two different proglucagon processing cell lines. FEBS Lett 344(1):65–68

Bollard J, Couderc C, Blanc M, Poncet G, Lepinasse F, Hervieu V, Gouysse G, Ferraro-Peyret C, Benslama N, Walter T, Scoazec J, Roche C (2013) Antitumor effect of everolimus in preclinical models of high-grade gastroenteropancreatic neuroendocrine carcinomas. Neuroendocrinology 97(4):331–340. doi:10.1159/000347063

Cordier-Bussat M, Bernard C, Haouche S, Roche C, Abello J, Chayvialle JA, Cuber JC (1997) Peptones stimulate cholecystokinin secretion and gene transcription in the intestinal cell line STC-1. Endocrinology 138(3):1137–1144

Dyer J, Salmon K, Zibrik L, Shirazi-Beechey S (2005) Expression of sweet taste receptors of the T1R family in the intestinal tract and enteroendocrine cells. Biochem Soc Trans 33(Pt 1):302–305. doi:10.1042/bst0330302

Foltz M, Ansems P, Schwarz J, Tasker MC, Lourbakos A, Gerhardt CC (2008) Protein hydrolysates induce CCK release from enteroendocrine cells and act as partial agonists of the CCK1 receptor. J Agric Food Chem 56(3):837–843

Geraedts M, Troost F, Saris W (2009) Peptide-YY is released by the intestinal cell line STC-1. J Food Sci 74(2):H79–H82. doi:10.1111/j.1750-3841.2009.01074.x

Geraedts M, Troost F, Fischer M, Edens L, Saris W (2011) Direct induction of CCK and GLP-1 release from murine endocrine cells by intact dietary proteins. Mol Nutr Food Res 55(3):476–484. doi:10.1002/mnfr.201000142

Geraedts M, Troost F, Saris W (2012) Addition of sucralose enhances the release of satiety hormones in combination with pea protein. Mol Nutr Food Res 56(3):417–424. doi:10.1002/mnfr.201100297

Glassmeier G, Herzig KH, Hopfner M, Lemmer K, Jansen A, Scherubl H (1998) Expression of functional GABAA receptors in cholecystokinin-secreting gut neuroendocrine murine STC-1 cells. J Physiol 510(Pt 3):805–814

Gribble FM (2012) The gut endocrine system as a coordinator of postprandial nutrient homoeostasis. Proc Nutr Soc 71(4):456–462

Hall WL, Millward DJ, Long SJ, Morgan LM (2003) Casein and whey exert different effects on plasma amino acid profiles, gastrointestinal hormone secretion and appetite. Br J Nutr 89(2):239–248

Hand K, Bruen C, O'Halloran F, Giblin L, Green B (2010) Acute and chronic effects of dietary fatty acids on cholecystokinin expression, storage and secretion in enteroendocrine STC-1 cells. Mol Nutr Food Res 54(Suppl 1):S93–s103. doi:10.1002/mnfr.200900343

Hand K, Bruen C, O'Halloran F, Panwar H, Calderwood D, Giblin L, Green B (2013) Examining acute and chronic effects of short- and long-chain fatty acids on peptide YY (PYY) gene expression, cellular storage and secretion in STC-1 cells. Eur J Nutr 52(4):1303–1313. doi:10.1007/s00394-012-0439-9

Kieffer TJ, Huang Z, McIntosh CH, Buchan AM, Brown JC, Pederson RA (1995) Gastric inhibitory polypeptide release from a tumor-derived cell line. Am J Physiol 269(2 Pt 1):E316–E322

Komarnytsky S, Cook A, Raskin I (2011) Potato protease inhibitors inhibit food intake and increase circulating cholecystokinin levels by a trypsin-dependent mechanism. Int J Obes (Lond) 35(2):236–243

Liou A, Chavez D, Espero E, Hao S, Wank S, Raybould H (2011a) Protein hydrolysate-induced cholecystokinin secretion from enteroendocrine cells is indirectly mediated by the intestinal oligopeptide transporter PepT1. Am J Physiol Gastrointest Liver Physiol 300(5):G895–G902. doi:10.1152/ajpgi.00521.2010

Liou AP, Lu X, Sei Y, Zhao X, Pechhold S, Carrero RJ, Raybould HE, Wank S (2011b) The G-protein-coupled receptor GPR40 directly mediates long-chain fatty acid-induced secretion of cholecystokinin. Gastroenterology 140(3):903–912

Mangel AW, Prpic V, Scott L, Liddle RA (1994) Inhibitors of ATP-sensitive potassium channels stimulate intestinal cholecystokinin secretion. Peptides 15(8):1565–1566

Miyata M, Kurogi M, Oda M, Saitoh O (2014) Effect of five taste ligands on the release of CCK from an enteroendocrine cell line, STC-1. Biomed Res 35(2):171–176

Palazzo M, Balsari A, Rossini A, Selleri S, Calcaterra C, Gariboldi S, Zanobbio L, Arnaboldi F, Shirai Y, Serrao G, Rumio C (2007) Activation of enteroendocrine cells via TLRs induces hormone, chemokine, and defensin secretion. J Immunol 178(7):4296–4303

Pasman WJ, Heimerikx J, Rubingh CM, van den Berg R, O'Shea M, Gambelli L, Hendriks HF, Einerhand AW, Scott C, Keizer HG, Mennen LI (2008) The effect of Korean pine nut oil on in vitro CCK release, on appetite sensations and on gut hormones in post-menopausal overweight women. Lipids Health Dis 7:10

Ratineau C, Plateroti M, Dumortier J, Blanc M, Kedinger M, Chayvialle J, Roche C (1997) Intestinal-type fibroblasts selectively influence proliferation rate and peptide synthesis in the murine entero-endocrine cell line STC-1. Differentiation 62(3):139–147. doi:10.1046/j.1432-0436.1997.6230139.x

Reimann F, Habib AM, Tolhurst G, Parker HE, Rogers GJ, Gribble FM (2008) Glucose sensing in L cells: a primary cell study. Cell Metab 8(6):532–539

Rindi G, Grant S, Yiangou Y, Ghatei M, Bloom S, Bautch V, Solcia E, Polak J (1990) Development of neuroendocrine tumors in the gastrointestinal tract of transgenic mice. Heterogeneity of hormone expression. Am J Pathol 136(6):1349–1363

Saitoh O, Hirano A, Nishimura Y (2007) Intestinal STC-1 cells respond to five basic taste stimuli. Neuroreport 18(18):1991–1995. doi:10.1097/WNR.0b013e3282f242d3

Tanaka T, Katsuma S, Adachi T, Koshimizu T, Hirasawa A, Tsujimoto G (2008) Free fatty acids induce cholecystokinin secretion through GPR120. Naunyn Schmiedebergs Arch Pharmacol 377(4–6):523–527. doi:10.1007/s00210-007-0200-8

Tolhurst G, Reimann F, Gribble F (2009) Nutritional regulation of glucagon-like peptide-1 secretion. J Physiol 587(Pt 1):27–32. doi:10.1113/jphysiol.2008.164012

Veldhorst MA, Nieuwenhuizen AG, Hochstenbach-Waelen A, van Vught AJ, Westerterp KR, Engelen MP, Brummer RJ, Deutz NE, Westerterp-Plantenga MS (2009) Dose-dependent satiating effect of whey relative to casein and to soy. Physiol Behav 96(4–5):675–682

Wang Y, Chandra R, Samsa LA, Gooch B, Fee BE, Cook JM, Vigna SR, Grant AO, Liddle RA (2011) Amino acids stimulate cholecystokinin release through the Ca2+-sensing receptor. Am J Physiol Gastrointest Liver Physiol 300(4):G528–G537

Chapter 20
NCI-H716 Cells

Jeffrey Gagnon and Patricia L. Brubaker

Abstract The endocrine response to nutrient ingestion is vital to the maintenance of energy homeostasis in the body. Glucagon like peptide-1 (GLP-1) is one such hormone that is released from L-cells of the distal small intestine and colon in response to meal ingestion. GLP-1 acts on various systems in the body to enhance glucose-stimulated insulin secretion, delay gastric emptying and promote satiety. As such, elevating the levels of active GLP-1 in the circulation, as well as enhancing GLP-1 bioactivity, is the basis of several recent anti-diabetic medications. Gaining an understanding of how GLP-1 secretion is regulated at the cellular level requires in vitro L-cell models. NCI-H716 is a cell line derived from ascites fluid of a colorectal adenocarcinoma from a 33 year old Caucasian male. This cell line is currently the only human model available for the in vitro study of GLP-1 regulation and is the topic of the following chapter. This chapter will cover the origin, characteristics and methods for using this model. Comparisons are then made between other available in vitro GLP-1 models.

Keywords GLP-1 • Endocrine hormone • Nutrient • Neurotransmitter • Receptor • Cell line • Cell culture • L-cell • Secretion • Enteroendocrine • Human

20.1 Introduction

Glucagon like peptide 1 (GLP-1) is a key hormone in the regulation of nutrient metabolism. Its release from intestinal enteroendocrine L-cells is stimulated by the ingestion of nutrients (both directly, through luminal nutrient sensing, and indirectly, via parasympathetic innervation). The L-cells are distributed primarily in the distal

J. Gagnon
Department of Physiology, University of Toronto, Toronto, ON, Canada

P.L. Brubaker (✉)
Department of Physiology, University of Toronto, Toronto, ON, Canada

Department of Medicine, University of Toronto, Toronto, ON, Canada
e-mail: p.brubaker@utoronto.ca

© The Author(s) 2015
K. Verhoeckx et al. (eds.), *The Impact of Food Bio-Actives on Gut Health*,
DOI 10.1007/978-3-319-16104-4_20

small intestine and colon. Once released into the circulation, GLP-1 acts to promote satiety as well as the storage of ingested nutrients. A well-established and pharma-cologically-targeted aspect of GLP-1 action is its role in potentiating glucose-stimulated insulin secretion from the β cells of the pancreas. This action creates a larger insulin response to ingested glucose as compared to isoglycemic levels of glucose delivered intravenously, and is known as the incretin effect. The GLP-1 responses to various dietary nutrients and their respective intracellular mechanisms of action on the L-cell have been, and continue to be, intensively studied using in vitro GLP-1-secreting cell models. One such model is the human colonic cell line, NCI-H716. This review will describe the origin, utility and associated methods of use of this cell line in the study of GLP-1 biology.

20.2   Origin

NCI-H716 cells were one of several colorectal cancer cell lines originally described in 1987 (Park et al. 1987). This particular cell line was developed from cells har-vested in ascites fluid of a 33-year old Caucasian male with colorectal cancer. Interestingly, of the 14 cell lines described in this study, NCI-H716 was the only line that contained dense-core granules, which are characteristic of endocrine cells. While it was unclear as to the contents in these granules at the time of discovery, later work in several laboratories elucidated the hormones produced by these cells. Early studies with these cells demonstrated the expression of both the secretory granule marker, chromogranin A, and mucin, as well as receptors for gastrin, soma-tostatin and serotonin (de Bruine et al. 1992). This not only confirmed that these cells had an endocrine phenotype but suggested that they were also sensitive to other gastrointestinal hormones. After several years of being used as model of tumour/endocrine differentiation, Reimer et al. demonstrated both expression and regulated secretion of glucagon-like peptide-1 (GLP-1) by these cells (Reimer et al. 2001). In that study, NCI-H716 cells were found to secrete high levels of active GLP-1 in amounts that could be easily measured by collecting cell culture medium and cell lysates. They also showed that nutrients, including fatty acids and protein hydrolysate, dose-dependently stimulated GLP-1 secretion by these cells (Reimer et al. 2001). Following this, work by our group elucidated roles for several hor-mones (Anini and Brubaker 2003b; Lim et al. 2009), neurotransmitters (Anini and Brubaker 2003a), fatty acids (Lauffer et al. 2009) and anti-diabetic medications (Mulherin et al. 2011) in the regulation of GLP-1 secretion. To date, the NCI-H716 cells remain the only characterized model of the human intestinal L-cell.

20.3   Features and Mechanisms

NCI-H716 cells are considered pseudo-diploid with an average chromosome count of 61 (Park et al. 1987). Electron micrographs shown in this initial study clearly demonstrate many dense-core granules. In culture, these cells appear spherical and undifferentiated, and they grow in suspension, occasionally forming clumps. Plating the cells on a basement membrane-coated surface allows the cells to attach and grow horizontally. Although this plating strategy was initially reported to allow the cells to become differentiated towards an endocrine lineage (Reimer et al. 2001), studies by our laboratory and other groups have indicated that plating does not change the levels of GLP-1 in these cells (Anini and Brubaker 2003a; Cao et al. 2003).

Anini and Brubaker demonstrated expression of the leptin receptor in NCI-H716 cells by western blot and fluorescent immunocytochemistry (Anini and Brubaker 2003b). This study further showed that leptin stimulates GLP-1 secretion by these cells, and that the hyperleptinemia that occurs during obesity leads to leptin resistance and impaired GLP-1 secretion in this condition. In that same year, Anini and Brubaker also demonstrated expression of the M1, M2 and M3 muscarinic receptors in the NCI-H716 cells by western blot and fluorescent immunocytochemistry (Anini and Brubaker 2003a). Herein, they also established a role for M1 muscarinic regulation of GLP-1 secretion. Shortly after, Lim et al. showed that NCI-H716 cells express the insulin receptor, by semi-quantitative reverse transcriptase PCR and fluorescent immunocytochemistry (Lim et al. 2009). They further demonstrated that, while insulin was able to stimulate GLP-1 secretion acutely, extended pre-incubation with insulin led to a loss in the ability of these cells to respond to other GLP-1 secretagogues. Hence, the hyperinsulinemia that occurs in patients with type 2 diabetes mellitus (T2DM) may also result in impaired GLP-1 secretion.

The ability of the L-cell to respond to luminal nutrients is a vital component of the incretin system. GLP-1 regulation by ingested nutrients through specific receptors has been demonstrated using the NCI-H716 cell line. In the initial NCI-H716 and GLP-1 secretion study, the authors showed that meat hydrolysate stimulated GLP-1 release (Reimer et al. 2001). Several years later, they also demonstrated mRNA expression of several amino acid transporters by these cells, including Y^+LAT2, ASCT2, and ATA-2 (Reimer 2006).

Luminal glucose sensing by the L-cells has also been demonstrated through the sweet taste receptor system. NCI-H716 cells express several of the type 1 taste receptors and α-gustducin, which is a key component of the taste transduction pathway (Jang et al. 2007). In this study, knockdown of α-gustducin prevented the glucose-stimulated GLP-1 response by NCI-H716 cells. While this study and others (Zhang et al. 2012) have demonstrated that glucose can stimulate GLP-1 secretion from NCI-H716 cells, the presence of a glucose transporter has yet to be demonstrated.

Lipid sensing by the NCI-H716 cell line has been demonstrated by several groups (Chen et al. 2012; Lauffer et al. 2009). Expression of the monoacylglycerol-sensing G-protein coupled receptor, GPR119, by RT-PCR in NCI-H716 cells was shown by Lauffer et al. (2009). While examining the effects of milk products on GLP-1 secretion, Chen et al. also demonstrated expression of fatty acid transport protein 4 in the NCI-H716 cells by quantitative RT-PCR (Chen and Reimer 2009).

The above studies clearly indicate that NCI-H716 cells possess many of the receptors and intracellular machinery required to sense hormones, neurotransmitters and nutrients, and to release GLP-1 in response to these agents.

20.4   Stability/Consistency/Reproducibility

The increasing use of the NCI-H716 cells as a model of the human intestinal L-cell necessitates an examination of the consistency of findings between different research groups using these cells. In terms of GLP-1 secretion, many groups now present secretion data relative to an untreated control, making interpretation of actual percent secretion difficult. However, early studies with the NCI-H716 cells demonstrated percent secretion to be in the range of 1–15 % of the total well/plate content of GLP-1 over a 2 h incubation period (Reimer et al. 2001; Anini and Brubaker 2003a; Lim et al. 2009; Lauffer et al. 2009). Regarding the error in replicates of GLP-1 secretion, these studies report standard error to be ~10 % (of the mean).

Some groups have examined the effect of similar treatments on GLP-1 secretion. Anini et al. found that the protein kinase C activator, phorbol 12-myristate 13-acetate (PMA) caused a 2.5-fold stimulation in GLP-1 secretion (Anini and Brubaker 2003a), whereas Reimer et al. saw a 4.3-fold stimulation with the same dose and duration of incubation (Reimer et al. 2001). While both studies showed a robust stimulation of GLP-1 secretion, the difference in magnitude may be due to differences in media/buffers used during the secretion experiment or in sample preparation (both subjects of further discussion below).

20.5   Relevance to the Human L-Cell In Vivo

Since NCI-H716 cells were derived from a human sample, they are of particular significance to the study of human GLP-1 secretion. However, while these cells do secrete GLP-1 in response to nutrients, as found in vivo, it must be recognized that they are derived from a tumour cell which was likely to have been relatively undifferentiated initially. Furthermore, one study examined the ability of NCI-H716 cells to regulate expression of reporter plasmids encoding human and rat *proglucagon* (i.e. the GLP-1 prohormone gene). Surprisingly, no basal promoter activity was observed even in the presence of several positive regulators of rodent *proglucagon* gene expression (Cao et al. 2003). In addition, compounds that are known to enhance *proglucagon* gene expression in rodent models failed to stimulate the endogenous human *proglucagon* gene in NCI-H716 cells. These findings suggest that either the NCI-H716 cells do not regulate *proglucagon* gene expression normally or that there are marked species-specific differences in the regulation of *proglucagon* expression.

20.6   General Protocol

20.6.1   Cell Maintenance Protocol

NCI-H716 cells are normally grown in suspension culture. Our group uses Roswell Park Memorial Institute (RPMI) media containing L-glutamine and Phenol Red. This media is further supplemented to contain 10 % fetal bovine serum (FBS) and

100 units/ml each of penicillin and streptomycin ('complete RPMI'). Cells can be grown in suspension in cell culture flasks and should be kept at a density of ~500,000 cells/ml to ensure a rapid growth rate; lower densities may cause stalled growth. Under these conditions, cells double every ~3 days and should be passaged once a week. To passage, cells do not require any enzymatic digestion; any small clumps of cells are simply dissociated using repeated serological pipetting.

20.6.2 Experimental Protocol for Test Compounds

As NCI-H716 cells require a basement membrane matrix coating to attach to cell culture plates, preparation must occur before seeding cells for an experiment. Matrigel (BD Biosciences) is provided as a concentrate with lot-dependent concentration. It is stored frozen and gels at 15 °C. Matrigel is therefore thawed on ice (stock aliquots of the concentrate can be thawed overnight on ice in a 4 °C fridge, as per manufacturer's directions). Plates or wells are then coated with 0.5 g/ml Matrigel diluted in ice-cold sterile Hanks Buffered Saline Solution (HBSS), with the area of the surface being coated dictating how much diluted Matrigel is required (generally ≥200 µl/well in a 24-well plate, ≥ 800 µl/well in a 6-well plate, etc.). After allowing the Matrigel to gel for 1 h at room temperature, the plates/wells are then rinsed with HBSS and allowed to dry for 30 min; the surface is then ready for seeding of the cells. Cells should be reconstituted in complete RPMI to ~200,000 cells/ml (1 ml for each well in a 24-well plate, 2 ml for each in a 12-well plate, etc.). This density is optimized to allow ~2 days of growth before experimental treatments. On the day of treatment, cells are rinsed with low-serum RPMI media (i.e. 0.5 % FBS) then incubated with treatments in low-serum RPMI. For GLP-1 secretion studies, 2-h of incubation time is ideal as any possible effects on de novo GLP-1 synthesis are unlikely to be observed in this relatively short time-frame. Over a 2-h incubation period, a 24-well plate treated with vehicle (control) will have ~300 pg of GLP-1 (total) per well in the collected media, with ~7,000 pg of GLP-1 in the cell lysate. Based on these values, the percent secretion (i.e. pg in media/(pg in media+cells)×100) is 4.1 %.

Similar plating densities can be used for RNA and protein extraction experiments. A larger surface area can be used for such studies (i.e. 6-well, 10-cm), depending on expression levels for the mRNA transcript or protein of interest. From a 6-well plate containing ~800,000 cell/well, ~20 µg of RNA and ~60 µg of protein can be collected.

20.7 Assess Viability

To ensure treatments do not affect cell health or viability, standard tests should be employed. These tests include the 3-(4,5-dimethylthiazol-2-yl)-2,5-diphenyltetrazolium (MTT) assay and neutral red uptake assay. Both of these tests can be done within a few hours of an experimental treatment, and should include a positive control, such as hydrogen peroxide, to ensure that the assay is reliable.

20.8 Experimental Readout

Sample processing from NCI-H716 cells is dependent on the nature of the experiment being conducted. Samples collected for GLP-1 secretion assay, as described in the experimental protocol above, are processed as follows.

Immediately after the 2-h incubation with treatments is complete, media is collected and centrifuged for 5 min at 1,000×g and 4 °C to remove any floating cells. This media is then decanted into a new tube containing 1 % trifluoroacetic acid (TFA) to make a final TFA concentration of 0.1 %. This acidic environment prevents protease activity and prepares the sample for the subsequent peptide purification step. The culture plate with the cells is simultaneously placed on ice and then 1 ml of ice-cold acidic cell lysis solution (i.e. 1 % TFA, 1 N hydrochloric acid, 5 % formic acid, and 1 % NaCl) is added. The cells are scraped and wells are rinsed with an additional volume of lysis buffer. These cell lysates are then sonicated, pelleted by centrifugation, and the supernatants collected on ice. However, of note, media and cells may be collected in a variety of different buffers, as indicated by the desired assay protocol, as long as proteolysis is prevented.

Depending on the method of hormone assay, samples may require additional purification. Using a total GLP-1 radio-immuno assay (RIA; i.e. Millipore), we have found that additional purification by C18 solid phase peptide extraction removes salts and proteases that may interfere with the assay. Peptides are thus purified using Sep Pak C18 Classic columns (Waters Associates) with 0.1 % TFA water (Solution A) and 0.1 % TFA/80 % isopropyl alcohol (Solution B), as follows: (1) 4 ml Solution B (wetting/clearing); (2) 4 ml Solution A (equilibrating); (3) sample loading; (4) 4 ml Solution A (washing); and (5) 4 ml Solution B (sample elution). A portion of the total 4 ml peptide eluate is then dried under vacuum in preparation for the RIA. Once samples have been purified using this method, they can be sealed and stored at −20 °C for later analysis.

Media and cell GLP-1 content determined from the assay can be used to calculate percent GLP-1 secretion, as described above. These percent secretion values may differ slightly from 1 week to the next due to inter-assay variability. To control for differences in percent secretion, treatment groups are often normalized to a vehicle-treated control; as such, data often appears "relative to control" in papers examining GLP-1 secretion, although absolute values should always be included in the publication to facilitate comparison between studies. The use of an appropriate control in each experiment is critical. The control treatment group must thus contain the solvent in which the treatment group is dissolved at an identical final concentration. This may include dimethyl sulfoxide, ethanol or a buffer solution. In addition, experiments should also be designed to include a positive control for GLP-1 secretion. Strong activators of cAMP signaling, such as forskolin, have thus been used by several groups to demonstrate cell responsiveness to a known secretagogue in parallel with experimental treatment groups.

In GLP-1 secretion experiments with NCI-H716 cells, at least three biological replicates should be produced for each treatment and control group. Additionally, any given experiment should be repeated at least twice to ensure statistical power for subsequent analysis, thereby constituting a minimum of six replicates.

For RNA and protein extraction, standard methods apply. For RT-PCR and qRT-PCR studies, the target gene of interest should be determined relative to an internal control, such as 18S ribosomal RNA. However, quantification of appropriate reference genes for a particular treatment must also be done to ensure that the internal control does not change in response to the experimental treatment. Similarly, protein expression analysis by immunoblot should also include a suitable validated control, such as actin, for protein loading.

20.9   Conclusions

The ability of the L-cell to respond to nutrient, hormonal and neurotransmitter cues is vital in the regulation of GLP-1 secretion and, ultimately, glucose homeostasis. Strategies targeting the GLP-1 system in humans have garnered great interest and application in the treatment of T2DM (Drucker 2011). Cell systems that enable a direct mechanistic examination of GLP-1 regulation have been and continue to be essential tools in developing novel GLP-1 therapies. The NCI-H716 cell line is the only human cell line currently available to study GLP-1. This places it in an important position to relate the effects of various treatments to human biology, although experimental effects should always be cross-examined through the use of additional in vitro and in vivo models to strengthen the hypothesis. The NCI-H716 cells are thus a useful tool to study the control of GLP-1 release and the underlying cell signaling pathways as well as, potentially, *proglucagon* gene expression. Overall, they are an excellent model to investigate the regulation of human GLP-1.

Acknowledgement JG was supported by a Post-doctoral Fellowship from the Canadian Institutes of Health Research and PLB by the Canada Research Chairs Program. Studies on GLP-1 in the Brubaker laboratory are supported by operating grants from the Canadian Diabetes Association (#2973OG-3-13-4024-PB) and the Natural Sciences and Engineering Research Council (RGPIN418).

References

Anini Y, Brubaker PL (2003a) Muscarinic receptors control glucagon-like peptide 1 secretion by human endocrine L cells. Endocrinology 144(7):3244–3250. doi:10.1210/en.2003-0143
Anini Y, Brubaker PL (2003b) Role of leptin in the regulation of glucagon-like peptide-1 secretion. Diabetes 52(2):252–259
Cao X, Flock G, Choi C, Irwin DM, Drucker DJ (2003) Aberrant regulation of human intestinal proglucagon gene expression in the NCI-H716 cell line. Endocrinology 144(5):2025–2033. doi:10.1210/en.2002-0049

Chen Q, Reimer RA (2009) Dairy protein and leucine alter GLP-1 release and mRNA of genes involved in intestinal lipid metabolism in vitro. Nutrition 25(3):340–349. doi:10.1016/j. nut.2008.08.012

Chen Y, Li ZY, Yang Y, Zhang HJ (2012) Uncoupling protein 2 regulates glucagon-like peptide-1 secretion in L-cells. World J Gastroenterol 18(26):3451–3457. doi:10.3748/wjg.v18.i26.3451

de Bruine AP, Dinjens WN, Pijls MM, vd Linden EP, Rousch MJ, Moerkerk PT, de Goeij AF, Bosman FT (1992) NCI-H716 cells as a model for endocrine differentiation in colorectal cancer. Virchows Arch B Cell Pathol Incl Mol Pathol 62(5):311–320

Drucker DJ (2011) Incretin-based therapy and the quest for sustained improvements in beta-cell health. Diabetes Care 34(9):2133–2135. doi:10.2337/dc11-0986

Jang HJ, Kokrashvili Z, Theodorakis MJ, Carlson OD, Kim BJ, Zhou J, Kim HH, Xu X, Chan SL, Juhaszova M, Bernier M, Mosinger B, Margolskee RF, Egan JM (2007) Gut-expressed gustducin and taste receptors regulate secretion of glucagon-like peptide-1. Proc Natl Acad Sci U S A 104(38):15069–15074. doi:10.1073/pnas.0706890104

Lauffer LM, Iakoubov R, Brubaker PL (2009) GPR119 is essential for oleoylethanolamide-induced glucagon-like peptide-1 secretion from the intestinal enteroendocrine L-cell. Diabetes 58(5):1058–1066. doi:10.2337/db08-1237

Lim GE, Huang GJ, Flora N, LeRoith D, Rhodes CJ, Brubaker PL (2009) Insulin regulates glucagon-like peptide-1 secretion from the enteroendocrine L cell. Endocrinology 150(2):580–591. doi:10.1210/en.2008-0726

Mulherin AJ, Oh AH, Kim H, Grieco A, Lauffer LM, Brubaker PL (2011) Mechanisms underlying metformin-induced secretion of glucagon-like peptide-1 from the intestinal L cell. Endocrinology 152(12):4610–4619. doi:10.1210/en.2011-1485

Park JG, Oie HK, Sugarbaker PH, Henslee JG, Chen TR, Johnson BE, Gazdar A (1987) Characteristics of cell lines established from human colorectal carcinoma. Cancer Res 47(24 Pt 1):6710–6718

Reimer RA (2006) Meat hydrolysate and essential amino acid-induced glucagon-like peptide-1 secretion, in the human NCI-H716 enteroendocrine cell line, is regulated by extracellular signal-regulated kinase1/2 and p38 mitogen-activated protein kinases. J Endocrinol 191(1): 159–170. doi:10.1677/joe.1.06557

Reimer RA, Darimont C, Gremlich S, Nicolas-Metral V, Ruegg UT, Mace K (2001) A human cellular model for studying the regulation of glucagon-like peptide-1 secretion. Endocrinology 142(10):4522–4528. doi:10.1210/endo.142.10.8415

Zhang H, Li J, Liang X, Luo Y, Zen K, Zhang CY (2012) Uncoupling protein 2 negatively regulates glucose-induced glucagon-like peptide 1 secretion. J Mol Endocrinol 48(2):151–158. doi:10.1530/JME-11-0114

Chapter 21
Murine GLUTag Cells

Manuel Gil-Lozano and Patricia L. Brubaker

Abstract The incretin hormone, glucagon-like peptide-1 (GLP-1) serves as a link between ingested nutrients and insulin secretion. Because of the anti-diabetic properties exerted by GLP-1, several derivative drugs are currently in the market for the treatment of patients with type 2 diabetes. Over the past few years, several cell lines have been established as useful models for the study of the GLP-1-secreting enteroendocrine L-cells. This review focuses on the murine GLUTag cell line, derived from colonic tumours of transgenic mice expressing large T antigen under the control of the *proglucagon* promoter. These cells are widely used to examine the effects of food components on GLP-1 secretion, as well as the signaling pathways underlying nutrient-induced GLP-1 release. The effects of different food components, as well as of nutrient-related signals, on GLP-1 secretion, and the protocols for the maintenance and use of the murine GLUTag L-cell line will be discussed.

Keywords Cell line • Cell culture • Endocrine hormone • Enteroendocrine • Food components • GLP-1 • L-cell • Murine • Nutrient • Neurotransmitter • Receptor • Secretion

21.1 Introduction

The potent insulinotrophic activity of the intestinal hormone, glucagon-like peptide-1 (GLP-1), in the perfused rat pancreas was first described almost 30 years ago. This finding was rapidly followed by the demonstration that physiological levels of the peptide also stimulate insulin secretion in a glucose-dependent manner in humans. Since then, our knowledge of the physiological effects of GLP-1 has vastly increased and several anti-diabetic properties such as suppression of glucagon

M. Gil-Lozano
Department of Physiology, University of Toronto, Toronto, ON, Canada

P.L. Brubaker (✉)
Department of Physiology, University of Toronto, Toronto, ON, Canada

Department of Medicine, University of Toronto, Toronto, ON, Canada
e-mail: p.brubaker@utoronto.ca

© The Author(s) 2015
K. Verhoeckx et al. (eds.), *The Impact of Food Bio-Actives on Gut Health*,
DOI 10.1007/978-3-319-16104-4_21

secretion, induction of satiety, reduction of gastric activity and potentiation of pancreatic β-cell mass have also been attributed to the peptide. Overall, these effects made GLP-1 a promising novel therapeutic tool for the treatment of type 2 diabetes mellitus, and GLP-1 derivative drugs have been successfully used to reduce glycemia in patients with type 2 diabetes since 2005. These anti-diabetic agents include agonists of the GLP-1 receptor long-lasting analogues of the native peptide and inhibitors of the enzyme responsible for the rapid degradation of the active forms of GLP-1 (Dong and Brubaker 2012). Interestingly, potentiation of the endogenous secretion of the peptide is another possible therapeutic approach which has gained much attention lately; however, a better understanding of the processes regulating GLP-1 secretion by the intestinal L-cell (an enteroendocrine type of cell which is the major source of circulating GLP-1 in the body) is still required.

Over the past few years, the establishment of different enteroendocrine cell lines has markedly improved our understanding of the different signals that modulate GLP-1 expression and secretion, as well as the intracellular pathways activated by these signals. The use of cell lines has also helped to troubleshoot the inherent difficulties derived from the use of intestinal primary cell cultures, such as low L-cell density and a high risk of bacterial contamination. Among all cell line models for the study of GLP-1 secretion, the murine GLUTag cells are the most widely used as well as the most specific. Thus, GLUTag cells not only have been demonstrated to appropriately respond to the same stimuli that trigger GLP-1 secretion in vivo, but they also have been shown to closely, although not fully, mirror the profile of genes and proteins expressed in non-immortalized L-cells. The present chapter will cover the modulation of GLP-1 secretion by the GLUTag cells, primarily focusing on the effect of food components, as well as the methodology for the use of this cell line.

21.2 Origin

GLUTag cells derive from a colonic tumour of a transgenic mouse expressing SV40 large T antigen under the control of approximately 2,300-base pairs of the promoter for *proglucagon*, the gene that encodes GLP-1 amongst other bioactive peptides. The mice developed endocrine carcinoma of the large bowel, the cells of which were positive for immunoreactive GLP-1 and cholecystokinin but were negative for other hormones, including adrenocorticotrophin, β-endorphin, calcitonin, corticotrophin-releasing hormone, gastrin, growth hormone-releasing hormone, insulin, pancreatic polypeptide, serotonin, somatostatin and vasoactive intestinal peptide, suggesting an L-cell lineage for these tumour cells. Fragments of the colonic tumours were subcutaneously propagated in nude mice, and individual cells from one of these tumours were isolated and single-cell cloned (Drucker et al. 1994). The cultured cells displayed pleomorphic nuclei with prominent nucleoli, and well-developed cytoplasmic organelles along with abundant secretory granules. GLUTag cells were found to express both *proglucagon* and cholecystokinin transcripts. The cells secrete GLP-1 in response to incubation with protein kinase A and protein kinase C activators, whereas *proglucagon* expression was only induced by activators of protein kinase A, confirming previous studies in primary cells (Drucker et al. 1994). As described for the normal intestinal L-cell, the predominant GLP-1 form produced by the GLUTag cells is GLP-1(7-36)-amide (Drucker et al. 1994). Importantly, similar profiles in the expression of genes that encode mediators of nutrient-induced GLP-1 secretion, including glucokinase, sodium-glucose linked transporter 1 (SGLT1) and G protein-coupled receptors such as GPR40, GPR41, GPR119, GPR120 and GPR131 (TGR5), have been found in GLUTag and non-immortalized L-cells (Reimann et al. 2008; Tolhurst et al. 2012). The GLUTag cells are thus widely considered as the best in vitro model of the murine L-cell.

21.3   Features and Mechanisms

21.3.1   Regulation of GLUTag Cell Secretory Activity by Nutrients

Nutrient ingestion is the most powerful mechanism for triggering GLP-1 secretion by the intestinal L-cell (Dong and Brubaker 2012). Since the L-cells are open-type enteroendocrine cells residing in the intestinal epithelium, with their apical surfaces and associated microvilli open into the gut lumen, they can directly sense and respond to luminal nutrients. The effects of the major groups of food components on GLUTag cell physiology are discussed below.

Carbohydrates GLUTag cells are highly sensitive to glucose, such that cells that are propagated at 5 mmol/L concentration of glucose become quiescent and hyper-polarized when this monosaccharide is removed from the medium. Accordingly, when glucose is added back to the bath solution, the membrane conductance decreases, and the frequency of action potentials in the cells, as well as the secretion of GLP-1 are strongly enhanced, even at very low glucose concentrations (Reimann and Gribble 2002). Initial studies demonstrated that glucose metabolism was associated with the closure of K^+ channels sensitive to ATP (K_{ATP}), resulting in increases in membrane conductance and enhanced GLP-1 release (Reimann and Gribble 2002). However, recent studies have indicated a more important role for a metabolism-independent glucose-sensing pathway in GLUTag cells (Parker et al. 2012a). Hence, GLUTag cells express two sodium-glucose co-transporters, SGLT1 and SGLT3a, which have been found to increase glucose uptake inducing an inward current of Na^+ ions that results in cell depolarization and stimulation of GLP-1 secretion. These effects are appropriately diminished by SGLT inhibitors (Parker et al. 2012a). SGLTs also appear to be involved in the effects of non-metabolisable sugars, such as methyl-α-glucopyranoside, on GLP-1 secretion by the GLUTag cell line, but not in the effects of dietary sugars such as fructose that are not substrates for these transporters. Expression of SGLT1 in the intestinal epithelium is enhanced by dietary carbohydrates, a process believed to be mediated by type 1 taste receptors and gustducin (Margolskee et al. 2007). Interestingly, these receptors are also expressed by the

GLUTag cells and they seem to mediate the effects of artificial sweeteners in this cell line, although they may not be relevant in vivo (Fujita et al. 2009). The facilitative glucose transporters, GLUT, are also involved in the stimulatory effects of glucose on GLP-1 secretion in the GLUTag cell line, while SGLT1 appears to be the most important glucose-sensing mechanism for the L-cells in vivo. The different glucose-sensing mechanisms exhibited by GLUTag and non-immortalized L-cells may explain why GLUTag cells respond to small changes in glucose levels when they are incubated at very low concentrations of glucose (<1 mmol/L) but they are not sensitive to additional increments in glucose concentration (5–25 mmol/L; Reimann and Gribble 2002), as opposed to the dose-dependent effect of glucose in primary L-cells (Reimann et al. 2008).

Lipids Free fatty acids are potent stimulators of GLUTag cell activity, the magnitude of the effects being proportional to their chain length (Iakoubov et al. 2007). Furthermore, GLP-1 secretion is stimulated by long-chain monounsaturated fatty acids, such as oleic acid, but not by similar length saturated fatty acids such as palmitic acid, in agreement with previous observations in primary L-cell cultures. The effects of oleic acid on GLP-1 secretion are known to be mediated by the atypical isozyme protein kinase C ζ in GLUTag cells, as also found in vivo (Iakoubov et al. 2007; Dong and Brubaker 2012), while a role for fatty acid transport protein 4 in the uptake of oleic acid by GLUTag cells has also been demonstrated (Poreba et al. 2012).

The activation of the Gq-coupled receptors, GPR40 and GPR120, by long-chain unsaturated fatty acids has also been associated with enhanced GLP-1 release in vivo (Edfalk et al. 2008; Hirasawa et al. 2005). However, although GLUTag cells have also been reported to express these receptors (Lauffer et al. 2009), their role in the physiology of this cell line has not been extensively studied to date. In contrast, the Gs-coupled receptor, GPR119, has been shown to mediate the stimulatory effects on GLP-1 secretion of oleoylethanolamide, an endogenously-occurring oleic acid derivate (Lauffer et al. 2009). Furthermore, synthetic GPR119 agonists have been found to stimulate GLP-1 secretion both in vivo and in GLUTag cells (Chu et al. 2008). Another G-protein-coupled receptor, TGR5, has also been recently demonstrated to play a role in the stimulatory effects of bile acids on GLP-1 secretion in GLUTag cells, as previously also found in vivo (Parker et al. 2012b; Dong and Brubaker 2012).

Finally, short-chain fatty acids (SCFAs), including acetate, butyrate, and propionate, are generated by the gut microbiota through the digestion of complex carbohydrates, such as dietary fibers. It has been recently reported that SCFAs stimulate GLP-1 secretion via the G-protein-coupled receptor GPR43 (Tolhurst et al. 2012), establishing a novel and exciting link between gut microflora and gastrointestinal hormones. Unfortunately, the GLUTag cell line appears to be an inappropriate model to study the effects of SCFAs, since they express barely detectable levels of GPR43 and do not release GLP-1 in response to 1 or 10 mmol/L SCFAs (Tolhurst et al. 2012).

Proteins Protein hydrolysates stimulate GLP-1 secretion and production from a variety of cell lines, including GLUTag cells (Hira et al. 2009). When the effects of individual amino acids were examined, several were found to stimulate GLP-1 secretion, with a particularly robust effect promoted by glutamine. Two major mechanisms have been invoked in the stimulatory effects of glutamine on GLUTag cells: electrogenic sodium-coupled amino acid uptake which leads to membrane depolarization and voltage-gated calcium entry, and an elevation of intracellular cAMP levels (Tolhurst et al. 2011). The synergy of these two mechanisms appears to be a particularly strong stimulus for GLP-1 release in vitro. Other amino acids, such as alanine and glycine, have been found to trigger GLP-1 release from GLUTag cells via a pathway involving ionotrophic glycine receptors (Gameiro et al. 2005), whereas the G protein-coupled receptor family C group 6 subtype A receptor has been implicated in the effects on GLP-1 secretion elicited by L-ornithine (Oya et al. 2013).

21.3.2 Regulation of GLUTag Cell Secretory Activity by Other Signals

Despite the predominant role of nutrients in the regulation of GLP-1 release, a number of other signals have been identified that exert stimulatory effects on the primary L-cell as well as on GLUTag cells. As many of these signaling mediators are induced by nutrient ingestion, this constitutes an indirect pathway that links food components to the release of GLP-1. Such indirect actions are particularly important to GLP-1 release, since L-cells are mainly concentrated in the distal ileum and colon (Dong and Brubaker 2012), and nutrients have not reached this area of the gastrointestinal tract when the initial phase of prandial GLP-1 secretion is triggered. Initially, a neurohormonal mechanism, involving the vagus nerve, was shown to mediate the very rapid rise in GLP-1 levels after nutrient ingestion (Dong and Brubaker 2012). Consistent with this hypothesis, a stimulatory effect of two muscarinic agonists, bethanechol and carbachol, has been found in GLUTag cells (Brubaker et al. 1998), although expression of muscarinic receptors by this cell line has not been reported to date. Notwithstanding, direct activation of downstream protein kinase C signaling does stimulate GLP-1 release from the GLUTag cells (Brubaker et al. 1998). Similarly, the incretin hormone, glucose-dependent insulinotrophic peptide (GIP), secreted by the enteroendocrine K cells in response to carbohydrates and fats, has been reported to trigger GLP-1 secretion at a concentration range of 30–100 nM (Brubaker et al. 1998). The effects of GIP on GLUTag cells are mediated through activation of the protein kinase A signaling pathway (Simpson et al. 2007). Other agents capable of raising the intracellular levels of cAMP, such as forskolin plus the cAMP phosphodiesterase inhibitor, 3-isobutyl-1-methylxanthine, have also been found to stimulate both GLP-1 secretion and production in GLUTag cells (Drucker et al. 1994) and are commonly used as positive controls when the secretory activity of this cell line is examined. Supporting the prominent role of intracellular cAMP levels in GLP-1 secretion, it has been reported that phosphodiesterase inhibition enhances the GLP-1 response to physiological secretagogues such as GIP (Simpson et al. 2007). Finally, other nutrient-dependent bioactive peptides, including leptin and insulin, have also been shown to stimulate GLP-1 secretion from GLUTag cells. The effects of leptin are believed to be produced via activation of the Janus-activated kinase/signal transducers and activators of transcription signaling pathway (Anini and Brubaker 2003), whereas the effects of insulin are mediated through the activation of the extracellular signal-regulated kinases and are dependent upon the presence of glucose (Lim et al. 2009).

21.4   Relevance to the Human Situation (L-Cell In Vivo)

Despite the rodent origin of GLUTag cells, they are the most extensively used in vitro model for the study of GLP-1 release. They have been demonstrated to respond to the same stimuli that trigger GLP-1 secretion in humans and they have been used to test potential therapeutic agents (i.e. GPR119 agonists). The difficulties in obtaining primary L-cell cultures from humans, as well as the homogeneity of the GLUTag cells (i.e. single-cell clone) as compared to other L-cell models, have made GLUTag cells an invaluable tool for the characterization of the intra-cellular pathways that govern GLP-1 secretion.

21.5   General Protocol

21.5.1   Maintenance of the GLUTag Cell Line

GLUTag cells are routinely grown in Dulbecco's Modified Eagle's Medium (DMEM) supplemented with 10 % fetal bovine serum (FBS), whereas supplementation with antibiotics, normally penicillin and streptomycin, is optional but not required. Cells must be trypsinized and passaged every 3–4 days when they reach 60–80 % confluence. The presence of floating cells may be indicative of contamination or overgrowth. There is no standard glucose concentration to grow the cells but a final concentration within the range 5–25 mmol/L is commonly used. It has recently been reported that cells grown with 25 mmol/L of glucose show increased reactive oxygen species production, upregulated *proglucagon*, prohormone convertase 1/3 and glucokinase content, and elevated basal secretion of GLP-1 as compared to cells maintained under 5 mmol/L conditions, suggesting enhanced metabolic activity. However, no difference in cell viability was found between the two conditions, and the cells grown with 25 mmol/L glucose were more resistant to a further metabolic insult (Puddu et al. 2014). Moreover, cells grown at either a low (5 mmol/L) or high (25 mmol/L) concentration of glucose have been demonstrated to respond to stimuli in a similar way, and no differences in GLP-1 secretion or production were found in cells exposed to low or high glucose for 2 h (Reimann and Gribble 2002), which is the standard incubation period to examine the GLUTag cells secretory response to stimuli.

21.5.2 Performance of Secretion and Expression Assays

To study the effects of different treatments on GLP-1 release (secretion assays), cells can be cultured in 6- or 24-well plates coated with either poly-D-lysine or Matrigel® to avoid detachment of cells during the washes. A 2-day period between plating and the secretion assay is recommended in order to allow the cells to recover from trypsin and to reach the proper size and confluence. When the cells are too confluent (over 80 %), they tend to grow on top of each other, forming clumps. This should be avoided since it is associated with higher basal levels of secretion (personal observations). During the test period of 2 h, GLUTag cells should be maintained in either FBS-free DMEM or DMEM with a very low concentration of FBS (0.5 %), containing the treatments tested at different doses; buffers such as Krebs–Ringer are also acceptable. To study the effects of treatments on gene expression or protein content, larger surface areas, such as 10 cm culture plates, may be required and longer incubation periods are recommended.

A recent study has shown that GLUTag cells possess an endogenous metabolic clock and that the activity of the cells can be synchronized after overnight incubation with a low concentration of FBS (0.5 %) followed by a 1-h shock with 10 % FBS and 10 µmol/L forskolin (Gil-Lozano et al. 2014). Synchronized GLUTag cells show a rhythm in clock genes expression as well as in their GLP-1 response to secretagogues, with higher responses at 4 h after synchronization and lack of effect at 16 h. This protocol may be adopted to obtain maximum secretory responses to a particular treatment.

21.6 Assess Viability

Some treatments can affect viability in GLUTag cells, inducing abnormally high levels of GLP-1 secretion that can lead to misinterpretation of the results. Both the 3-(4,5-dimethylthiazol-2-yl)-2,5-diphenyltetrazolium bromide (MTT) and the neutral red uptake assays have been routinely performed in our laboratory to test the effects of new treatments on cell viability. Both are colorimetric assays and a reduction in absorbance corresponds with loss of cell viability. These tests are not quantitative and the addition as a reference of a positive control, such as hydrogen peroxide, is required.

21.7 Experimental Readout of the System

In secretion assays, GLP-1 content in medium and cells should be determined independently. Both medium and cell lysate should be collected and acidified to ensure preservation of peptides and facilitate the subsequent purification step. Our laboratory routinely extracts peptides from the medium and cell lysate by reversed-phase adsorption to C18 silica columns after acidification with trifluoroacetic acid. GLP-1 levels can then be determined by radioimmunoassay, ELISA or alternative quantification methods. GLP-1 secretion must be calculated as the total GLP-1 content of the medium divided by the total GLP-1 content of medium plus cells. A ratio of secretion is then obtained, which can be normalized to the percentage of secretion of the control group. A two- to fourfold increment in secretion can be found for the most potent secretagogues, such as the combination of forskolin and 3-isobutyl-1-methylxanthine, which are often included in the study as positive controls.

In expression assays, medium is discarded and cells are scraped in a solution containing lysis buffer and β-mercaptoethanol to inactivate RNAses. Our laboratory has found that extraction of RNA using RNeasy Plus Mini Kit with QIAshredder (Qiagen, MD) permits collection of high quality RNA with a 260/280 nm ratio over 2 and 30–100 µg of RNA per 10 cm plate. In real-time PCR analysis involving cDNA derived from GLUTag cells, *H3f3a* amplicon (H3 histone gene) is strongly recommended as an internal control, since it shows much better consistency than other primers commonly used as housekeeping genes, such as *18S* amplicons (personal observations). For protein extraction, medium is discarded and cells are scraped in a lysis buffer such as RIPA (radioimmunoprecipitation assay) and then sonicated. As much as 100 µg of protein per well can be collected from a 12-well plate.

21.8   Conclusions

Since the original generation of the GLUTag cell line in 1994 by Drucker, Brubaker and colleagues, these cells have been instrumental in the studies addressing the role of nutrients in L-cell physiology, and particularly in GLP-1 secretion. The easy maintenance of the cells, their close similarity in phenotype with non-immortalized L-cells and the ability to apply advanced techniques such as patch-clamp recording and gene silencing are clear advantages for the use of this cell line. Yet, as is common for many tumoural cell lines, some phenotypic differences are found between GLUTag cells and the normal L-cell, notably, the former presenting a higher diversity of glucose-sensing mechanisms and, apparently, insensitivity to SCFAs. Nevertheless, the use of GLUTag cells has been crucial to develop potential novel therapies for diabetes, such as GPR119 agonists. It is highly likely that GLUTag cells will continue to be intensively used as a model of the enteroendocrine L-cell until the isolation of primary L-cells becomes an easier and more standardized procedure.

Acknowledgements MGL was supported by Post-doctoral Fellowships from the Canadian Institutes of Health Research Sleep and Biological Training Program, University of Toronto, and PLB by the Canada Research Chairs Program. Studies on GLP-1 in the Brubaker laboratory are supported by operating grants from the Canadian Diabetes Association (OG-3-13-4024-PB) and the Natural Sciences and Engineering Research Council (RGPIN418).

References

Anini Y, Brubaker PL (2003) Role of leptin in the regulation of glucagon-like peptide-1 secretion. Diabetes 52:252–259

Brubaker PL, Schloos J, Drucker DJ (1998) Regulation of glucagon-like peptide-1 synthesis and secretion in the GLUTag enteroendocrine cell line. Endocrinology 139:4108–4114

Chu ZL, Carroll C, Alfonso J et al (2008) A role for intestinal endocrine cell-expressed g protein-coupled receptor 119 in glycemic control by enhancing glucagon-like peptide-1 and glucose-dependent insulinotropic peptide release. Endocrinology 149:2038–2047

Dong CX, Brubaker PL (2012) Ghrelin, the proglucagon-derived peptides and peptide YY in nutrient homeostasis. Nat Rev Gastroenterol Hepatol 9:705–715

Drucker DJ, Jin T, Asa SL et al (1994) Activation of proglucagon gene transcription by protein kinase-A in a novel mouse enteroendocrine cell line. Mol Endocrinol 8:1646–1655

Edfalk S, Steneberg P, Edlund H (2008) Gpr40 is expressed in enteroendocrine cells and mediates free fatty acid stimulation of incretin secretion. Diabetes 57:2280–2287

Fujita Y, Wideman RD, Speck M et al (2009) Incretin release from gut is acutely enhanced by sugar but not by sweeteners in vivo. Am J Physiol Endocrinol Metab 296:E473–E479

Gameiro A, Reimann F, Habib AM et al (2005) The neurotransmitters glycine and GABA stimulate glucagon-like peptide-1 release from the GLUTag cell line. J Physiol 569:761–772

Gil-Lozano M, Mingomataj EL, Wu WK et al (2014) Circadian secretion of the intestinal hormone, glucagon-like peptide-1, by the rodent L-cell. Diabetes 63:3674-3685. doi:10.2337/db13-1501

Hira T, Mochida T, Miyashita K et al (2009) GLP-1 secretion is enhanced directly in the ileum but indirectly in the duodenum by a newly identified potent stimulator, zein hydrolysate, in rats. Am J Physiol Gastrointest Liver Physiol 297:G663–G671

Hirasawa A, Tsumaya K, Awaji T et al (2005) Free fatty acids regulate gut incretin glucagon-like peptide-1 secretion through GPR120. Nat Med 11:90–94

Iakoubov R, Izzo A, Yeung A et al (2007) Protein kinase Czeta is required for oleic acid-induced secretion of glucagon-like peptide-1 by intestinal endocrine L cells. Endocrinology 148:1089–1098

Lauffer LM, Iakoubov R, Brubaker PL (2009) GPR119 is essential for oleoylethanolamide-induced glucagon-like peptide-1 secretion from the intestinal enteroendocrine L-cell. Diabetes 58:1058–1066

Lim GE, Huang GJ, Flora N et al (2009) Insulin regulates glucagon-like peptide-1 secretion from the enteroendocrine L cell. Endocrinology 150:580–591

Margolskee RF, Dyer J, Kokrashvili Z et al (2007) T1R3 and gustducin in gut sense sugars to regulate expression of Na+-glucose cotransporter 1. Proc Natl Acad Sci U S A 104:15075–15080

Oya M, Kitaguchi T, Pais R et al (2013) The G protein-coupled receptor family C group 6 subtype A (GPRC6A) receptor is involved in amino acid-induced glucagon-like peptide-1 secretion from GLUTag cells. J Biol Chem 88:4513–4521

Parker HE, Adriaenssens A, Rogers G et al (2012a) Predominant role of active versus facilitative glucose transport for glucagon-like peptide-1 secretion. Diabetologia 55:2445–2455

Parker HE, Wallis K, le Roux CW et al (2012b) Molecular mechanisms underlying bile acid-stimulated glucagon-like peptide-1 secretion. Br J Pharmacol 165:414–423

Poreba MA, Dong CX, Li SK et al (2012) Role of fatty acid transport protein 4 in oleic acid-induced glucagon-like peptide-1 secretion from murine intestinal L cells. Am J Physiol Endocrinol Metab 303:E899–E907

Puddu A, Sanguineti R, Montecucco F et al (2014) Glucagon-like peptide-1 secreting cell function as well as production of inflammatory reactive oxygen species is differently regulated by glycated serum and high levels of glucose. Mediators Inflamm 2014:923120

Reimann F, Gribble FM (2002) Glucose-sensing in glucagon-like peptide-1-secreting cells. Diabetes 51:2757–2763

Reimann F, Habib AM, Tolhurst G et al (2008) Glucose sensing in L cells: a primary cell study. Cell Metab 8:532–539

Simpson AK, Ward PS, Wong KY et al (2007) Cyclic AMP triggers glucagon-like peptide-1 secretion from the GLUTag enteroendocrine cell line. Diabetologia 50:2181–2189

Tolhurst G, Zheng Y, Parker HE et al (2011) Glutamine triggers and potentiates glucagon-like peptide-1 secretion by raising cytosolic Ca2+ and cAMP. Endocrinology 152:405–413

Tolhurst G, Heffron H, Lam YS et al (2012) Short-chain fatty acids stimulate glucagon-like peptide-1 secretion via the G-protein-coupled receptor FFAR2. Diabetes 61:364–371

Part V
In Vitro Intestinal Tissue Models: General Introduction

A reliable prediction and thorough understanding of the food-intestine interactions and oral bioavailability in humans is crucial for the food industry. For this purpose different in silico methods, in vitro cell lines, ex vivo intestinal tissue and/or in vivo animal studies can be used. However, their predictive value for the human situation is often limited. In vitro intestinal models are used to study many different biological processes and are used to minimize the use of animals. The optimal conditions and the choice of in vitro model depends on the research questions. For example, for mechanistic studies aiming to elucidate cellular processes, a simple cell-line model may be accurate. However when crosstalk and possible feedback mechanisms have to be studied, these single cell type culture models are not satisfactory and either co-cultures of multiple cell types and/or intact tissue models or functional organs are needed. In that case, often ex vivo experiments are used which include procedures with viable functional tissues or organs isolated from an organism and incubated outside the organism in an artificial environment under highly controlled conditions.

The simplest in vitro model is one containing a monolayer of intestinal epithelial cells from either human or animal origin, which can be exposed to bacteria, food or drug compounds. However, a major drawback of these cell models are, that they are composed of a single cell type (e.g. absorptive or secretive cells), whereas the intestinal epithelium is a conglomerate of absorptive enterocytes and other cells such as goblet cells, endocrine cells, and Paneth cells (Madara and Trier 1994). For this reason, more complicated models using intestinal tissue are preferable to study the effect of food and food compounds. In the following section, the most common in vitro intestinal tissue models are described and an overview of the major characteristics of these models are presented in Table 1.

Table 1 Characteristics of the most common in vitro intestinal models

	Intestinal rings	Intestinal segments	Everted sac	Ussing system	Gut organoids
System	Fresh ex vivo intestinal tissue from animal and human origin and different parts of the intestine	Fresh ex vivo intestinal tissue from animal and human origin and different parts of the intestine	Fresh ex vivo intestinal tissue from animal and human origin and different parts of the intestine	Fresh ex vivo intestinal tissue from animal and human origin and different parts of the intestine	Isolated crypt cells from mouse, pig or human origin, cultured and differentiated in Matrigel to multiple organoids per well
Tissue	Epithelial, enteroendocrine goblet and paneth cells, blood and lymph vessels, M-cells, Peyers Patches, immune cells	Epithelial, enteroendocrine goblet and paneth cells, blood and lymph vessels, M-cells, Peyers Patches, immune cells	Epithelial, enteroendocrine goblet and paneth cells, blood and lymph vessels, M-cells, Peyers Patches, immune cells	Epithelial, enteroendocrine goblet and paneth cells, blood and lymph vessels, M-cells, Peyers Patches, immune cells	Epithelial cells, enteroendocrine cells, paneth cells, goblet cells
Muscle layer	Yes	No	Yes	No	No
Basolateral and apical compartment	Rings float in culture medium	Segment floats in culture medium	Differentiation between lumen (apical) and blood side (basolateral)	Differentiation between lumen and blood side	Intestinal lumen within the organoid
Exposure	To both luminal and serosal side	To both luminal and serosal side	To either luminal or blood side via individual compartment	To either luminal or blood side via individual compartment	To either basolateral (media) or luminal exposure via microinjection in organoids
Viability	30–60 min	0–2 h	0–2 h	0–2 h	Continuous culturing
Special skills	Relatively simple	Relatively simple	Relatively simple	Special skills and laborious	Special skills and laborious
Throughput	High	High	Low	Intermediate	Intermediate
Applications	Regional absorption mechanisms	Regional absorption, GI hormone release, endocannabinoids	Regional absorption mechanisms. Intestinal metabolism of drugs. Roles of transporter in drug absorption. Role of intestinal enzymes during drug transport	Regional absorption, translocation proteins/bacteria, permeability and restoration of barrier. GI hormone release, transcriptomics, early immune response	GI hormone release, transcriptomics, early immune response

Intestinal Rings and Intestinal Segments

The use of intestinal rings as an in vitro intestinal model is first described in 1954 by Agar et al. (1954) and is mainly used to measure the accumulation of compounds into the enterocytes rather than transport through the enterocytes (Hillgren et al. 1995). In the intestinal ring model, a part of the intestine is isolated and after thorough washing everted (inside out) over a glass rod and cut into small rings. The rings are then submerged in high oxygenated buffer containing the compound of interest. The intestinal rings are generally viable for 30–60 min, probably due to bad oxygenation caused by the presence of the muscle layers.

In case of intestinal segments, the outer muscle layers are removed and segments are punched out the intestinal tissue. The intestinal segments are viable for approx. 2 h. Like the intestinal rings, the segments are also submerged in high oxygenated buffer. Both systems are easy to use and many samples can be obtained from one animal. The main disadvantage of these models is that the serosal as well as the mucosal layer are exposed to the buffer containing the compound of interest. The use of radiolabelled compounds is preferable, since this simplifies the analysis of absorbed compound in the tissue.

Everted Sac

The everted gut sac system was first reported for the study of the transport of glucose and amino acids across rat and hamster intestinal barrier (Wilson and Wiseman 1954). In the everted sac method, the intestinal section is everted and both ends are tied after filling the sac with buffer. The 'sausage' is incubated into buffer with the compound of interest (Mariappan and Singh 2004). The technique can be used to study the absorptive properties of food compounds by measuring the amount ending up on the serosal side and in the epithelial cells, provided that the used analytical method is sensitive enough. The use of radiolabelled compounds is highly recommended. A downside of this type of experiment is that in everted sac experiments, the speed of absorption might be unrealistically slow as, in contrast to the in vivo situation, since the compound has to cross the muscle layers and the fluid is stagnant. Under optimal conditions the tissue is viable for up to 2 h, and the system is predominantly used for studying drug absorption mechanisms, intestinal metabolism of drugs, roles of transporter in drug absorption, and for investigating the role of intestinal enzymes during drug transport through the intestine (Acra and Ghishan 1991).

Ussing Chamber

Ussing chambers were initially developed to measure the transport of ions, nutrients, and drugs across various epithelial tissues (Ussing and Zerhan 1951, see Chap. 23). With the Ussing chamber technique, the intestinal segment is mounted in an Ussing chamber where one side will be exposed to buffer with the compound of interest (apical or lumen side) and the other side to buffer without the compound of interest (basolateral or mucosal side). The major advantage of this system is that different parts of the intestine (from duodenum to colon) can be used (Smith et al. 1992). However the method is laborious and has a relatively low-throughput, and short-term survival in culture. Recently, TNO described a more high throughput system, InTESTine™ which allow up to 96 incubations simultaneously (Roeselers et al. 2013, see Chap. 24).

Isolated and Perfused Intestinal Segments

Isolated and perfused intestinal segments were first described by Baker et al. (1968) who prepared these segments from dog ileum. During the last decade, a wide range of isolated organ systems, either from whole organs or resected intestinal tissue, have been developed for biomedical and pharmaceutical research. For this system a part of the intestine including the vascular bed is isolated and both ends of the intestine is sealed before the tissue is mounted in to the perfusion system, where the artery is perfused with a well-defined buffer. The compound of interest can be administered as a bolus injection into the gut lumen and samples can be withdrawn at several times points form the recirculating perfusate (Wei et al. 2009). A major advantage of the system is that the model displays a normal morphology, histology and physiology. However, the difficulty of obtaining sufficient quality and quantity of organs and the limited duration of experiments are a major drawback.

3-D Culture Systems

The development of 3D cell culture models have been increased in the past few years (Salerno-Goncalves et al. 2011; Yu et al. 2012). 3-D culture systems are prepared from cell lines, primary cells and/or organ cultures using various methods, including spontaneous aggregation in a suspension culture, implantation onto 3D scaffolds (e.g. collagen or synthetic materials) and culture in Transwell® systems or in rotating culture bioreactors (Schmeichel and Bissell 2003). These systems range in complexity and carry distinct advantages and disadvantages including experimental costs, level of expertise, optimization, reproducibility and validation.

The power of 3-D culture systems is that they permit cells to change shape and form cell–cell connections that are prohibited on rigid conventional culture substrates. The models can be applied for instance to study cell attachment, migration and proliferation. One example of a 3-D intestinal system is the gut organoid. Organoids are cultured from adult intestinal stem cells and have self-renewing capacity. Their structure and hierarchy highly resemble the in vivo intestinal epithelium. The major advantage of the gut organoid system over other GI models is the long-term culturing, next to the broad range host species and GIT compartments (stomach, ileum, colon) and gene manipulation possibilities (Koo et al. 2012). So far, development and application of the intestinal organoid system focused mainly towards the use of organoids for regenerative medicine purposes. However, further expansion of this system will lead to broader application range and expansion of the read out possibilities.

Most ex vivo intestinal tissue models make use of tissue obtained from different animal species (e.g. rabbit, piglets or rats), since the availability of healthy human intestinal tissue is limited. The disadvantage of using animal tissue, is the inter-species differences in anatomy, physiology, metabolism, diet and micro-biota, which complicates the extrapolation of data to humans (Nejdfors et al. 2000; Deferme et al. 2008). Pigs share more physiological and immunological similarities to humans than rodents, and the use of (mini)pigs is becoming increasingly common in nutritional research (Patterson et al. 2008, see Chap. 24).

The majority of the in vitro intestinal tissue models are used for absorption studies using drug or food compounds. Since intestinal tissue consist of multiple cell types (e.g. immune cells, enteroendocrine cells) it can be envisioned that these models could also be applicable to study innate immune responses or satiety. Unfortunately hardly any evidence was found to use these models to study the effect of food on innate immune response, this is most likely due to viability issues. Intestinal tissue is viable for approx. 2 h due to intestinal oedema and disruption of the villi (Plumb et al. 1987). Two hours is too short to measure the production of excreted compounds like cytokines or chemokines. Efforts to elongate viability of tissue incubations are on-going and promising (Tsilingiri et al. 2012). However measurements of these compounds on mRNA level should be possible. Recently, intestinal segments were used to study the effect of sugars, proteins and fatty acids on the release of satiety hormones (Voortman et al. 2012, see Chap. 23). The advantage of this model is that the hormone release can be studied in different parts of the intestine. This is very important, since it has been shown that the secretion of satiety hormones is site specific. A novel method to study effects of foods and drug compounds on intestinal tissue is the organoid model (see Chap. 22). This is a promising model, since the lifetime of this, from stem cells derived tissue, is much longer than excised intestinal tissue. However the model is still being in its infancy and the wide applicability still has to be proven.

In the following chapters we will focus more in depth on the applicability of three intestinal models namely, Ussing chamber, intestinal segments and organoids.

References

Acra SA, Ghishan FK (1991) Methods of investigating intestinal transport. JPEN J Parent Enteral Nutr 15:93S–98S

Agar WJ, Hird FJR, Sidhu GS (1954) The uptake of amino acids by the intestine. Biochem Biophys Acta 14:80–84

Baker RR, Merz T, Kim ST, Tolo VT (1968) Evaluation of isolated perfused segments of dog ileum as a means of growing tumor. J Surg Res 8:458–461

Deferme S, Annaert P, Augustijns P (2008) In vitro screening models to assess intestinal drug absorption and metabolism. In: Ehrhardt C, Kim K (eds) Drug absorption studies: in situ, in vitro and in silico models. Springer, New York, pp 182–215

Hillgren KM, Kato A, Borchardt RT (1995) In vitro systems for studying intestinal drug absorption. Med Res Rev 15:83–109

Koo B-K, Stange DE, Sato T, Karthaus W, Farin HF, Huch M et al (2012) Controlled gene expression in primary Lgr5 organoid cultures. Nat Methods 9:81–83

Madara JL, Trier JS (1994) Functional morphology of epithelium of the small intestine. In: Johnson LR (ed) Handbook of physiology, the gastrointestinal system, intestinal absorption and secretion, vol Supplement 19, 2nd edn. Raven Press, New York, pp 1577–622

Mariappan TT, Singh S (2004) Evidence of efflux-mediated and saturable absorption of rifampicin in rat intestine using the ligated loop and everted gut sac techniques. Mol Pharm 1:363–367

Nejdfors P, Ekelund M, Jeppsson B, Westrom BR (2000) Mucosal in vitro permeability in the intestinal tract of the pig, the rat, and man: species- and region-related differences. Scand J Gastroenterol 35:501–507

Patterson JK, Lei XG, Miller DD (2008) The pig as an experimental model for elucidating the mechanisms governing dietary influence on mineral absorption. Exp Biol Med 233:651–664

Plumb JA, Burston D, Baker TG, Gardner ML (1987) A comparison of the structural integrity of several commonly used preparations of rat small intestine in vitro. Clin Sci 73:53–59

Roeselers G, Ponomarenko M, Lukovac S, Wortelboer HM (2013) Ex vivo systems to study host-microbiota interactions in the gastrointestinal tract. Best Pract Res Clin Gastroenterol 27:101–113

Salerno-Goncalves R, Fasano A, Sztein MB (2011) Engineering of a multicellular organotypic model of the human intestinal mucosa. Gastroenterology 141:e18–e20

Schmeichel KL, Bissell MJ (2003) Modeling tissue-specific signaling and organ function in three dimensions. J Cell Sci 116:2377–2388

Smith PL, Wall DA, Gochoco CH, Wilson G (1992) (D) Routes of delivery: case studies: (5) oral absorption of peptides and proteins. Adv Drug Deliv Rev 8:253–290

Tsilingiri K, Barbosa T, Penna G, Caprioli F, Sonzogni A, Viale G, Rescign M (2012) Probiotic and postbiotic activity in health and disease: comparison on a novel polarised ex-vivo organ culture model. Gut 61:1007–1015

Ussing HH, Zerhan K (1951) Active transport of sodium as the source of electric current in the short-circuited isolated frog skin. Acta Physiol Scand 23:110–121

Wei Y, Neves LA, Franklin T, Klyuchnikova N, Placzek B, Hughes HM et al (2009) Vascular perfused segments of human intestine as a tool for drug absorption. Drug Metab Dispos Biol Fate Chem 37:731–736

Wilson TH, Wiseman G (1954) The use of sacs of everted small intestine for the study of the transference of substances from the mucosal to the serosal surface. J Physiol 123:116–125

Yu J, Peng S, Luo D, March JC (2012) In vitro 3D human small intestinal villous model for drug permeability determination. Biotechnol Bioeng 109:2173–2178

Chapter 22
Intestinal Crypt Organoids as Experimental Models

Sabina Lukovac and Guus Roeselers

Abstract When it comes to studying the effect of food bioactives on gut health, one of the essential steps that needs to be assessed is characterizing specific effects of the bioactives on the physical barrier of the lumen, the gastrointestinal tissue. In addition to studying the effects on transport function (e.g. by using Ussing chambers or cell culture systems), it is of great interest to evaluate the effects on morphology, cell biology, gene expression, and relevant functions of different cell types that are resident in the gastrointestinal (GI) tract. An ideal near-physiological model should contain a mixture of different GI epithelial cells (e.g. Paneth cells, goblet cells, absorptive and hormone secretive epithelial cells), which can be cultured indefinitely. Recently, the culture and applications of long-term primary multi-cellular cluster structures gastrointestinal organoids (or enteroids) have been demonstrated, and within the last 5 years the number of researchers that commonly use this tissue culture model has increased rapidly. This multi-cellular system may be a promising addition for existing ex vivo and alternative for animal models for testing effects of food bioactives on the intestinal tissue, and could provide a model for pre-screening of compounds prior to moving to the large scale testing systems. Moreover, intestinal organoids can be cultured from different species (e.g. human, pig and mouse). In this chapter we will focus on organoids cultured from mouse and pig crypt cells. We will give a short overview on how to isolate, culture, incubate, and apply them in different research fields.

Keywords Ex vivo • Intestinal epithelium • Paneth cells • Crypts • Lgr5 • Stem cells

S. Lukovac (✉)
TNO, Microbiology & Systems Biology Group,
Utrechtseweg 48, 3704 HE Zeist, The Netherlands

TMC, Herculesplein 44, 3584 AA Utrecht, The Netherlands
e-mail: Sabina.Lukovac@tmc.nl

G. Roeselers
TNO, Microbiology & Systems Biology Group,
Utrechtseweg 48, 3704 HE Zeist, The Netherlands
e-mail: guus.roeselers@tno.nl

© The Author(s) 2015   245
K. Verhoeckx et al. (eds.), *The Impact of Food Bio-Actives on Gut Health*,
DOI 10.1007/978-3-319-16104-4_22

22.1 Gastrointestinal Organoids

Several attempts have been made to isolate the proliferative regions of the intestinal epithelium from different species (e.g. mouse, porcine, human) (Pageot et al. 2000; Agopian et al. 2009). Proliferative regions of the intestinal epithelium consist of the transit amplifying cells, which are undifferentiated progenitor cells that can differentiate into all different epithelial cells of the intestinal epithelium. Initially cultured intestinal tissue contained crypt and villus domains, resembling the intestinal epithelial architecture present in vivo. However, these cultured intestinal tissue pieces from the intestinal epithelial could be maintained in culture for up to several days at maximum, and thus were not self-renewing .

The need for a self-sustaining, long-term GI model remained. The big breakthrough came when Dr. Hans Clevers and colleagues, for the first time, identified the intestinal stem cells by thorough lineage tracing experiments (Barker et al. 2007). These experiments were performed over a period of 60 days in mice, and confirmed the presence of Lgr5 (marker for adult stem cells)-positive stem cells in small intestinal and colonic crypts. These stem cells are able to differentiate in to all intestinal epithelial cells (enterocytes, Paneth cells, Goblet cells, enteroendocrine cells, but also stem and progenitor cells), as shown in vivo and in culture where these Lgr5-positive cells formed so-called 'mini-guts' or 'organoids' (Sato et al. 2009) (Fig. 22.1).

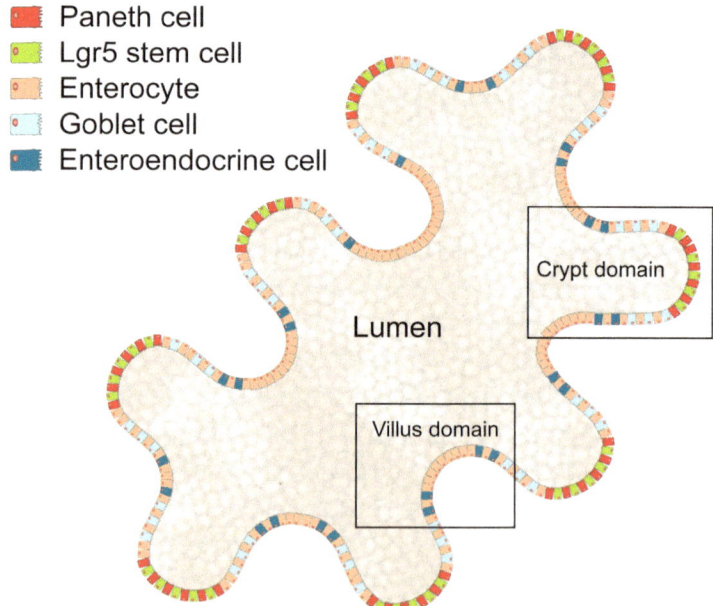

Fig. 22.1 Schematic image of an intestinal organoid. All epithelial cell types normally present in vivo are also present in the cultured intestinal organoids, as indicated by different colored cells. Figure adapted from Roeselers et al. (2013)

Secretion of mucins into the luminal space and lysozyme production by Panneth cells illustrate the functionality of this epithelium (McCracken et al. 2011).

The discovery of these stem cells opened a new world of the possibilities for GI culture methods, by manipulation of these cells. The great potential of these now self-sustaining 'mini-guts' in the field of transplantation and regenerative medicine was immediately evident, and soon engraftment of intestinal organoids in mice was successfully performed (Yui et al. 2012; Shaker and Rubin 2012). So far, the human organoid transplantation has not been performed in humans yet, although human intestinal organoids can be cultured and maintained in culture for some time now (Sato et al. 2011a).

Recently human organoids cultured from induced pluripotent stem cells (iPSCs) have been transplanted into mice (in vivo), where intestinal adaptive capacity as well as epithelial function (peptide uptake and permeability) have been demonstrated in the transplants (Watson et al. 2014). In addition, Clevers and colleagues (University Medical Center Utrecht, the Netherlands), proposed treatment for children suffering from a rare gastrointestinal disease (microvillus inclusion disease) by transplantation of stem-cell derived organoids (Wiegerinck et al. 2014), but the outcomes of these studies have not been published yet. Intestinal organoids have been applied to study cystic fibrosis by using primary organoids from cystic fibrosis patients. This model facilitates diagnosis, functional studies, drug development and personalized medicine approaches in cystic fibrosis (Dekkers et al. 2012).

Recently, we have demonstrated that intestinal organoids provide a powerful model for host–microbiome interaction studies (Lukovac et al. 2014). In this study we observed that specific gut bacteria and their metabolites differentially modulate epithelial transcription in mouse organoids (Lukovac et al. 2014).

22.2 General Protocol

Isolation of mouse intestinal crypts and organoid culture, adapted from (Sato and Clevers 2013) is schematically represented in Fig. 22.2. A detailed protocol of the culture protocol is described below.

22.2.1 Small Intestinal Crypt Isolation

Thaw aliquots of Matrigel™ Basement Membrane Matrix (BD Biosciences) on ice before isolation and pre-incubate a 24-well plate in a CO_2-incubator (5 % CO_2, 37 °C). Matrigel™ is a solubilized gelatinous protein mixture and is derived from the Engelbreth–Holm–Swarm (EHS) mouse sarcoma cells. The extract resembles the complex extracellular environment found in many tissues and is commonly used as a substrate for culturing cells. At room temperature, Matrigel™ polymerizes into a three dimensional structure that is useful for both cell culture and studying cellular processes in three dimensions, including cell migration.

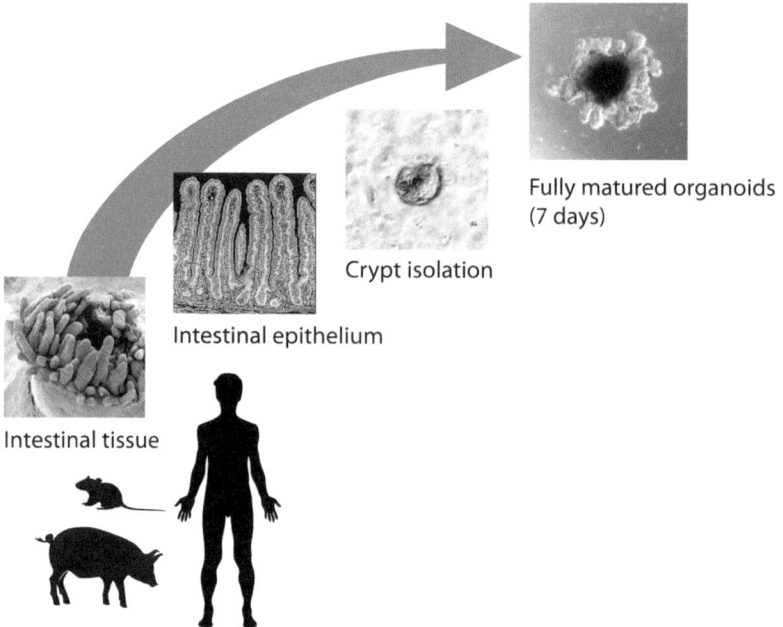

Fig. 22.2 Schematic summary of intestinal epithelial crypt isolation and subsequent organoid cultures

Isolate a piece of mouse small intestine (at least 5 cm) and open longitudinally. Wash the intestine with ice cold PBS until most of the luminal contents are cleared and scrape off the villi using a coverslip. After another washing with ice-cold PBS, cut the intestine into 2–4 mm pieces with scissors and transfer them to a 50-mL tube. Add 30 mL ice-cold PBS and wash the fragments by gently pipetting the fragments up and down with a 10 mL pipette and discard supernatant after settling down. Repeat this step 5–10 times until the supernatant is almost clear. Next, add 30 mL of ice-cold crypt isolation buffer (5 mM EDTA in PBS; one can also use higher EDTA concentration, up to 10 mM if yields are low) and gently rock the tube at 4 °C for 30 min. Keep pipetting up and down the tissue in solution every 10 min in order to release as much epithelium as possible. After settling down the fragments, the supernatant can be removed and 20 mL of ice-cold PBS can be added to wash the fragments using a pipet. Repeat this procedure when the fragments are settled down, in order to release most of the crypts, that will also settle down after some time. Villous fractions present in the supernatant are discarded and crypts fractions are passed through a 70-mm cell strainer and collected into 50 mL tube(s), after adding 5 % FBS to a crypt solution. Spin down the crypt fractions at $300 \times g$ 5 min and resuspend the pellets in 10 mL of ice-cold basal culture medium (advanced Dulbecco's modified Eagle medium/F12 supplemented with penicillin/streptomycin, 10 mmol/L HEPES, Glutamax, $1 \times N2$, $1 \times B27$, and 1 mmol/L N-acetylcysteine). The suspension is transferred to a 15 mL Falcon tube and centrifuged at $150–200 \times g$ for 2 min, to remove single cells (mostly lymphocytes), which end up in the supernatant.

This washing step should be repeated 2–3 times until most single cells are cleared. After the washing steps, the number of crypts can be calculated (take 10 µL of the crypts suspension and count numbers of crypts using a heamocytometer by inverted microscopy; the total number of crypts = the number of crypts counted × 1,000). It should be noted that at this point it is difficult to distinguish true crypts from other epithelial debris, hence the count is only a rough estimate.

22.2.2 Small Intestinal Organoid Culture

Centrifuge small intestinal crypts, and remove supernatant as much as possible to avoid dilution of Matrigel™ in the next step. Keep the tube at 4 °C and resuspend the crypts pellet in Matrigel™ (200–500 crypts/50 µL matrigel). Pipet 50 µL of the crypt-Matrigel™ suspension into the pre-warmed 24-well plate. The suspension should be applied on the center of the well so a hemispherical droplet can be formed. Transfer the plate back into the CO_2 incubator (5 % CO_2, 37 °C) as soon as possible after the seeding. Allow the Matrigel™ to solidify for 5–10 min and add 500 µL complete organoid culture medium per well. For preparation of the organoid culture medium (e.g. amount of growth factors) check Table 22.1. Place the 24-wells plate in a CO_2-incubator and refresh the culture medium every 4–6 days.

22.2.3 Small Intestinal Organoid Passage

Crypt organoids can be passaged 7–14 days after seeding. Place the plate with organoids on ice to allow the matrigel to dissolve. Remove the culture medium and gently break up the Matrigel™ by pipetting with a 100–1,000 µL-pipette. Resuspend the organoids in 1–2 mL basal culture medium and transfer them to a 15 mL-tube. Gently disrupt the organoids using a fire-polished Pasteur pipette to remove released dead cells and single cells. Pellet everything by centrifuging at 300×g for 5 min, and wash the organoids afterwards with 10 mL basal culture medium. Pellet again at the same speed and add the correct amount of Matrigel™, namely 50 µL per well that you would like to plate. Count the "crypts/immature organoids" as described before (by hemocytometer) and pipet the Matrigel™ suspension with crypts (50 µL per well) into a new 24 culture plate and place the plate back in CO_2 incubator (5 % CO_2, 37 °C) to culture the organoids.

22.3 Monitoring Viability

Organoid culture and development across time after isolation, and post-passage can be verified using microscopy. Mainly, the dead cells of the organoid tissue will be visible in the central part of the organoid, corresponding to the intestinal lumen

Table 22.1 Concentrations of growth factors and other compounds added to the basal culture medium used for mouse and pig organoid culture

Reagent (supplier)	Mouse[a]	Pig
Advanced DMEM/F12 (ADF) (Invitrogen, Cat #11320-082)	1×	1×
HEPES buffer (Sigma-Aldrich, Cat #83264-100ML-F)	10 mM	10 mM
GlutaMax (Invitrogen, Cat #35050-061)	2 mM	2 mM
Penicillin/Streptomycin (Invitrogen, Cat #15070-063)	0.5 U/mL (100×)	0.5 U/mL (100×)
N2 Supplement (Invitrogen, Cat #17502-048)	1× (stock: 100×)	1× (stock: 100×)
B-27 Supplement Minus Vitamin A (Invitrogen, Cat #12587-010)	1× (stock: 50×)	1× (stock: 50×)
Mouse Recombinant EGF (Invitrogen, Cat #PMG8041)	50 ng/mL	50 ng/mL
Mouse Recombinant Noggin (Peprotech, Cat #250-38)	100 ng/mL	100 ng/mL
Human Recombinant RSPO1 (R&D Systems, Cat #4645-RS-025)[b]	500 ng/mL	500 ng/mL
Wnt3a (Cell Guidance Systems, Cat # GFM77)	–	100 ng/mL
Gastrin (Sigma-Aldrich, Cat # SCP0152)	–	15 nM
Nicotinamide (Sigma-Aldrich, Cat # N0636)	–	10 mM
P38 Mapk inhibitor SB202190 (Sigma-Aldrich, Cat # S7067)	–	10 µM
TGFb tyoe Itrecepror inhibitor A83-01 (Tocris, Cat # 2939)	–	600 nM
N-Acetylcysteine (Sigma-Aldrich, Cat # A9165-5G)	–	12.5 mM
Primocin (Invitrogen, Cat # ant-pm-1)	–	100 µg/mL

For the protocol used for the isolation and cultivation of human intestinal organoids we refer to Sato et al. (2011a, b)
[a]The mouse organoid culture protocol was adapted from Sato et al. (2009)
[b]Instead of using the commercially available RSPO1, one can culture transfected HEK293 cells, which carry a plasmid encoding mRSPO1-Fc (construct developed by Dr. Calvin Kuo at Stanford University, La Jolla, CA). For the protocol for HEK293 RSPO1 production, we refer to Carmon et al. (2011) and the ATCC website: http://www.lgcstandards-atcc.org/en.aspx

in vivo (these aggregates of dead cells can be visualized by microscopy as shown in Fig. 22.3 at day 5–7). Figure 22.3 shows a very simplified summary of a mouse intestinal crypt development into a mature organoid structure. Additional conformation of the viability of the organoids can be performed upon by the use of the LDH toxicity assay or the immunofluorescent staining procedures for viability markers (e.g. apoptotic caspase 3 staining), but the organoids have to be sampled destructively in order to perform this assay, thus this step can be only applied at the final stages of specific experiments. It should be noted that once growth, normal cell proliferation and differentiation stops, and organoids start to die, that this will immediately become visible, thus it is essential to monitor the organoid growth and proliferation on a daily basis.

Culture and maintenance of intestinal organoids, relies on the presence and activity of mature (intestinal) stem cells. Even though intestinal organoids can be isolated

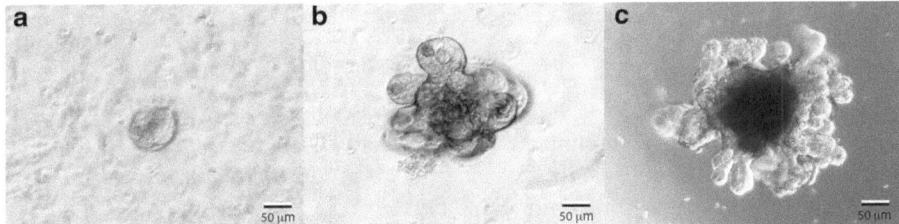

Fig. 22.3 Phase contrast images of a developing mouse intestinal crypt into a organoid in culture. (**a**) immature organoid in culture 0.5–1 day after crypt isolation; (**b**) mature organoid in culture 3–5 days after crypt isolation; (**c**) mature organoid in culture 5–14 days after crypt isolation, ready to be passaged (note the accumulated debris of dead cells in the lumen of the organoid). This morphology is representative for other species as well [image adapted from Lukovac et al. (2014)]

and cultured from human induced pluripotent stem cells (iPPS), as described by McCracken et al. (2011), most studies culture organoids from mature stem cells from the intestinal crypt regions. The reason for this is that this method described is relatively elaborative and it takes much longer to mature the intestinal organoids in culture compared to the other method based on the organoid culture from adult intestinal cells. For a detailed protocol of iPPS-based culture of the intestinal tissue we refer to McCracken et al. (2011). As mentioned previously, lineage studies confirmed the stem cell hierarchy in intestinal organoids, either by fast cycling Lgr5+- or slowly cycling Bmi1+-stem cell activity (Sangiorgi and Capecchi 2008; Sato et al. 2009). Stem cell-based organoid cultures are an attractive model due to the presence of the self-renewing capacity; they can be passaged every 5–7 days. In addition, the structure and hierarchy of the mature organoids (after 5–7 days of culturing), highly resembles the in vivo intestinal epithelium (Fig. 22.1). Similar to the intestinal epithelium, stem cells and highly proliferative transit amplifying cells reside in the crypt-like domain of the organoids (Fig. 22.1). Differentiated cell types reside in the villus-region of the organoids and include enterocytes, goblet cells, Paneth cells, and enteroendocrine cells. These cells can be identified using immunohistochemistry with intestinal epithelial cell-specific markers.

22.4 General Advantages and Limitations of Intestinal Organoid Systems

Despite the described significant advances in culturing organoids, this system has some limitations. A major advantage of intestinal models in which an apical to basolateral polarity is preserved is the possibility to characterize epithelial transport mechanisms and barrier integrity upon exposure to microorganisms and microbial components. Indeed intestinal organoids consist of polarised, columnar epithelia. However, the inwards orientation of the epithelia (i.e. villi protrude into the lumen of the organoids) makes the apical side relatively inaccessible for direct experimental stimulation (Fig. 22.1).

Also, the organoids lack several essential components of the living digestive tract, such as the enteric nervous system, the vascular system, lymphatic systems and functional adaptive and innate immune systems. Additionally, the 3D architecture is not as regular as seen in vivo, and the crypt-villus structures are variable in size and shape from one organoid to another.

Regardless of these drawbacks, this model offers great experimental utility for understanding and modeling human intestinal development, physiology, and host–microbe interactions. Indeed, this model has several other advantages, such as a long-term culturing, broad range host species and GI tract compartments (stomach, ileum, colon) and gene manipulation possibilities (Koo et al. 2011). Moreover, this system provides a viable starting point for the development of screening platforms for pharmaceutical and food compounds.

Further expansion of this system will lead to broader application range and expansion of the read out possibilities. One example is transition of the existing system towards a structure with the apical/luminal compartment exposed on the outer site of the organoids, which will make the apical site of the organoids experimentally more accessible. Furthermore, this system might be a promising tool in studies on epithelial interactions with food bioactive in individual subjects, as it has been demonstrated that organoids can be cultured from human intestinal biopsy samples (Sato et al. 2011b).

References

Agopian VG, Chen DC, Avansino JR, Stelzner M (2009) Intestinal stem cell organoid transplantation generates neomucosa in dogs. J Gastrointest Surg 13:971–982

Barker N, van Es JH, Kuipers J, Kujala P, van den Born M, Cozijnsen M, Haegebarth A, Korving J, Begthel H, Peters PJ, Clevers H (2007) Identification of stem cells in small intestine and colon by marker gene Lgr5. Nature 449:1003–1007

Carmon KS, Gong X, Lin Q, Thomas A, Liu Q (2011) R-spondins function as ligands of the orphan receptors LGR4 and LGR5 to regulate Wnt/beta-catenin signaling. Proc Natl Acad Sci U S A 108:11452–11457

Dekkers JF, Wiegerinck CL, de Jonge HR, de Jong NWM, Bijvelds MJC, Nieuwenhuis EES, van den Brink S, Clevers H, van der Ent CK, Middendorp S (2012) WS14.5 a functional CFTR assay using primary cystic fibrosis intestinal organoids. J Cyst Fibros 11:S32

Koo B-K, Stange DE, Sato T, Karthaus W, Farin HF, Huch M, van Es JH, Clevers H (2011) Controlled gene expression in primary Lgr5 organoid cultures. Nat Methods 9:81–83

Lukovac S, Belzer C, Pellis L, Keijser BJ, de Vos WM, Montijn RC, Roeselers G (2014) Differential modulation by *Akkermansia muciniphila* and *Faecalibacterium prausnitzii* of host peripheral lipid metabolism and histone acetylation in mouse gut organoids. mBio 5(4):e01438-14

McCracken KW, Howell JC, Wells JM, Spence JR (2011) Generating human intestinal tissue from pluripotent stem cells in vitro. Nat Protoc 6:1920–1928. doi:10.1038/nprot.2011.410

Pageot LP, Perreault N, Basora N, Francoeur C, Magny P, Beaulieu JF (2000) Human cell models to study small intestinal functions: recapitulation of the crypt-villus axis. Microsc Res Tech 49:394–406

Roeselers G, Ponomarenko M, Lukovac S, Wortelboer HM (2013) Ex vivo systems to study host–microbiota interactions in the gastrointestinal tract. Best Pract Res Clin Gastroenterol 27:101–113

Sangiorgi E, Capecchi MR (2008) Bmi1 is expressed in vivo in intestinal stem cells. Nat Genet 40:915–920

Sato T, Clevers H (2013) Primary mouse small intestinal epithelial cell cultures. Methods Mol Biol 945:319–328. doi:10.1007/978-1-62703-125-7_19

Sato T, Vries RG, Snippert HJ, van de Wetering M, Barker N, Stange DE, van Es JH, Abo A, Kujala P, Peters PJ, Clevers H (2009) Single Lgr5 stem cells build crypt-villus structures in vitro without a mesenchymal niche. Nature 459:262–265

Sato T, Stange DE, Ferrante M, Vries RGJ, Van Es JH, Van Den Brink S, Van Houdt WJ, Pronk A, Van Gorp J, Siersema PD, Clevers H (2011a) Long-term expansion of epithelial organoids from human colon, adenoma, adenocarcinoma, and Barrett's epithelium. Gastroenterology 141:1762–1772. doi:10.1053/j.gastro.2011.07.050

Sato T, van Es JH, Snippert HJ, Stange DE, Vries RG, van den Born M, Barker N, Shroyer NF, van de Wetering M, Clevers H (2011b) Paneth cells constitute the niche for Lgr5 stem cells in intestinal crypts. Nature 469:415–418

Shaker A, Rubin DC (2012) Stem cells: one step closer to gut repair. Nature 485:181–182. doi:10.1038/485181a

Watson CL, Mahe MM, Múnera J, Howell JC, Sundaram N, Poling HM, Schweitzer JI, Vallance JE, Mayhew CN, Sun Y, Grabowski G, Finkbeiner SR, Spence JR, Shroyer NF, Wells JM, Helmrath MA (2014) An in vivo model of human small intestine using pluripotent stem cells. Nat Med. doi:10.1038/nm.3737

Wiegerinck CL, Janecke AR, Schneeberger K, Vogel GF, Van Haaften-Visser DY, Escher JC, Adam R, Thöni CE, Pfaller K, Jordan AJ, Weis CA, Nijman IJ, Monroe GR, Van Hasselt PM, Cutz E, Klumperman J, Clevers H, Nieuwenhuis EES, Houwen RHJ, Van Haaften G, Hess MW, Huber LA, Stapelbroek JM, Müller T, Middendorp S (2014) Loss of syntaxin 3 causes variant microvillus inclusion disease. Gastroenterology. doi:10.1053/j.gastro.2014.04.002

Yui S, Nakamura T, Sato T, Nemoto Y, Mizutani T, Zheng X, Ichinose S, Nagaishi T, Okamoto R, Tsuchiya K, Clevers H, Watanabe M (2012) Functional engraftment of colon epithelium expanded in vitro from a single adult Lgr5⁺ stem cell. Nat Med 18:618–623

Chapter 23
Porcine Ex Vivo Intestinal Segment Model

D. Ripken and H.F.J. Hendriks

Abstract This chapter describes the use of the porcine ex vivo intestinal segment model. This includes the advantages and disadvantages of the segment model and a detailed description of the isolation and culture as well as the applications of the porcine ex vivo intestinal segment model in practice. Compared to the Ussing chamber (Chap. 24) the porcine ex vivo small intestinal segment model is a relatively simple to use intestinal tissue model. The main difference being that the tissue segment is not mounted in a chamber, but is freely floating in a solution. Therefore the ex vivo intestinal segment model does not distinguish between the apical and basolateral side of the tissue. The intestinal segments can be obtained from various anatomical regions of the small intestine (e.g. duodenum, jejunum, ileum or even the colon) and the segments consist of various cell types (e.g. epithelial cells, paneth cells, goblet cells, enterochromaffin cells and enteroendocrine cells). The intestinal segment model has been shown to be a suitable tool to study compound and location specific effects on the release of gastrointestinal hormones and gastrointestinal metabolism of endocannabinoids and related compounds.

Keywords Ex vivo • Porcine ex vivo intestinal segment model • Gastro-intestinal hormone release

D. Ripken (✉)
Top Institute of Food and Nutrition, Nieuwe Kanaal 9A, 6709 PA Wageningen, The Netherlands

TNO, Utrechtseweg 48, 3704 HE Zeist, The Netherlands

Division of Human Nutrition, Wageningen University, Bomenweg 2, 6703 HD Wageningen, The Netherlands
e-mail: dina.ripken@tno.nl

H.F.J. Hendriks
Top Institute of Food and Nutrition, Nieuwe Kanaal 9A, 6709 PA Wageningen, The Netherlands

© The Author(s) 2015
K. Verhoeckx et al. (eds.), *The Impact of Food Bio-Actives on Gut Health*,
DOI 10.1007/978-3-319-16104-4_23

23.1 The Porcine Ex Vivo Intestinal Segment Model

Although it is possible to obtain human intestinal material for example by taking biopsies using endoscopic techniques, the availability of human tissue is often a bottleneck. Despite human biopsies from the duodenum can be obtained, biopsies from the more distal parts of the small intestine, such as the ileum, are difficult to obtain. This makes it difficult to study region specific effects. The use of tissue segments from other species (e.g. rat, mice, dogs and pigs) can provide a solution for these limitations. It is possible to obtain intestinal tissue segments from various regions of the intestine e.g. duodenum, jejunum, ileum or even the colon. Since the pig intestine shows a high degree of macroscopic and microscopic resemblance with that of humans the best alternative for human tissue is pig (Clouard et al. 2012; Miller and Ullrey 1987; Patterson et al. 2008). Despite the difference in dietary quantity between the two species, pigs and humans have the same dietary habits. Furthermore they have comparable organ sizes, and especially of importance are the similarities in the anatomy, morphology and physiology of the gastrointestinal tract (Clouard et al. 2012; Miller and Ullrey 1987; Patterson et al. 2008). The model was therefore optimized using pig tissue. The porcine ex vivo intestinal segment model will be further referred to as intestinal segment model.

The development of the intestinal segment model was necessary to obtain a more high throughput screening system compared to in vivo studies and the Ussing chamber technique. Moreover a model resembling the human in vivo situation more than single cell cultures was needed to study processes in the gastrointestinal tract, such as nutrient transport and/or hormone release. Currently different cell lines are used for this purpose; Caco-2 and HT-29 cell lines for nutrient uptake and enteroendocrine cells such as murine STC-1 cells (Abello et al. 1994; Geraedts et al. 2009; McLaughlin et al. 1998), GLUTag cells (Reimer et al. 2001) and the human NCI-H716 cell line (Brubaker et al. 1998) to study satiety hormone release. These cell lines are already discussed extensively in the previous chapters. Since intestinal segments consist of many different cell types (e.g. epithelial cells, paneth cells, goblet cells, enterochromaffin cells and enteroendocrine cells), it can be envisioned that this model can be used to study many different biological processes (e.g. nutrient uptake, immune response, hormone release). However up till now only a few applications were found and the model has to compete with a more established model, the Ussing chamber. The main difference of this model compared to the Ussing chamber is that the tissue segment is not mounted in a chamber and thus not distinguishes between the apical and basolateral side of the tissue.

Up till now the model has been used predominantly to study gastrointestinal hormone release due to its short viability (150 min). In this chapter we will therefore focus on gastrointestinal hormone release following interaction with nutrients or related compounds. The presence of nutrients and other molecules in the small intestine could be sensed and consequently stimulate the release of hormones such as Glucagon Like Peptide-1 (GLP-1), Cholecystokinin (CCK) and Peptide YY (PYY). These hormones are released following interaction of nutrients and other compounds with G-protein coupled receptors and solute carrier transporters located on enteroendocrine cells. CCK, GLP-1 and PYY are involved in generating satiety and satiation (Gribble 2012; Tolhurst et al. 2012). Enteroendocrine cells are scattered throughout the epithelial layer of the gastrointestinal tract (Tolhurst et al. 2012). Enteroendocrine I-cells mainly release CCK and are relatively abundant in the duodenum, whereas GLP-1 and PYY are released by L-cells mainly present in the distal jejunum, ileum and colon (Holst 2007; Iwasaki and Yada 2012). Satiety hormone responses depend on the type of nutrient and anatomical region of the small intestine. Nutrient type and anatomical region can both be varied when the intestinal segment model is being used.

23.2   General Protocol

Small intestinal tissue can be obtained from sacrificed pigs. The reproducibility of the results will improve when intestine is used from pigs with comparable health status, age and diet. This is certainly true for satiety studies, however the exact effect of varying health status, age and diet on the physiology of the intestinal segment system is never studied. Furthermore, it is important to collect the intestine as quick as possible; within 5 min after sacrificing the animal. During the procedure the tissue has to be kept on ice. Various anatomical regions of the small intestine can be collected, including duodenum (starting 10 cm distal of the pylorus, total length ± 0.5 m), proximal, mid and distal jejunum (1.5, 5 and 10 m distal of the pylorus, respectively), proximal, mid and distal ileum (1.5, 1 and 0.5 m proximal from the ileal cecal valve), and colon. The intestine is stored in ice cold Krebs-Ringer Bicarbonate (KRB) buffer, which is bubbled with a O_2/CO_2 (95 %/5 %) gas mixture to prevent ischemia, and transported as quickly as possible to the laboratory. Upon arrival the intestinal tissue is carefully flushed and rinsed with ice cold KRB buffer, to remove luminal debris and put in a beaker with KRB on ice. The intestine is cut open in a longitudinal direction and the outer muscle layers are carefully stripped off with the basolateral side upwards. The mucosal tissue has to be placed on gauze (pores $= 250$ µm, Sefar Nitex 03-250/50, Sefar Heiden Switzerland) with the apical side upward, then circles of tissue (tissue segments) can be punched out (payers patches are excluded) using a biopsy punch with a diameter of 8 mm (Miltex, York, USA) (Fig. 23.1b). Tissue segment size may be varied depending on research question and amount of tissue required for analysis. After collecting all segments in the 24-wells plate filled with 500 µl ice cold KRB buffer/well (Fig. 23.1c, d), the tissue is brought to room temperature in 30 min, followed by a pre-incubation at 37 °C for 1 h in a humidified incubator at 5 % v/v CO_2. To study the effects of nutrients on intestinal hormone release, incubations can be initiated by replacing the KRB buffer with 500 µl pre-warmed solution without D-glucose (0 mM) containing the test compounds and incubate for 1 h at 37 °C at 5 % v/v CO_2. Add phenylmethanesulfonyl fluoride (PMSF) (final concentration 100 µM) to the collected media to inactivate serine protease activity and to avoid cleavage of active GLP-1 (7-36 and

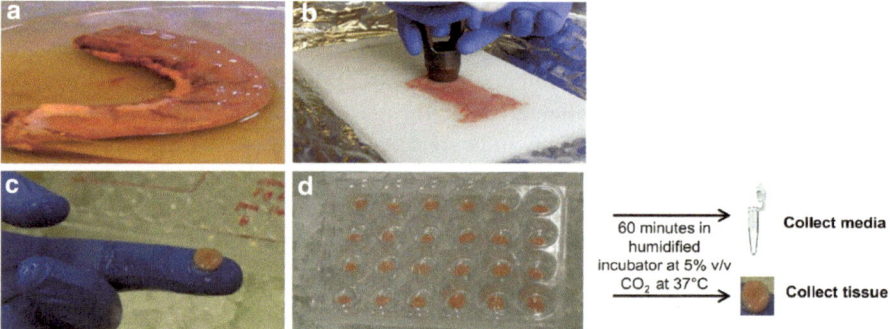

Fig. 23.1 Porcine ex vivo intestinal segment model

7-37) by dipeptidyl peptidase IV (DPP-IV). The collected media as well as the tissue can be stored at 4 °C, or −80 °C until further analysis. For lactate dehydrogenase (LDH) analysis, store media samples at 4 °C for maximal 48 h, for GLP-1, PYY and CCK at −80 °C. Tissue sample has to be snap frozen immediately after the incubation and stored at 80 °C until further analysis (e.g. mRNA, or fatty acid and metabolites thereof (endocannabinoids (ECs), *n*-acyl ethanolamines (NAEs) or *n*-acyl serotonins (NAs)).

23.3 Sample Analysis

Satiety hormones can be analyzed using commercially available ELISA kits according to the manufacturer's instructions. Porcine GLP-1 can be measured using the human GLP-1 kit from Millipore (EGLP-35K, Billerica, MA, USA), since the GLP-1 hormone gene sequence is highly preserved. GLP-2 can be determined using a competitive human GLP-2 (1-34) ELISA kit from Phoenix Pharmaceuticals Inc. (Belmont, CA, USA) according to the manufacturer's instructions. There is a 92 % homology of GLP-2 protein between humans and pigs. The cross reactivity of this kit with GLP-1 is 6 %.

PYY release can be measured with a commercial available ELISA kit for total PYY (Bachem, Peninsula Laboratories, San Carlos, CA, USA). This kit measures porcine PYY, which is identical to human PYY (Adrian et al. 1987, 11).

CCK concentrations can be determined using a human EURIA-CCK radioimmunoassay (RIA) kit (Euro-diagnostica, Malmö, Sweden) according to the manufacturer's instructions. An identical sequence of CCK-8 has been found form most mammals, among them pigs and human. The kit measures CCK-8 sulfate (CCK 26-33). Tissue concentrations of ECs (AEA and 2-AG) and NAEs (DHEA, EPEA, DLE, OEA, PEA and SEA) can be analyzed according to a method described previously (Balvers et al. 2012). For this analysis 50–100 mg freeze-dried tissue is required to be within the detection limit of LC–MS method. Although the paper describes the extraction of ileal mouse tissue, this method can also be used for porcine pig tissue (unpublished data). Tissue concentrations of NAEs can be analyzed according to the method described previously (Verhoeckx et al. 2011). For this method 50–300 mg intestinal tissue is required.

23.4 Monitoring Viability

The viability of the tissue segments is important to analyze, especially when secretagogue effects are studied. It has to be avoided that the release of satiety hormones is secondary to cell lysis or other nonspecific toxic effects. To analyze tissue viability macroscopic and microscopic tissue checks could be done after staining with markers for proliferation and apoptosis (e.g. hematoxylin–eosin staining). Another method to check the viability is measuring leakage of intracellular lactate dehydrogenase (LDH). LDH is stable enzyme common in all cells which can be readily detected when cell membranes are no longer intact for instance due to active proteases or mechanical forces during the preparation of the biopsies. The enzyme activity is determined with respect to the total intracellular lactate dehydrogenase. The total LDH activity of the tissue can be determined by homogenize the tissue in ice cold KRB buffer, using a Potter Elvehjem-type teflon pestle tissue grinder (Braun, Melsungen Germany) for 5 min at 200 rpm. It is also possible to use Triton-x as a positive control. Triton-x dissolves membranes and makes the tissue leaky. Samples are excluded when the LDH leakage is more than 10 % of the total intracellular LDH. The incubation time of the segments should not exceed 1 h, since at longer incubation times, LDH leakage will be >10 % (Voortman et al. 2012).

23.5 Incubation with Food Components and Digested Food Samples

The intestinal segment model can be used to study the effect of digested and un-digested food components (e.g. sugars, fatty acids and proteins) as long as the components are dissolved and diluted with KRB buffer (Voortman et al. 2012). When diluting compounds in the KRB buffer, care should be taken that the levels of alcohol and DMSO are below 1 %. It has been shown that the use of undiluted digested food samples from the TIM model (including bile, pancreatic enzymes and/or food components) can be applied onto the apical side of the tissue segments if mounted in a chamber (Westerhout et al. 2014). However, the use of undiluted digested food samples is not yet tested for the intestinal segment system. In this system tissue is exposed on both the apical and the basolateral side and therefore the use of digestion fluid has to be tested first.

23.6 Readout of the Porcine Ex Vivo Intestinal Segment Model

The intestinal segment model is used to study the effects of fatty acids, carbohydrates and proteins on the release of gastrointestinal hormones, such as GLP-1, PYY and CCK (Table 23.1). Hormone concentrations can be expressed as both the relative release compared to basal release after the negative control set to 100 % (KRB buffer without D-glucose), or as the absolute hormone release (pmol/l) per surface area (cm²). The model is also used to study the tissue concentrations of both ECs and NAEs and NAs. The concentration can be expressed per gram tissue, or as relative concentration compared to the negative control set to 100 % (KRB buffer without D-glucose).

The model can also be used to study the uptake of food compounds after labelling these compounds radioactively. The radioactivity in the tissue and in the media can be analyzed using liquid scintillation counting. However since this model does not distinguish between the apical and basolateral side, it is not possible to study transport of molecules from the apical side to the basolateral side with this model. To study this transport, the Ussing technology, or Caco-2 cells might be better systems.

23.7 General Advantages and Disadvantages of the Model

The advantages and disadvantages of the model are summarized in Table 23.2. The advantage of the current intestinal segment model compared to cell cultures is that this intestinal segment model consist of multiple cell types (e.g. epithelial cells, enterochromaffin cells, paneth cells, goblet cells and enteroendocrine cells) instead of one cell type only. This makes it possible to study effects which depend on cell–cell communication. Another advantage of this intestinal segment model is that it can be used to study region specific effects (duodenum, jejunum, ileum, etc.), since the anatomical region of the small intestine is an important parameter to study

Table 23.1 Different studies using the ex vivo porcine intestinal segment model

References	Compound tested	Analysis in tissue	Analysis in culture media
Voortman et al. (2012)	Casein hydrolysate Long chain fatty acids Short chain fatty acids	GLP-1 GLP-2 PYY CCK-8	GLP-1 GLP-2 PYY
Verhoeckx et al. (2011)	Fatty acids N-Acyl serotonins Serotonin	N-Acyl serotonins	GLP-1
Ripken et al. (2014)	Casein (intact protein) Sucrose Safflower oil Rebaudioside A	Endocannabinoids	GLP-1 PYY

Table 23.2 Advantages and disadvantages

Advantages	Disadvantages
Multi-cell system	Biological variations
All intestinal regions can be used	No distinction between apical and
Less labor intensive than Ussing chamber technology	basolateral side
More incubations possible (from the same intestine),	Not studied whether the effect of
since the surface area is smaller compared to the	undiluted digested food samples can be
Ussing chamber technology	studied
	Influence of microbiota not studied

compound specific effects on satiety hormone release. Furthermore the intestinal segment model has a higher throughput since it is a simplified version of the Ussing chamber model. Since it is not required to mount the tissue of the intestinal segment model in a Ussing chamber, the practical skills required for the intestinal segment model are relatively simple compared to the Ussing chamber technology.

An important aspect which might be different from other systems such as organoids and Caco-2 cells is the presence of microorganisms in the intestinal segment model (Roeselers et al. 2013). Although the presence of microorganisms in the intestinal segment system was never tested, it is likely that the natural flora of the host is still present since the intestine is directly isolated from the animal. The influence of microorganisms on nutrient specific effects is still not known and has to be investigated.

One of the main disadvantages of this model is the biological variation in the release of satiety hormones per intestinal segment. The enteroendocrine cells are not homogenously distributed over the tissue and account for only 1 % of the total epithelial cells. Therefore the number of enteroendocrine cells per surface area (cm^2) tissue may vary. For this reason the amount of replicates and the use of different pigs are important when using this model. The number of intestinal replicates can be determined using a power calculation, which depend on the variation of the measured parameter within the population. To study the effects of a specific compound the statistical method should test the effect within the animal (compared to the KRB buffer). The inter-tissue coefficient of variation (CV) for GLP-1 release is 33 ± 7 %, based on the average CV of incubations in triplicate from 10 pigs. The intra-animal CV for basal GLP-1 release is 78 % based on the basal GLP-1 release (after KRB buffer) from 10 pigs. For this reason the effects should be studied within the animal, and not between animals.

Another issue is the tissue source. It is recommended to use tissue from healthy animals who were kept on a controlled diet. The controlled diet is especially important when the effects of nutrients or digested compounds have to be studied. However the effects of various diets, age, weight or health status on the release on satiety hormones using the intestinal segment model are never studied.

References

Abello J, Ye F, Bosshard A, Bernard C, Cuber JC, Chayvialle JA (1994) Stimulation of glucagon-like peptide-1 secretion by muscarinic agonist in a murine intestinal endocrine cell line. Endocrinology 134:2011–2017

Adrian TE, Bacarese-Hamilton AJ, Smith HA, Chohan P, Manolas KJ, Bloom SR (1987) Distribution and postprandial release of porcine peptide YY. J Endocrinol 113:11–14

Balvers MG, Verhoeckx KC, Bijlsma S et al (2012) Fish oil and inflammatory status alter the n-3 to n-6 balance of the endocannabinoid and oxylipin metabolomes in mouse plasma and tissues. Metabolomics 8:1130–1147

Batterham RL, Cowley MA, Small CJ et al (2002) Gut hormone PYY(3-36) physiologically inhibits food intake. Nature 418:650–654

Brubaker PL, Schloos J, Drucker DJ (1998) Regulation of glucagon-like peptide-1 synthesis and secretion in the GLUTag enteroendocrine cell line. Endocrinology 139:4108–4114

Clouard C, Meunier-Salauen MC, Val-Laillet D (2012) Food preferences and aversions in human health and nutrition: how can pigs help the biomedical research? Animal 6:118–136

Geraedts MC, Troost FJ, Saris WH (2009) Peptide-YY is released by the intestinal cell line STC-1. J Food Sci 74:H79–H82

Gribble FM (2012) The gut endocrine system as a coordinator of postprandial nutrient homoeostasis. Proc Nutr Soc 71:456–462

Habib AM, Richards P, Cairns LS et al (2012) Overlap of endocrine hormone expression in the mouse intestine revealed by transcriptional profiling and flow cytometry. Endocrinology 153:3054–3065

Holst JJ (2007) The physiology of glucagon-like peptide 1. Physiol Rev 87:1409–1439

Iwasaki Y, Yada T (2012) Vagal afferents sense meal-associated gastrointestinal and pancreatic hormones: mechanism and physiological role. Neuropeptides 46:291–297

McLaughlin JT, Lomax RB, Hall L, Dockray GJ, Thompson DG, Warhurst G (1998) Fatty acids stimulate cholecystokinin secretion via an acyl chain length-specific, Ca2+-dependent mechanism in the enteroendocrine cell line STC-1. J Physiol 513(Pt 1):11–18

Miller ER, Ullrey DE (1987) The pig as a model for human nutrition. Annu Rev Nutr 7:361–382

Patterson JK, Lei XG, Miller DD (2008) The pig as an experimental model for elucidating the mechanisms governing dietary influence on mineral absorption. Exp Biol Med (Maywood) 233:651–664

Reimer RA, Darimont C, Gremlich S, Nicolas-Metral V, Ruegg UT, Mace K (2001) A human cellular model for studying the regulation of glucagon-like peptide-1 secretion. Endocrinology 142:4522–4528

Ripken D, van der Wielen N, Wortelboer H, Meijerink J, Witkamp R, Hendriks H (2014) Stevia glycoside rebaudioside A induces GLP-1 and PYY release in a porcine ex vivo intestinal model. J Agric Food Chem 62:8365–8370. doi:10.1021/jf501105w

Roeselers G, Ponomarenko M, Lukovac S, Wortelboer HM (2013) Ex vivo systems to study host–microbiota interactions in the gastrointestinal tract. Best Pract Res Clin Gastroenterol 27:101–113

Tolhurst G, Reimann F, Gribble FM (2012) Intestinal sensing of nutrients. Handb Exp Pharmacol (209):309–335

Verhoeckx KC, Voortman T, Balvers MG, Hendriks HF, Wortelboer HM, Witkamp RF (2011) Presence, formation and putative biological activities of N-acyl serotonins, a novel class of fatty-acid derived mediators, in the intestinal tract. Biochim Biophys Acta 1811:578–586

Voortman T, Hendriks HF, Witkamp RF, Wortelboer HM (2012) Effects of long- and short-chain fatty acids on the release of gastrointestinal hormones using an ex vivo porcine intestinal tissue model. J Agric Food Chem 60:9035–9042

Westerhout J, van de Steeg E, Grossouw D et al (2014) A new approach to predict human intestinal absorption using porcine intestinal tissue and biorelevant matrices. Eur J Pharm Sci 63:167–177

Chapter 24
Ussing Chamber

Joost Westerhout, Heleen Wortelboer, and Kitty Verhoeckx

Abstract The Ussing chamber system is named after the Danish zoologist Hans Ussing, who invented the device in the 1950s to measure the short-circuit current as an indicator of net ion transport taking place across frog skin (Ussing and Zerahn, Acta Physiol Scand 23:110–127, 1951). Ussing chambers are increasingly being used to measure ion transport in native tissue, like gut mucosa, and in a monolayer of cells grown on permeable supports. However, the Ussing chamber system is, to date, not often applied for the investigation of the impact of food bioactives (proteins, sugars, lipids) on health.

An Ussing system is generally comprised of a chamber and a perfusion system, and if needed, an amplifier and data acquisition system. The heart of the system lies in the chamber with the other components performing supporting roles. The classic chamber design is still in wide use today. However, several newer designs are now available that optimize for convenience and for diffusion- or electrophysiology-based measurements. A well designed Ussing chamber supports an epithelia membrane or cell monolayer in such a way that each side of the membrane is isolated and faces a separate chamber-half. The chambers are then filled with a physiologically relevant solution, such as Ringer's solution. This configuration allows the researcher to make unique chemical and electrical adjustments to either side of the membrane with complete control. The Ussing chamber technique has its strengths and limitations, which will be explained in more detail in this chapter.

Keywords Ex vivo • Intestine • Intestinal barrier • Intestinal transport • Ussing chamber

J. Westerhout (✉) • H. Wortelboer • K. Verhoeckx
TNO, Utrechtseweg 48, 3704 HE Zeist, The Netherlands
e-mail: joost.westerhout@tno.nl

© The Author(s) 2015
K. Verhoeckx et al. (eds.), *The Impact of Food Bio-Actives on Gut Health*,
DOI 10.1007/978-3-319-16104-4_24

24.1   The Ussing Chamber

A commercially available 'classic' Ussing chamber system (Fig. 24.1) is machined from solid acrylic into two halves and has vertical and horizontal ports in each half for connection to the circulation system and for making electrical connections. The face of one chamber-half is imbedded with sharp stainless-steel pins which mate with corresponding holes in the other chamber-half face. These pins allow puncturing and positioning of an epithelium membrane within the chamber. Each chamber-half also has a separate air/gas inlet to drive the circulation system. Gas, commonly a 95 % O_2/5 % CO_2 mixture, is forced under low pressure into these inlets and allows contact with the buffer solution. The rising bubbles drive circulation and oxygenate the buffer. Each chamber-half also has an inlet or outlet port for access to the water jacket. Water at the desired temperature is pumped through the jacket and allowing thermoregulation of the perfusion solutions. Each chamber is supplied with 2 Ag-AgCl pellet electrodes (for voltage) and 2 Ag wire electrodes (for current). Electrodes are connected into the chamber by means of an agar salt bridge.

One of the major limitations of the classic Ussing chamber model is that it does not allow simultaneous preparation and analyses of a large set of segments of epithelial tissues thereby having a relatively low throughput. The original Ussing chamber system incorporates the intestinal tissue segments in a vertical position, whereas other systems also allow horizontal placement (e.g. TNO's InTESTine™ system or Warner Instruments' NaviCyte™). This horizontal setup can be used for transport or toxicology studies using cells or tissue exposed to liquids and semi-solid compounds, such as digested food samples, while the basolateral surface remains exposed to the medium.

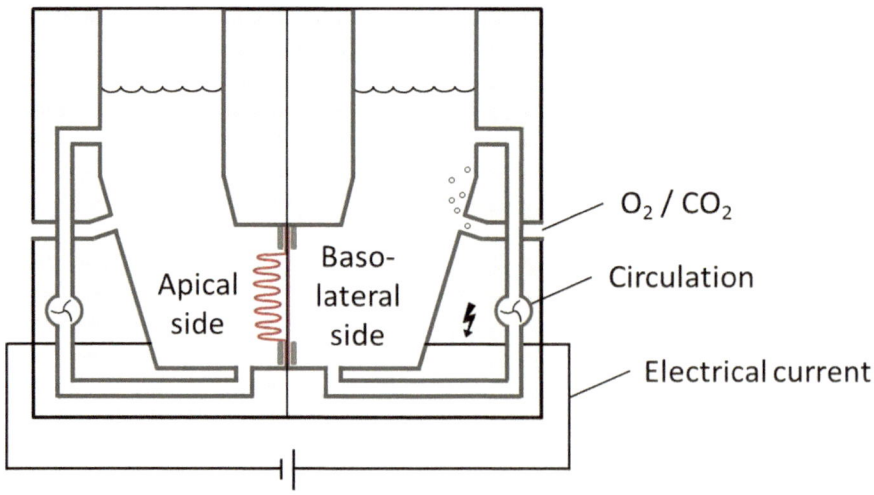

Fig. 24.1 Schematic representation of a small piece of intestinal epithelial tissue mounted in the Ussing chamber

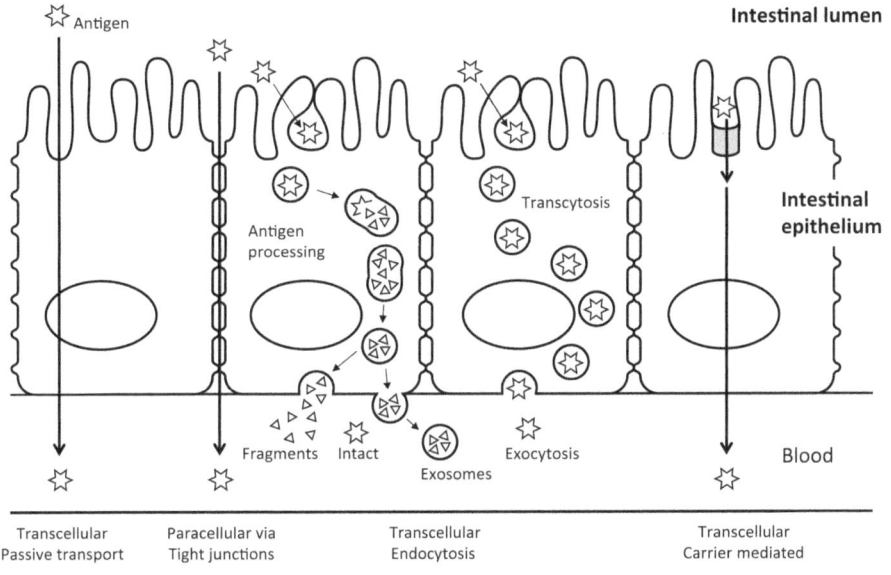

Fig. 24.2 Different transport routes across the intestinal wall (Reitsma et al. 2014, copyright Wiley-VCH Verlag GmbH & Co. KGaA. Reproduced with permission)

Further development of the Ussing chamber system has led to various multichannel systems capable of supporting up to 24 independent chambers, all operating in parallel. TNO recently developed a medium throughput system (InTESTine™) in which up to 96 ex vivo intestinal tissue segments can be used per system per day to investigate intestinal absorption (Westerhout et al. 2014).

Food compounds can be transported across the intestinal barrier by different transport mechanisms (Fig. 24.2). The transport route and velocity of absorption of these compounds are influenced by different factors that are related to physiological factors of the tissue (e.g. composition and thickness of the mucus layer, disease state or membrane permeability, intestinal pH, concentration bile acids, composition of pancreatic juice, surface area, enzyme activity, lipid and protein content of the enterocyte membrane and the amount of Peyer's patches in the tissue), physicochemical factors of the compound (e.g. water solubility, molecular weight, aggregation state, charge, H-bonding capacity, and hydrophobicity) and formulation factors (e.g. food matrix) (Deferme et al. 2008).

In general compounds are transported via one of the following mechanisms:

(1) Paracellular (e.g. small hydrophilic and polar compounds)
(2) Transcellular

- passive transport (lipophilic compounds)
- endocytosis (e.g. proteins and small nucleotides)
- carrier mediated transport (e.g. glucose and amino acids)

(3) Cell mediated transport

In contrast to single cell-culture models, such as the epithelial Caco-2 cell model, intestinal epithelium consists of absorptive enterocytes and other cells such as goblet cells, endocrine cells, and M cells, with the mucus secreting goblet cells representing the second most frequent cell type, ranging from 10 % in the small intestine and 24 % in the distal colon (Madara and Trier 1987). The transport of food compounds across the intestinal barrier can also be facilitated by the other cell types present in the lamina propria (Fig. 24.3), including M cells, Peyer's patches, dendritic cells and macrophages.

As the Ussing chamber technique uses intestinal tissue segments, they still contain the morphological and physiological features of the intestine, including interplay of many complex processes, such as interaction between the multicellular environment (Rozehnal et al. 2012). Using ex vivo intestinal tissue segments gives a better representation of the complex in vivo morphology (multicellular conglomeration, presence of the mucus layer) and thereby a better representation of the various possible processes involved in the in vivo situation. Furthermore, ex vivo intestinal tissue segments from the different regions of the small and large intestine allow the investigation of regional absorption and immune responses, which is not possible when using single cell-culture models.

The major limitation of using ex vivo intestinal tissue segments is the limited availability of (healthy) human tissue. Therefore, animal tissue is often used. Porcine intestinal tissue could possibly be used as alternative based on its high similarity with that from humans (Patterson et al. 2008; Walters et al. 2011; Groenen et al. 2012). When using human tissue, often a small segment of non-diseased biopsy is

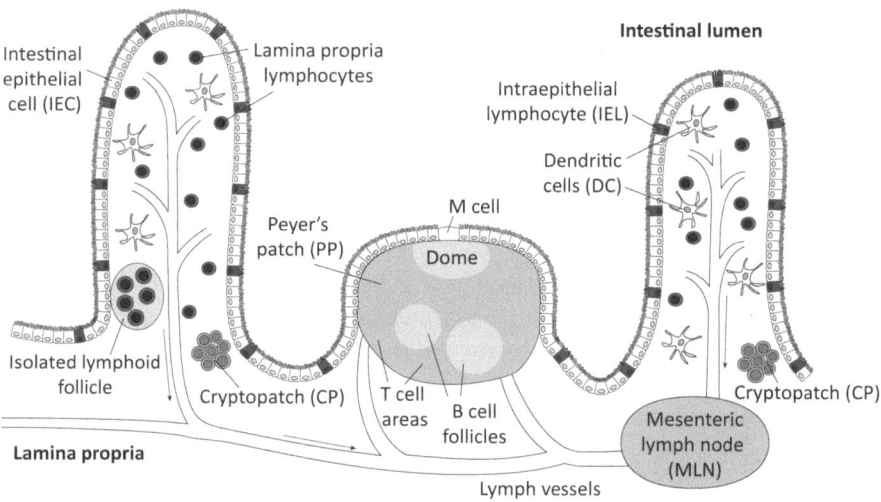

Fig. 24.3 The intestine at a cellular level (Reitsma et al. 2014, copyright Wiley-VCH Verlag GmbH & Co. KGaA. Reproduced with permission)

used, which could be more-or-less influenced by the adjacent diseased tissue. However, besides the effects of the possible side effects of diseased tissue and/or historic drug treatment, the inter-individual variability among humans or any other non in-bred animal model also in a healthy stage is high, making it difficult to directly compare results.

Another major limitation of using ex vivo intestinal tissue segments is the limited viability of the tissue. Previous studies by Haslam et al. (2011), Rozehnal et al. (2012) and Sjöberg et al. (2013) have shown that the intestinal tissue remains viable only for up to 150 min (on average 120 min), which is long enough to study absorption and/or secretion of satiety hormones. For the investigation of the impact of food bioactives on other health effects 2 h may be too short for any clear response, such as release of cytokines, reversible disruption of tight junctions, extensive metabolism. However, changes in mRNA levels within 2 h should be measurable by using reverse transcription polymerase chain reactions (RT-PCR).

24.2 General Protocol

Human intestinal tissue segments are often obtained via biopsies or by surgical resection, whereas intestinal tissue segments from animals are often obtained from section of sacrificed animals. Collect the intestine, directly after opening the abdominal cavity, in a cold Krebs' Ringer Bicarbonate buffer (pH 7.4), supplemented with 25 mM HEPES (KRB/HEPES), which is bubbled with a O_2/CO_2 (95/5 %) gas mixture, prior to use to prevent ischemia (e.g. during transport). The intestine can be divided in different segments; duodenum can be found directly after the stomach and is approx. 20 cm long, followed by the jejunum (approx. 1.0 m), ileum (approx. 1.6 m) and after the cecum you will find the colon (approx. 1.1 m) (DeSesso et al. 2001).

Upon arrival in the laboratory, flush and rinse the intestinal tissue segments carefully with ice cold physiologically relevant buffer, such as KRB/HEPES buffer, to remove luminal debris and put in a beaker with KRB/HEPES on ice. The tissue segments are then opened longitudinally next to the juncture to the peritoneum and the serosa and muscle layers are removed (stripping) while the tissue is submerged in chilled and oxygenated KRB/HEPES buffer. Removal of muscle layers is either performed using scissors keeping the feather scissor at an angle towards the muscles, or using a pair of curved tweezers to gently strip off the muscle layers while holding the rim of the tissue segment with a Kocher. After the muscle layer is gently removed, the tissue is cut clean from fat, connective tissue and parts of the submucosa to reach the blood vessel level. Depending on the system, intestinal tissue segments with a surface area of 0.1–1.8 cm^2 are prepared using for instance a punch. Keep the tissue and segments in cold buffer during the whole procedure. Stripped intestinal segment is then mounted in the Ussing chamber system (Fig. 24.4). Since epithelial tissue is polarized, it contains an apical (mucosal) and basolateral (serosal) side, so pay attention on how to mount the tissue in the system.

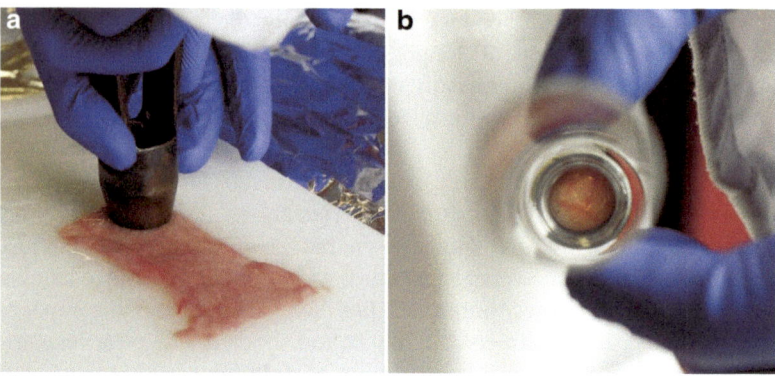

Fig. 24.4 (**a**) Photograph of obtaining a small segment from stripped porcine intestinal tissue by using a hollow punch. (**b**) The smaller segments are then mounted in the InTESTine™ system

Both chamber halves are then filled with ice cold physiologically relevant buffer, such as KRB/Hepes buffer. The basolateral KRB/HEPES buffer is sometimes supplemented with 0.1 % bovine serum albumin (BSA) to enhance intestinal transport. The solution is then oxygenated (95 % O_2:5 % CO_2), circulated, and warmed to 37 °C. Once at 37 °C, carefully flush the mucus layer from the segments and replace the apical buffer by fresh, pre-warmed buffer containing the compound of interest, to investigate the permeability. In most cases compounds are labelled radioactively (e.g. ^{14}C) or with a fluorescent label. During the permeability studies small (100 μl) samples are taken over time from the basolateral and apical compartment for up to 150 min. After 150 min, the viability of the tissue and the integrity of the intestinal barrier starts to decrease (Sjöberg et al. 2013). If desired, the basolateral and apical buffer can be supplemented with fresh, pre-warmed, blank buffer, to complement for the samples taken. Usually, an apparent permeability value (Papp value, unit is cm/s) is calculated according to Eq. (24.1).

$$Papp = \frac{dQ/dt}{SA * C_{Api,0}} \tag{24.1}$$

Here, dQ/dt (in disintegrations per minute (dpm)/s or mg/s) indicates the rate of transport of the compound from the apical side to the basolateral side over time when the rate of transport is linear (usually between T=60 and T=120 min), SA is the surface area of the exposure area (cm^2) and $C_{Api,0}$ is the initial donor concentration of the compound (concentration at apical side at T=0 in dpm/ml or mg/ml).

To study the absorption of compounds, the presence of an unstirred water layer upon the enterocytes should be prevented by continuous moving of the incubation buffers, either by gassing of the media (Ussing system) or placement of the complete system on a rocker platform (InTESTine™ system, Westerhout et al. 2014). Care must also be taken to prevent nonspecific binding of the test compound, evaporation of the media during incubation, cross-contamination of the test compound to other chambers (especially in case of gassing of the incubation media). The reusable Ussing system therefore requires thorough cleaning and fine adjustment of the gassing rate prior to performing experiments. The InTESTine™ system inserts are generated from disposable glass material, which reduces nonspecific binding of the compound of interest and risk of contamination of the mounted device. Furthermore, the set-up and horizontal mounting enables the InTESTine™ system to be incubated in a carbogenated, humidified incubator at 37 °C on a rocker platform thereby reducing the unstirred water layer, evaporation and possible foaming (Westerhout et al. 2014).

24.3 Monitoring Viability

Previous studies by Haslam et al. (2011), Rozehnal et al. (2012) and Sjöberg et al. (2013) have used the potential difference (>4 mV), short-circuit current (>100 $\mu A/cm^2$), trans epithelial electrical resistance (TEER; >20 Ω cm^2) for continuous monitoring of the viability. Low resistance indicates tissue leakage. Due to the set-up of the Ussing chamber system, these measurements can be performed for each experiment. Other markers that can be studied in parallel to a typical transport study include the intestinal metabolism of testosterone and midazolam, by measuring the presence of metabolites apically, basolaterally and intracellularly by means of, for instance, High Pressure Liquid Chromatography (HPLC) (Sjöberg et al. 2013). Transport via the paracellular pathway can be assessed by measuring the permeability of Lucifer yellow (Papp value <6×10^{-6} cm/s) (Rozehnal et al. 2012) leakage of fluorescein isothiocyanate dextran to the basolateral side (FD4; <0.5 %) (Westerhout et al. 2014), or the linearity of transport of atenolol (R^2>0.995 with at least 3 datapoints) (Haslam et al. 2011). Values higher than the ones mentioned above, indicate leakage of the intestinal tissue. Ideally, one would monitor multiple parameters to evaluate the viability and integrity of the intestinal mucosa in parallel to the transport of the compound of interest, but in some cases the compound of interest has the same properties as the marker compounds (e.g. when using fluorescently labeled or radio-labeled compounds).

24.4 Use of Digested Food Samples

While we have shown that undiluted digested food samples (including bile, pancreatic enzymes and/or food components), obtained by using TNO's dynamic computer-controlled in vitro gastrointestinal model of the stomach and small intestine (TIM-1, Minekus et al. 1995) can be applied directly onto intestinal tissue segments (Westerhout et al. 2014), it must be noted that the digested food samples can also be presented in a different formulation. Care should be taken that the levels of alcohol and DMSO are below 1 % and other conditions, such as bile salt concentration, osmolarity and pH, are in concordance to what has been presented by Bergström et al. (2014). In general, bile salt concentrations should be between 1.4 and 24 mM, the osmolarity should be between 137 and 416 mOsm and the pH should be between 5.4 and 7.8.

24.5 Readout of the Ussing Chamber System

Since the Ussing chamber system is not often applied for the investigation of the impact of food bioactives (proteins, sugars, lipids) on health it is difficult to identify typical readout parameters. To date, the Ussing chamber has been applied to investigate the general transport of proteins, sugars and lipids across intestinal tissue (Table 24.1). The readout for these types of studies is the percentage of the applied dose that can be detected in the basolateral compartment intact using for instance radiolabel counting (e.g. ^{3}H, ^{125}I, ^{14}C), HPLC coupled to a mass spectrometer or UV

Table 24.1 Different absorption studies using intestinal tissue segments of different species mounted in the Ussing chamber system

References	Species	Protein	Sugar	Lipid	Food-related	Analysis
Ducroc et al. (1983)	Rabbit	HRP			No	Enzymatic activity, ^{3}H-radioactive labeling
Gabler (2009)	Pig		Glucose		Yes	Potential difference
Hardin et al. (1999)	Rabbit	BSA			No	Immunoblotting, ELISA and ^{125}I-radioactive labeling
Heyman et al. (1984)	Human (0.5–13 years)	HRP			No	Enzymatic activity and ^{3}H-radioactive labeling
Herrmann et al. (2012)	Pig		Glucose		Yes	Potential difference
Keljo and Hamilton (1983)	Pig	HRP			No	Enzymatic activity
Larsen et al. (2001)	Human		Glucose		Yes	Potential difference
Majamaa and Isolauri (1996)	Human (0.5–8 years)	HRP			No	Enzymatic activity and ^{125}I-radioactive labeling
Schulthess et al. (1996)	Rabbit			Cholesterol	Yes	Merckotest
Sjöberg et al. (2013)	Human (adults)		Glucose		Yes	^{14}C-radioactive labeling
Sutas et al. (1997)	Wistar rat pups	HRP			No	^{125}I-radioactive labeling and HPLC gel filtration

Table 24.2 General advantages, disadvantages and limitations of the Ussing chamber system

Advantages	• Use of ex vivo (human) intestinal tissue, which contain the morphological and physiological features of the intestine • Possibility to study regional (duodenum, jejunum, ileum or colon) absorption and immune responses • Animal tissue allows the possibility to investigate the effect of sensitization on the intestinal absorption of proteins by immunizing the tissue in vivo prior to the in vitro studies (Walker et al. 1974)
Disadvantages	• Viability of the tissue segments up to 150 min • Limited availability of healthy human tissue. Alternatively, animal tissue can be used, but the extrapolation from animals to humans remains to be investigated • Large inter-individual variability, making it difficult to interpret results
Limitations	• Relatively low throughput • Labor intensive

detector or enzymatic activity. While horseradish peroxidase (HRP) and BSA are not food-related, at least it shows that the Ussing chamber technique can be applied to investigate the transport of proteins across intestinal tissue. At TNO we are currently investigating the possibility to use porcine intestinal tissue segments for studying the transport of different allergens across intestinal tissue. Furthermore, we also use porcine intestinal tissue segments to monitor cytokine release at mRNA levels, since cytokine release at protein level is limited due to viability issue's mentioned before.

24.6 General Advantages, Disadvantages and Limitations of the Ussing Chamber System

The Ussing chamber system is, to date, not often applied for the investigation of the impact of food bioactives (proteins, sugars, lipids) on health. This indicates that still a lot of research can be done in this field. The general advantages and disadvantages of the Ussing chamber system are summarized in Table 24.2.

References

Bergström CAS, Holm R, Jørgensen SA, Andersson SBE, Artursson P, Beato S, et al (2014) Early pharmaceutical profiling to predict oral drug absorption: current status and unmet needs. Eur J Pharm Sci 57:173–199
Deferme S, Annaert P, Augustijs P (2008) In vitro screening models to assess intestinal drug absorption and metabolism in drug absorption studies. In: Biotechnology: pharmaceutical aspects, vol VII. Springer, New York, pp 182–215

DeSesso JM, Jacobson CF (2001) Anatomical and physiological parameters affecting gastrointestinal absorption in humans and rats. Food Chem Toxicol 39:209–228

Ducroc R, Heyman M, Beaufrere B, Morgat JL, Desjeux JF (1983) Horseradish peroxidase transport across rabbit jejunum and Peyer's patches in vitro. Am J Physiol 245:G54–G58

Gabler NK, Radcliffe JS, Spencer JD, Webel DM, Spurlock ME (2009) Feeding long-chain n-3 polyunsaturated fatty acids during gestation increases intestinal glucose absorption potentially via the acute activation of AMPK. J Nutr Biochem 20:17–25

Groenen MAM, Archibald AL, Uenishi H, Tuggle CK, Takeuchi Y, Rothschild MF et al (2012) Analyses of pig genomes provide insight into demography and evolution. Nature 491: 393–398

Hardin JA, Kimm MH, Wirasinghe M, Gall DG (1999) Macromolecular transport across the rabbit proximal and distal colon. Gut 44:218–225

Haslam IS, O'Reilly DA, Sherlock DJ, Kauser A, Womack C, Coleman T (2011) Pancreatoduodenectomy as a source of human small intestine for Ussing chamber investigations and comparative studies with rat tissue. Biopharm Drug Dispos 32:210–221

Herrmann J, Schröder B, Klinger S, Thorenz A, Werner AC, Abel H, Breves G (2012) Segmental diversity of electrogenic glucose transport characteristics in the small intestines of weaned pigs. Comp Biochem Physiol A Mol Integr Physiol 163:161–169

Heyman M, Boudraa G, Sarrut S, Giraud M et al (1984) Macromolecular transport in jejunal mucosa of children with severe malnutrition: a quantitative study. J Pediatr Gastroenterol Nutr 3:357–363

Keljo DJ, Hamilton JR (1983) Quantitative determination of macromolecular transport rate across intestinal Peyer's patches. Am J Physiol 244:G637–G644

Larsen R, Mertz-Nielsen A, Hansen MB, Poulsen SS, Bindslev N (2001) Novel modified Ussing chamber for the study of absorption and secretion in human endoscopic biopsies. Acta Physiol Scand 173:213–222

Madara JL, Trier JS (1987) Functional morphology of the mucosa of the small intestine. In: Johnson R (ed) Physiology of the gastrointestinal tract, 2nd edn. Raven, New York, pp 1209–1249

Majamaa H, Isolauri E (1996) Evaluation of the gut mucosal barrier: evidence for increased antigen transfer in children with atopic eczema. J Allergy Clin Immunol 97:985–990

Minekus M, Marteau P, Havenaar R, Huis in't Veld JHJ (1995) A multi compartmental dynamic computer-controlled model simulating the stomach and small intestine. Altern Lab Anim 23:197–209

Patterson JK, Lei XG, Miller DD (2008) The pig as an experimental model for elucidating the mechanisms governing dietary influence on mineral absorption. Exp Biol Med (Maywood) 233:651–664

Reitsma M, Westerhout J, Wichers HJ, Wortelboer HM, Verhoeckx KC (2014) Protein transport across the small intestine in food allergy. Mol Nutr Food Res 58:194–205

Rozehnal V, Nakai D, Hoepner U, Fischer T, Kamiyama E, Takahasi M, Yasuda S, Mueller J (2012) Human small intestinal and colonic tissue mounted in the Ussing chamber as a tool for characterizing the intestinal absorption of drugs. Eur J Pharm Sci 46:367–373

Schulthess G, Compassi S, Boffelli D, Werder M, Weber FE, Hauser H (1996) A comparative study of sterol absorption in different small-intestinal brush border membrane models. J Lipid Res 37:2405–2419

Sjöberg Å, Lutz M, Tannergren C, Wingolf C, Borde A, Ungell A-L (2013) Comprehensive study on regional human intestinal permeability and prediction of fraction absorbed of drugs using the Ussing chamber technique. Eur J Pharm Sci 48:166–180

Sutas Y, Autio S, Rantala I, Isolauri E (1997) IFN-gamma enhances macromolecular transport across Peyer's patches in suckling rats: implications for natural immune responses to dietary antigens early in life. J Pediatr Gastroenterol Nutr 24:162–169

Ussing HH, Zerahn K (1951) Active transport of sodium as the source of electric current in the short-circuited isolated frog skin. Acta Physiol Scand 23:110–127

Walker WA, Isselbacher KJ, Bloch KJ (1974) Immunologic control of soluble protein absorption from the small intestine: a gut-surface phenomenon. Am J Clin Nutr 27:1434–1440

Walters EM, Agca Y, Ganjam V, Evans T (2011) Animal models got you puzzled?: think pig. Ann N Y Acad Sci 1245:63–64

Westerhout J, van de Steeg E, Grossouw D, Zeijdner EE, Krul CAM, Verwei M, Wortelboer HM (2014) A new approach to predict human intestinal absorption using porcine intestinal tissue and biorelevant matrices. Eur J Pharm Sci 63:167–177

Part VI
In Vitro Fermentation Models: General Introduction

The human colonic microbiota is a dense and highly diverse microbial community that contributes in many ways to host health, including through the recovery of metabolic energy from non-digestible dietary components and the maintenance of intestinal homeostasis (Bäckhed et al. 2012; Hooper et al. 2012). The evaluation of gut microbial composition and diversity, and its impact on food digestion, by either culture dependent or culture independent (e.g. metagenomics) methodologies has focused primarily on fecal samples, which are considered to be representative of the distal large intestine. These studies, however, do not provide insights into dynamic microbial processes and functionality or digestion at their locations in the gut. In vitro fermentative models are considered excellent tools to allow the screening of a large number of substances, ranging from dietary ingredients to pathogens, drugs and toxic or radioactive compounds, to assess how they alter and are altered by gastrointestinal environments and microbial populations without ethical constraints. In vitro models vary from batch incubations using anaerobic conditions and dense fecal microbiota to more complex continuous models involving one or multiple connected, pH-controlled chemostats inoculated with fecal microbiota and representing different parts of the human colon (Venema and Van den Abbeele 2013).

The common purpose of in vitro gut fermentation models is to cultivate a complex intestinal microbiota under controlled environmental conditions for carrying out microbial modulation and metabolism studies. Selection of the appropriate model depends on the study objectives and the advantages and limitations exhibited by each type of system (Payne et al. 2012). Part VI describes a number of these in vitro gut fermentation models.

Batch Fermentation Models

Batch-type simulators represent models ranging from closed bottles used to grow defined (single or mixed) bacterial strains to controlled reactors inoculated with fecal microbiota suspensions. These models usually operate under anaerobic conditions and studies proceed over short-term periods. They are used to study the effects of substrates on the physiology and biodiversity of intestinal microorganisms, and vice versa. The impact on the microbiota is evaluated by molecular quantitative and qualitative techniques and the impact on metabolic activity is assessed by evaluating the formation of short chain fatty acids (SCFA) or other metabolites. These models are appropriate to check for inter-individual variability in the response to a particular bioactive or agent or for comparison of the consequences of exposure to different sources or doses of compounds. In addition, as noted, they provide a first assessment of the types of microbial metabolites formed and helps to elucidate the pathways involved (see Chap. 25). On the other hand, they are limited by substrate depletion and the accumulation of the end products of microbial metabolism that alter the conditions in the batch away from the microbial balanced starting point, thus affecting the in vivo relevance in longer simulations.

Dynamic Fermentation Models

Several dynamic fermentation models have been developed in recent years with the purpose to establish in vitro a relatively stable microbial ecosystem under physiologically relevant colon conditions. Continuous multistage and single-fermentation models are useful for longer-term experiments needed to evaluate the spatial and temporal adaptation of the colonic microbiota to dietary compounds and the microbial metabolism of dietary ingredients.

Most of the dynamic multistage-fermentation models are based on the *Reading model* firstly described by Gibson et al. (1988). The model consists of three vessels of increasing size, aligned in series such that a sequential feeding of growth medium occurs, mimicking the conditions of proximal, transverse and distal colon. The operating conditions included setting of pH for the three vessels at 5.5, 6.2, and 6.8, respectively; values which were calculated using measurements made on colonic contents taken from sudden death victims (Macfarlane et al. 1992). Thus, the first vessel has a high availability of substrate, representing a rapid bacterial growth rate and is operated at an acidic pH, similar to events in the proximal colon. In contrast, the final vessel resembles the neutral pH, slow bacterial proliferation rate and low substrate availability which is characteristic of more distal regions. The authors also developed a nutritious medium that has been extensively used in both batch and dynamic fermentation models. The medium consists of protein substrates (casein and peptone), complex carbohydrates (pectin, xylan, arabinogalactan and resistant starches) that are not digested by gastrointestinal enzymes, and a mixture of salts and vitamins (Gibson et al. 1988). The viability of the in vitro gut microbiota is dependent on the continuous replenishment of nutrients and the control of physiological temperature and anaerobic conditions. The control of these factors allows the establishment of steady-state conditions, with respect to both microbial composition and metabolic activity, whereas the control of defined pH values, downstream nutrient limitations and retention times in the different vessels allows a region-specific differentiation of the colonic microbiota in terms of metabolic activity and microbial communities. Anaerobic conditions are usually reached through continuous CO_2 or N_2 flushing of the headspace. Two representative three-stage fermentation models are described in Chaps. 27 (SHIME) and 28 (SIMGI).

Continuous single-stage fermentation models are often designed to simulate the proximal colon conditions and to reproduce its metabolic activity. The most sophisticated dynamic model is represented by the multiple-component continuous system TIM-2 (Minekus et al. 1999). The model reproduces the peristaltic mixing of proximal colonic luminal content as well as the absorption of water and fermentation products (see Chap. 26). A comparison of the TIM-2 and the multistage-fermentation model SHIME (see Chap. 27) to study long-chain arabinoxylans as potential prebiotics has demonstrated that both models similarly revealed a compound-specific modulation of prebiotics in terms of short chain fatty acid production and stimulation of specific *Bifidobacterium* species (Van den Abbeele et al. 2013). The simulation of the human distal gut environment by maintaining neutral pH conditions and a constant culture volume has been recently described by McDonald et al. (2013) using a twin-vessel single-stage chemostat model. Moreover, recent developments are addressing the construction of single-stage models simulating the ileum microbiota (Venema and Van den Abbeele 2013). The ileum is the region harboring the highest bacterial population within the small intestine due to its slower peristalsis and lower concentrations of digestive enzymes and bile. This type of model can be of special utility to demonstrate bacterial competition against food-borne pathogens and toxin production (Ceuppens et al. 2012).

Inoculation of Gut Fermentation Models

The inoculation and colonization of in vitro fermentation systems influences the reproducibility of the studies and constitute a challenge of the models (Payne et al. 2012). Fermenters are usually inoculated with a liquid fecal suspension from an individual or pooling stools from several subjects. In the last case, the fecal samples are used to inoculate a fed-batch fermenter to produce a standardized inoculum that is stored frozen. Cinquin et al. (2004) developed an immobilization process for the entrapment of fecal microbiota in mixed xanthan–gellan gum gel beads to reach high cell densities and to maintain the microbial diversity over long time continuous colonic fermentations. The model has recently been updated into the model PolyFermS where it is possible to stably and reproducibly cultivate complex intestinal communities in multiple reactors allowing studying in parallel the impact of different treatments compared to a control reactor (Zihler Berner et al. 2013). In order to facilitate the reproducibility of experiments, recent developments are addressing the inoculation of fermentation models with defined populations of human gut microorganisms represented by common saccharolytic and amino acid-fermenting populations in the large intestine (Newton et al. 2013).

Host-Gut Microbiota Interaction

A major drawback of gut fermentation models is the limitations for simulating the host functionality. The combination of fermentation models and intestinal cell cultures represent a common approach to reproduce in vitro the host responses. These studies have been usually performed with colon epithelial cell cultures and/or immune cells to evaluate adherence, cytokine production and gene expression, among others (Venema and Van den Abbeele 2013). Additional tools that have improved modeling of the physiological colonic conditions are the incorporation in the reactors of a mucosal environment able to differentiate between the luminal microbiota and the microbial biofilms adhering to the colonic epithelium (Macfarlane et al. 2005; Van den Abbeele et al. 2012). The Host-Microbiota Interaction (HMI) module is a recently developed device adapted to the SHIME model that allows long-term studies of a complex microbial community colonizing a mucus layer, while being co-cultured microaerophilically (up to 48 h) in the presence of shear forces and a monolayer of enterocyte human cells (Marzorati et al. 2014). This combination of in vitro models represents an approximation to the conceptualized ideal experimental model for the study of host-gut microbiota interactions described by Fritz et al. (2013). These authors claim that the ideal model should include epithelial cells, complex gut microbiota, anaerobic/microaerophilic conditions, a mucus layer and physiological conditions of pH, fluid retention times and dissolved O_2 concentrations.

The progress made in developing in vitro fermentation models able to closely mimic the gut microbial environment and the increasing knowledge regarding microbial populations and host-gut microbiota interactions can offer remarkable insights into gut microbiota functions. Moreover, ongoing studies would allow the development of well-defined mixtures of microorganisms that retain a high level of diversity and encompass key functions required for healthy intestinal homeostasis in an approach coined "microbial ecosystem therapeutics" by Petrof et al. (2013).

References

Bäckhed F, Fraser CM, Ringel Y, Sanders ME, Sartor RB, Sherman PM, Versalovic J, Young V, Finlay BB (2012) Defining a healthy human gut microbiome: current concepts, future directions, and clinical applications. Cell Host Microbe 12:611–622

Ceuppens S, Uyttendaele M, Drieskens K, Heyndrickx M, Rajkovic A, Boon N, Van de Wiele T (2012) Survival and germination of *Bacillus cereus* spores without outgrowth or enterotoxin production during in vitro simulation of gastrointestinal transit. Appl Environ Microb 78:7698–7705

Cinquin C, Le Blay G, Fliss I, Lacroix C (2004) Immobilization of infant fecal microbiota and utilization in an *in vitro* colonic fermentation model. Microb Ecol 48:128–138

Fritz JV, Desai MS, Shah P, Schneider JG, Wilmes P (2013) From meta-omics to causality: experimental models for human microbiome research. Microbiome 1:14

Gibson GR, Cummings JH, Macfarlane GT (1988) Use of a three-stage continuous culture system to study the effect of mucin on dissimilatory sulfate reduction and methanogenesis by mixed populations of human gut bacteria. Appl Environ Microb 54:2750–2755

Hooper LV, Littman DR, Macpherson AJ (2012) Interactions between the microbiota and the immune system. Science 336:1268–1273

Macfarlane GT, Gibson GR, Cummings JH (1992) Comparison of fermentation reactions in different regions of the human colon. J Appl Bacteriol 72:57–64

Macfarlane S, Woodmansey EJ, Macfarlane GT (2005) Colonization of mucin by human intestinal bacteria and establishment of biofilm communities in a two-stage continuous culture system. Appl Environ Microb 71:7483–7492

Marzorati M, Vanhoecke B, De Ryck T, Sadaghian Sadabad M, Pinheiro I, Possemiers S, Van den Abbeele P, Derycke L, Bracke M, Pieters J, Hennebel T, Harmsen HJ, Verstraete W, Van de Wiele T (2014) The HMI™ module: a new tool to study the host-microbiota interaction in the human gastrointestinal tract in vitro. BMC Microb 14:133

McDonald JA, Schroeter K, Fuentes S, Heikamp-Dejong I, Khursigara CM, De Vos WM, Allen-Vercoe E (2013) Evaluation of microbial community reproducibility, stability and composition in a human distal gut chemostat model. J Microb Methods 95:167–174

Minekus M, Smeets-Peeters M, Bernalier A, Marol-Bonnin S, Havenaar R, Marteau P, Alric M, Fonty G, Huis in't Veld JH (1999) A computer-controlled system to simulate conditions of the large intestine with peristaltic mixing, water absorption and absorption of fermentation products. Appl Microb Biotechnol 53:108–114

Newton DF, Macfarlane S, Macfarlane GT (2013) Effects of antibiotics on bacterial species composition and metabolic activities in chemostats containing defined populations of human gut microorganisms. Antimicrob Agents Chemother 57:2016–2025

Payne AN, Zihler A, Chassard C, Lacroix C (2012) Advances and perspectives in in vitro human gut fermentation modeling. Trends Biotechnol 30:17–25

Petrof EO, Claud EC, Gloor GB, Allen-Vercoe E (2013) Microbial ecosystems therapeutics: a new paradigm in medicine? Benefic Microbes 4:53–65

Van den Abbeele P, Roos S, Eeckhaut V, MacKenzie DA, Derde M, Verstraete W, Marzorati M, Possemiers S, Vanhoecke B, Van Immerseel F, Van de Wiele T (2012) Incorporating a mucosal environment in a dynamic gut model results in a more representative colonization by lactobacilli. Microb Biotechnol 5:106–115

Van den Abbeele P, Belzer C, Goossens M, Kleerebezem M, De Vos WM, Thas O, De Weirdt R, Kerckhof FM, Van de Wiele T (2013) Butyrate-producing Clostridium cluster XIVa species specifically colonize mucins in an in vitro gut model. ISME J 7:949–961

Venema K, Van den Abbeele P (2013) Experimental models of the gut microbiome. Best Pract Res Clin Gastroenterol 27:115–126

Zihler Berner A, Fuentes S, Dostal A, Payne AN, Vazquez Gutierrez P, Chassard C, Grattepanche F, de Vos WM, Lacroix C (2013) Novel polyfermentor intestinal model (PolyFermS) for controlled ecological studies: validation and effect of pH. PLoS One 8:e77772

Chapter 25
One Compartment Fermentation Model

Anna-Marja Aura and Johanna Maukonen

Abstract In vitro colon model was first applied in an inter-laboratory dietary fibre (DF) fermentation study and adapted at VTT for whole foods and beverages, isolated dietary phenolic compounds and pharmaceuticals. The application of the models includes strict anaerobiosis, which ensures active anaerobic microbial community. Pooling of faecal samples from several donors ensures reproducibility between the experiments. The correlation of in vitro data with in vivo data is quantitatively challenging, but is qualitatively highly relevant. In this chapter we explain the applicability of the one compartment fermentation model, including the general protocol as well as the advantages and disadvantages of the system.

Keywords One compartment colon in vitro model • Human faecal microbiota • Anaerobic conditions • Microbial metabolites • Dietary fibre • SCFA

25.1 Description of the VTT One Compartment Fermentation Model

25.1.1 History of the Model

In 1995, VTT developed an enzymatic digestion model to simulate the digestion of DF in the upper intestine. The model was optimized for maximal starch digestion and to obtain non-digestible DF residue using physiological conditions (Aura et al. 1999). In the same year an inter-laboratory study of in vitro colon fermentation models was published by Barry et al. (1995). The physiological relevance of the model was ensured by a multidisciplinary approach involving specialists in nutrition and gastroenterology. The fermentation was expressed as the difference between the faecal control (inoculum, no substrates) and substrates under investigation (e.g. plant foods or other DF ingredients). Later the model was adapted for pure

A.-M. Aura (✉) • J. Maukonen
VTT Technical Research Centre of Finland Ltd., P.O. Box 1000, Tietotie 2, Espoo, Finland
e-mail: anna-marja.aura@vtt.fi

© The Author(s) 2015
K. Verhoeckx et al. (eds.), *The Impact of Food Bio-Actives on Gut Health*,
DOI 10.1007/978-3-319-16104-4_25

phenolic compounds (Aura 2005), fruit matrices and beverages (Bazzocco et al. 2008; Aura et al. 2013). The fermentation model from Barry et al. has been used in combination with the enzymatic digestion model from Aura et al. (1999) for cereal studies (Nordlund et al. 2012, 2013). Most recently the model has been coupled to a platform with bioinformatics tools to obtain non-targeted metabolite profiling of plant foods (Aura et al. 2013).

25.1.2 Special Features

Venema and van den Abbeele (2013) have pointed out recently a non-physiological slow rate of conversion in "static cultures". In reference to the unit operation the authors describe the "static" culture conditions (in contrast to the continuous "dynamic" culture conditions) (Venema and van den Abbeele 2013) often in context of one compartment cultures. When one compartment in vitro colon models are performed in well-buffered non-nutrient media with strictly anaerobic microbiota, accurately timed sampling and monitoring of pH changes, they appear to show also the distinctively higher rate of conversion. The time course studies under agitation bring out the dynamic processes, which occur from food components also in "batch" unit operations. Static, standing cultures without agitation cannot be used in conversions related to DF components or isolated phenolic compounds, because the suspension would sediment and enzyme–substrate interaction would be suppressed, especially in incubations with solid substrates such as DF components.

A distinctive characteristic of the VTT one compartment in vitro colon model is the use of a strong buffer with minerals as the main matrix of the medium (Barry et al. 1995) instead of nutritive medium with additional carbon sources (Hughes et al. 2008). The faecal material brings additional carbohydrates and a matrix to support the microbiota together with the buffer and mineral solution. It has been shown in previous studies (McBurney and Thompson 1989; Mortensen et al. 1991) that use of faeces from at least three donors secures the diversity of the microbiota and enables the reproducibility of the results. Therefore in the one compartment in vitro model applied at VTT a pooled human faecal suspension is prepared from samples from at least 3 but usually from 4 to 6 donors.

The faecal inoculum is dense (10–16.7 %, w/v), especially, when non-digestible carbohydrate fermentation is performed. The microbiota acts as a source of enzymes, which degrade and convert released components such as sugars to short-chain fatty acids (SCFA) or phenolic compounds to microbial phenolic acid or lactone metabolites.

One of the positive features of the VTT model is that the anaerobic conditions can be easily maintained as the system is closed and air-tight. This anaerobic environment is needed to avoid damaging of the strictly anaerobic microbiota during the preparation of the inoculum (homogenization, dilution and filtration) and its incubation with substrates. It also ensures that the microbial enzymes can perform degradation of DF constituents under the same conditions as in the human gastrointestinal tract.

The quality of the anaerobic conditions in the VTT model was tested by cultivating samples from the fresh faecal suspension in aerobic and in anaerobic conditions and counting the microbial cells after the cultivation. The initial (0 h) anaerobic cell count (1.6×10^9 CFU/ml) exceeded that from the corresponding aerobic cell count (3.1×10^7 CFU/ml). After 4 h of anaerobic incubation at 37 °C, cell counts were changed to 3.2×10^9 CFU/ml and 2.3×10^7 CFU/ml, for the anaerobic and aerobic cell counts, respectively. The corresponding log 10-anaerobic-to-aerobic ratios were 1.2 and 1.3 for 0 h and 4 h, respectively, indicating that anaerobic conditions were not disturbed during the incubation. When frozen faeces were used as an inoculum in three experiments, the log 10-anaerobic-to-aerobic ratio was 1.5 ± 0.2, but cell counts were 10 % of the fresh inoculum. After 4 h in corresponding conditions both the cell counts showed fourfold increase, and the log 10-anaerobic-to-aerobic-ratio was still 1.3 ± 0.04. Based on these results it was concluded that fresh inoculum was the most vital for DF fermentations and that the inoculum preparation conditions need to be strict enough for maintaining the adequate difference between the anaerobic and aerobic microbiota (2–3 log difference). When the cultivated log 10-anaerobic-to-aerobic ratio is between 1.2 and 1.6, anaerobic microbiota dominates the population. If the ratio were 1, strictly anaerobic strains would be dead and only microaerophilic strains would stay alive. This would be non-physiological situation in terms of DF fermentations.

Faecal samples from each donor and the corresponding faecal suspension were obtained from the in vitro colon model and kept frozen at −70 °C before microbiological analysis. Partial 16S rRNA gene was amplified for the denaturing gradient gel electrophoresis (DGGE) analysis of predominant bacteria, *Eubacterium rectale–Blautia coccoides* group (Erec-group, *Lachnospiraceae*), *Clostridium leptum* group (Clept-group, *Ruminococcaceae*), *Bacteroides*, *Bifidobacterium* and *Lactobacillus* as previously described (Maukonen et al. 2006, 2012). Above mentioned bacterial profiles of individual donors were compared to the bacterial profiles of the corresponding faecal suspension. Even though the inter-individual similarity between the donors was low (Fig. 25.1), and partly different people were used as donors in different in vitro colon model experiments, the bacterial profiles of faecal suspensions were rather similar (Fig. 25.2). In predominant bacteria, Clept-group, lactobacilli and bifidobacterial profiles clear differences between faecal suspensions were observed. However, the observed differences corresponded to intra-individual temporal variation previously observed in healthy adults (Maukonen et al. 2012). Moreover, in the Erec-group (Fig. 25.2b) and in *Bacteroides* profiles there were no major differences between the different faecal suspensions. Erec-group bacterial populations have also previously been shown to be temporally rather stable (Maukonen et al. 2006).

In conclusion, the microbiological reproducibility of the in vitro model was good: the microbiological variation observed between the different in vitro colon model experiments was equivalent to intra-individual temporal variation. Since DNA-based methods were used, which target both living and dead bacteria, it cannot be reliably stated that the active microbiota has been the same in each experiment. The used DGGE methods target only the rDNA amplicons obtained from the population that exceeds 1 % of the species present in the analysed community

Fig. 25.1 Bifidobacterial
profiles of samples used for
in vitro fermentation
(BM = bifidobacterial marker;
A–C = donors; 3 = faecal
slurry)

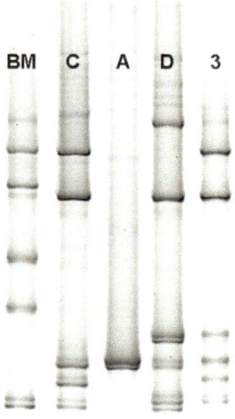

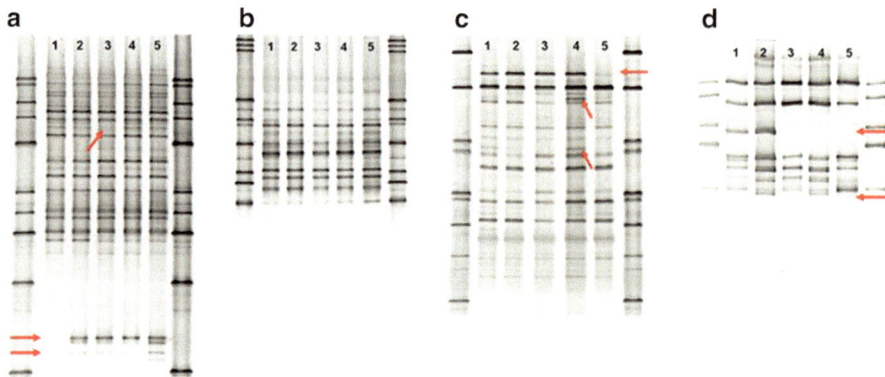

Fig. 25.2 Predominant bacterial (**a**), *Eubacterium rectale–Blautia coccoides* group (**b**), *Clostridium leptum* group (**c**) and *Bifidobacterium* (**d**) DGGE-profiles of faecal slurries used for five (1–5) different in vitro fermentations. *Red arrows* denote clear differences between samples

(Muyzer et al. 1993), therefore our results show that the most dominant bacteria within each bacterial group were similar. This study of microbiological reproducibility was a quality measure, which ensures that the metabolites are less dependent on the microbiota than from the substrates. Other microbial ecological studies are not recommended with the one compartment in vitro colon model, since the model is designed to study microbial metabolism of various substrates and for comparison of their metabolite dynamics.

Table 25.1 Variation of different microbial metabolites (mean±standard deviation) measured from the faecal control between experiments (n: number of experiments) in the beginning (0 h) and in the end (24 h) of the incubation

Response	0 h	24 h	n	References
Total SCFA (mM)	11.7±2.6	32.6±5.7	5	1
Acetic acid (mM)	7.6±1.5	20.5±4.5	5	1
Propionic acid (mM)	1.7±0.3	5.9±1.1	5	1
Butyric acid (mM)	2.4±0.9	6.2±1.3	5	1
pH (initial 6.9) (range)	6.99–7.07	6.67– 6.68	2	1
pH (initial 5.5)	6.07±0.05	5.83±0.12	3	
3-Hydroxyphenylpropionic acid (µM)	7.7±3.5	7.3±2.4	3	2
3-Phenylpropionic acid (µM)	96±15	235±71	3	2
3-Hydroxyphenylacetic acid (µM)	5.4±4.4	8.7±1.7	3	2

1: Nordlund et al. (2012), Bazzocco et al. (2008), Aura et al. (2013) and Nordlund et al. (2013)
2: Aura et al. (2013) and Bazzocco et al. (2008)

25.2   Validation of the System

The validation of the model was performed by comparing the variability of meta-bolite concentrations in the faecal control (inoculum, no substrates) and in the sub-strate incubation in experiments performed on different days. The faecal control measurement provided information on the variation of the background fermentation of the remaining fermentable carbohydrates or phenolic compounds from the diet of the faecal donors. Table 25.1 shows comparison of 4 or 5 experiments in respect to major microbial metabolites and pH in the faecal control at 0-h- and 24-h-time points. The smaller the response, the higher is the variation of the response. The variation is smallest (14–25 %) for total SCFA, acetic acid, propionic acid and 3-phenylpropionic acid, whereas butyric acid varies 21–37 % between the experiments. pH shows only 1–2 % variation, for the initial pH 5.5 of the buffer; and 3-hydroxyphenylpropionic and -acetic acids show 65–114 % (0 h) and 23–27 % (24 h) variations. Furthermore, when the concentrations of the metabolites in the faecal control are compared between the experiments, the differences are dependent on the metabolite and its dynamics and particularly the diets of the donors of the faeces. The differences are balanced by pooling the inoculum from several donors. Therefore, the major phenolic metabolites in the in vitro model show similar variation between experiments (65–114 %) in phenolic microbial metabolites than corre-sponding urinary excretion profiles shown between different human individuals (70–113 %) (Vetrani et al. 2014).

The true validation requires the same reference substrate for all experiments under investigation. Usually the space in the experiment is limited and the reference substrate is chosen according to the hypotheses of the specific projects. Rye bran and flaxseed meal were used in several experiments, the responses were measured as SCFA production and were summarized in Aura (2005). The total SCFA formation at the end of the fermentation showed 10 % and 18 % variations for enzymatically pre-digested rye bran (83.3 ± 8.5 mM, n = 4) and pre-digested flaxseed meal (95.3 ± 17.5 mM, n = 3), respectively. Taking into account the different inoculum concentrations (10 % or 16.7 %) used in these studies, the variation was surprisingly low for SCFA production. In contrast, the 97 % variation of the enterolactone formation from predigested flaxseed meal (43 ± 42 nM, n = 3) was highly dependent on the inoculum concentration. This could be explained by the minor population of enterolactone-converting bacteria (Clavel et al. 2005) and therefore the extent of conversion was susceptible to the depletion by the low concentration of inoculum. Furthermore, the resilient structure of flaxseed meal and low nano-molar concentrations may have contributed to the high variation (Aura 2005).

25.3 Relevance to Human In Vivo Situation

The one compartment in vitro colon model measures only those conversions, which occur in anaerobic conditions by faecal microbiota. These reactions take place in caecum prior to absorption. The model does not take into account the membrane functions which occur in the colon epithelia, the intestinal epithelia and in the liver. The major metabolites from the DF intake are SCFA and phenolic microbial metabolites. The major SCFA formed in the colon is acetic acid, which does not have distinctive structural characteristics, which would be needed for follow-up of its route without a radiolabel. For example a correlation was sought for SCFA production between a fibre blend and its components in vivo and in vitro (Koecher et al. 2014). In vitro experiments showed SCFA formation and analysis of SCFA from faeces revealed an increase of SCFA content in faeces after consumption of the fibre blend versus fibre free diet, however, significant differences between the ingredients of the blend were not observed in vivo (Koecher et al. 2014). The authors speculated about the balance between production and absorption rates of microbial metabolites and concluded that the in vivo interventions and in vitro studies may not be directly correlated, but in vitro models are additionally informative (Koecher et al. 2014).

Phenolic compounds and their microbial metabolites may be better biomarkers for plant food and dietary fibre intake than SCFA. The metabolites in urine include also membrane-derived metabolites, such as glucuronidated and sulfated derivatives of phenolic compounds. In addition to this urine contains the microbial metabolites derived from the diet (Aura 2008). The closest and most non-invasive measure is the analysis of 24-h-urine, which describes the excretion of metabolites from the food and beverage intakes (Vetrani et al. 2014). The first challenge is that the urinary analysis describes the excretion of the metabolites from the whole diet from several precursors (Vetrani et al. 2014) which share the microbial metabolites (Aura 2008). If the diet is not controlled the background can be disturbed by the excretion of phenolic metabolites from the non-controlled components, and statistical differences are not obtained (Lappi et al. 2013).

Another issue is that a single component does not change the excreted metabolites in a significant way (Lappi et al. 2013), whereas a fully controlled diet high or low in polyphenols can show significant metabolite profiles typical for the polyphenols (Vetrani et al. 2014). Therefore, the metabolites cannot be connected to a single food or component, but refer to a whole diet. The third challenge is that the comparison of in vivo responses match qualitatively well with the microbial metabolites expected to be excreted from the food phenolic components (Aura 2008; Lappi et al. 2013; Vetrani et al. 2014). However, quantitative comparisons are problematic because of the high individual variation of microbial metabolites in vivo (Vetrani et al. 2014) and the diversity of hepatic metabolites, the quantification of which requires unavailable authentic standards for most of them for significant correlation.

25.4 Quality in Relation to Other Models with the Same Applicability

When microbial metabolism of rutin (quercetin-rhamnoglucoside) and chlorogenic acid was studied in a one compartment fermentation model using four individual donors (Rechner et al. 2004), the patterns of the metabolite formations varied according to the subject. The formation of 3,4-dihydroxyphenylacetic or -propionic acid from rutin and chlorogenic acid, respectively, showed maxima at 8–10 h using 10 % (w/v) suspension (Rechner et al. 2004). In contrast, using the diluted suspension (5 %, w/v) pooled and homogenized under strict anaerobic conditions, the corresponding value was shown in 2 h and complete dehydroxylation was observed in 8 h (Aura et al. 2002). The inoculum in VTT one compartment system has to be diluted 1 % in order to observe the deglycosylation of quercetin derivatives or anthocyanins in different experiments (Aura 2005). It is likely that strict application of anaerobic conditions enables faster conversions, when vulnerable anaerobic microbial suspension is used as a source of activity.

When in vitro colon models using monogastric pig and human faeces as inocula were compared, human inoculum showed more efficient fermentation for several sources of DF, whereas pig inoculum was more efficient in cellulose digestion. The pig inoculum produced less SCFA and more gas compared with the human inoculum. These differences were attributed to an adaptation of human microbiota to a more diverse diet compared to the pig feed and an adaptation of pig microbiota to a cellulose-rich diet (Jonathan et al. 2012).

25.5 General Protocol

The latest description of the model is from Nordlund et al. (2012). In vitro colon model for measurement of SCFA and phenolic acids was started by weighing 100 mg per 10 ml of incubation suspension (dry w/v) of plant foods or their fractions to the bottles, and hydrating with 20 % of the volume of medium 1 day before inoculation. Human faeces were collected from at least 3 (usually from 4–6) healthy volunteers, who had not received antibiotics for at least 6 months and had given a written consent. Freshly passed faeces were immediately taken in an anaerobic chamber or closed in a container with an oxygen consuming pillow (Anaerocult Mini; Merck, Darmstadt, Germany) and a strip testing the anaerobiosis (Anaerotest; Merck, Darmstadt, Germany). Faecal suspension was prepared under strictly anaerobic conditions. Equal amounts of faecal samples were pooled and diluted to a 12.5 % (w/v) or 20.8 % (w/v) suspension, depending on the application and 80 % of the volume of the incubation suspension was dosed to the fermentation bottles to obtain a 10 % or 16.7 % (w/v) final faecal concentrations as described previously (Aura 2005). Lower concentrations can be used. For example 5 % suspension is suitable for isolated components, because the metabolite responses from faecal control are high in more dense inoculum and the substrate concentration should be below the saturation point of the substance to keep the substrate in the solution and not to suppress the activity of the microbiota.

The fermentation experiments were performed in triplicate and a time course of 0, 2, 4, 6, 8 and 24 h is followed using the same inoculum for all the substrates. Incubation is performed at 37 °C in tightly closed bottles and in magnetic stirring (250 rpm). Faecal background is incubated without addition of the supplements (Aura 2005). Headspace is sampled for the measurement of gas pressure (Nordlund et al. 2012) and the liquid space is sampled for SCFA and phenolic acid metabolite analyses (Nordlund et al. 2012; Aura et al. 2013).

25.6   Controls: Positive and Negative

Monitoring of metabolite background in the faecal control without substrate is crucial. In addition to this it is recommendable to use inactive microbiota as a negative control (Aura 2005). Relevant reference compounds/ingredients should be used. For instance when fermentation rate is the focus of the study, reference substrates such a "rapid" or "slow" standards can be used to make comparisons between different experiments. The controls should also be performed at least in three replicates for adequate statistical evaluation.

25.7   Read Out of the System

The read out of the in vitro colon model is caused by the interaction between microbial enzymes and the precursors and expressed as time course of metabolite formation. The precursors can be carbohydrates and the products are then SCFA or gas formation (Nordlund et al. 2012). Other precursors can be flavonoids, proanthocyanidins or plant-derived phenolic acids (ferulic acid, chlorogenic acid), which lead to the formation of benzoic acid derivatives or of hydroxylated phenylpropionic, -acetic or -valeric acids (Aura 2008; Aura et al. 2013; Nordlund et al. 2012). Plant lignans are the precursors for enterodiol and enterolactone (the enterolignans) (Heinonen et al. 2001) and isoflavonoids for equol or O-desmethylangolensin production (Heinonen et al. 2004; Possemiers et al. 2007). Furthermore ellagitannin conversion by intestinal microbiota results in formation of urolithins (Cerda et al. 2004).

The responses should be compared always within the experiment, using the adequate number of intermediary time points, at least in triplicates and in respect to the inoculum, the responses of which is dependent on the diet of the donor. Moreover, the quantitative results should not be extrapolated between experiments and the responses should be related to the clearly indicated faecal control read out within the experiment. Thus the accepted microbial metabolite can only be one which shows at least two- to fivefold higher responses than the background metabolites from the faecal control. Therefore the results show specific metabolite profiles from each substrate. These semi-quantitative results can be obtained from non-targeted metabolomics platform for identification of new metabolites. Quantitative measures are achieved using the targeted approach, which limits the profiling to those metabolites, which are available as authentic standards. A unit to be measured should be on molar basis, because structural transformations affect the molar masses and responses based on weight are not comparable.

The one compartment colon model can be applied to comparison of fermentation rates of different sources of DF or ingredients having different characteristics (Kaur et al. 2011; Koecher et al. 2014), whereas changes in microbial population requires a semi-continuous unit operation (Hughes et al. 2008).

25.8 Summary of Advantages, Disadvantages and Limitations of the System and Contingency Plan

Table 25.2 summarizes the advantages, disadvantages and limitations of the one compartment in vitro colon model and is divided into characteristics of the operation of the model (unit operation, microbiota, anaerobiosis, pH monitoring and control, timing and sampling, stirring) and into those of the outcome (data analysis, comparison of substrates, prediction of human metabolism, and absorption). It is important to judge the limitations in context of the hypotheses that are studied in the one compartment model.

In conclusion, the one compartment fermentation model experiments perhaps do not mimic human pH changes or include absorption as the computer-controlled continuous models do, but the one compartment incubations are very suitable when the perspective is in the food matrix or its chemistry. In vitro colon models do not have membrane functions required for full mimicking of the xenobiotic metabolism. However, in vitro one compartment digestion models explain the effects of food matrix on release of components or bioconversion of food components to their microbial metabolites and they can elucidate factors affecting these processing and predict in vivo bioavailability. This field of food biochemistry is essential to elucidate phenomena, by which food chemistry turns into nutrition.

Table 25.2 Advantages, disadvantages and limitations of the one compartment in vitro colon model

Characteristics	Advantages	Disadvantages/limitations
Unit operation: batch	Metabolic studies: targeted and non-targeted measures with the same background	Microbial ecological studies are not applicable
Microbiota	Anaerobic human microbiota can stay alive: authentic activities for DF fermentation	
Anaerobic environment	Easy to operate and maintain	Requires accuracy at all stages of the process
pH monitoring/control	Monitoring gives the effect of food on pH of the incubation	Inconvenient to regulate manually since the use of strong buffer is inadequate. Low pH can suppress conversions. Labour intensive
Timing/sampling	Accurate in respect to replicates and samples in comparison	Requires resources
Stirring	Stirring enhances enzyme–substrate interactions	
Data analysis	Replicates with the same control enable comparison with substrates. Non-targeted metabolomics coupled with bioinformatics and compound libraries makes the statistics and data filtration easy	Semi-quantitative. Identification of new components is labour intensive
Comparison of substrates	Time course studies reveal sensitively initial rates of fermentation and dynamic metabolite profiles within one experiment	Not applicable for comparison between experiments without reference substrates. Requires significant structural/chemical differences between substrates. Extent of fermentation is a robust measure
Prediction for human metabolism	Predicts qualitatively bioavailable metabolites with distinctive structural changes from dietary precursors. Time course studies reveal dynamic processes and intermediary products: role of microbiota in the nutritional metabolomics	Quantitative prediction is difficult, because membrane-bound converting enzymes are missing and individual variation is high in vivo. Product inhibition can occur at later time points
Absorption		Missing

References

Aura A-M (2005) In vitro digestion models for dietary phenolic compounds. PhD dissertation, Helsinki University of Technology

Aura A-M, O'Leary KA, Williamson G, Ojala M, Bailey M, Puupponen-Pimiä R, Nuutila AM, Oksman-Caldentey K-M, Poutanen K (2002) Quercetin derivatives are deconjugated and converted to hydroxyphenylacetic acids but not methylated by human fecal flora *in vitro*. J Agric Food Chem 50:1725–1730

Aura A-M, Härkönen H, Fabritius M, Poutanen K (1999) Development of an *in vitro* enzymic digestion method for removal of starch and protein and assessment of its performance using rye and wheat breads. J Cereal Sci 29:139–152

Aura A-M, Mattila I, Seppänen-Laakso T, Miettinen J, Oksman-Caldentey K-M, Orešič M (2008) Microbial metabolism of catechin stereoisomers by human faecal microbiota: comparison of targeted analysis and a non-targeted metabolomics method. Phytochem Lett 1:18–22

Aura A-M, Mattila I, Hyötyläinen T, Gopalacharyulu P, Cheynier V, Souquet J-M, Bes M, Le Bourvellec C, Guyot S, Orešič M (2013) Characterization of microbial metabolism of Syrah grape products in an *in vitro* colon model using targeted and non-targeted analytical approaches. Eur J Nutr 52:833–846

Barry JL, Hoebler C, Macfarlane GT, Macfarlane S, Mathers JC, Reed KA, Mortensen PB, Norgaard I, Rowland IR, Rumney CJ (1995) Estimation of the fermentability of dietary fibre *in vitro*: a European interlaboratory study. Br J Nutr 74:303–322

Bazzocco S, Mattila I, Guyot S, Renard CMGC, Aura A-M (2008) Factors affecting the conversion of apple polyphenols to phenolic acids and fruit matrix to short-chain fatty acids by human faecal microbiota *in vitro*. Eur J Nutr 47:442–452

Cerda B, Espin JC, Parra S, Martinez P, Tomas-Barberan FA (2004) The potent *in vitro* antioxidant ellagitannins from pomegranate juice are metabolised into bioavailable but poor antioxidant hydroxy-6H-dibenzopyran-6-one derivatives by the colonic microflora of healthy humans. Eur J Nutr 43:205–220

Clavel T, Henderson G, Alpert C-A, Philippe C, Rigottier-Gois L, Doré J, Blaut M (2005) Intestinal bacterial communities that produce active estrogen-like compounds enterodiol and enterolactone in humans. Appl Environ Microbiol 71:6077–6085

Heinonen S, Nurmi T, Liukkonen K, Poutanen K, Wähälä K, Deyama T, Nishibe S, Adlercreutz H (2001) *In vitro* metabolism of plant lignans: new precursors of mammalian lignans enterolactone and enterodiol. J Agric Food Chem 49:3178–3186

Heinonen S-M, Wähälä K, Liukkonen K-H, Aura A-M, Poutanen K, Adlercreutz H (2004) Studies of the *in vitro* intestinal metabolism of isoflavones aid in the identification of their urinary metabolites. J Agric Food Chem 52:2640–2646

Hughes SA, Shewry PR, Gibson GR, McCleary BV, Rastall RA (2008) *In vitro* fermentation of oat and barley derived beta-glucans by human faecal microbiota. FEMS Microbiol Ecol 64:482–493

Jonathan MC, van den Borne JJGC, van Wiechen P, Souza da Silva C, Schols HA, Gruppen H (2012) *In vitro* fermentation of 12 dietary fibres by faecal inoculum from pigs and humans. Food Chem 133:889–897

Kaur A, Rose DJ, Rumpagaporn P, Patterson JA, Hamaker BR (2011) *In vitro* batch fecal fermentation comparison of gas and short-chain fatty acid production using "slowly fermentable" dietary fibres. J Food Sci 76:H137–H142

Koecher KJ, Noack JA, Timm DA, Klosterbuer AS, Thomas W, Slavin JL (2014) Estimation and interpretation of fermentation in the gut: coupling results from a 24 h batch *in vitro* system with fecal measurements from a human intervention feeding study using fructo-oligosaccharides, inulin, gum acacia, and pea fibre. J Agric Food Chem 62:1332–1337

Lappi J, Aura A-M, Katina K, Nordlund E, Kolehmainen M, Mykkänen H, Poutanen K (2013) Comparison of postprandial phenolic excretions and glucose responses after digestion of breads with bioprocessed or native rye bran. Food Funct 4:972–981

Maukonen J, Mättö J, Satokari R, Söderlund H, Mattila-Sandholm T, Saarela M (2006) PCR DGGE and RT-PCR DGGE show diversity and short-term temporal stability in the *Clostridium coccoides-Eubacterium rectale* group in the human intestinal microbiota. FEMS Microbiol Ecol 58:517–528

Maukonen J, Simões C, Saarela M (2012) The currently used commercial DNA extraction methods give different results of clostridial and actinobacterial populations derived from human fecal samples. FEMS Microbiol Ecol 79:697–708

McBurney MI, Thompson LU (1989) Effect of human faecal donor in *in vitro* fermentation variables. Scand J Gastroenterol 24:359–367

Mortensen PB, Hove H, Clausen MR, Holtug K (1991) Fermentation to short-chain fatty acids and lactate in human faecal batch cultures. Intra- and inter-individual variations versus variations caused by changes in fermented saccharides. J Scand Gastroenterol 26:1285–1294

Muyzer G, de Waal EC, Uitterlinden AG (1993) Profiling of complex microbial populations by denaturing gradient gel electrophoresis analysis of polymerase chain reaction-amplified genes coding for 16S rRNA. Appl Environ Microbiol 59:695–700

Nordlund E, Aura A-M, Mattila I, Kössö T, Rouau X, Poutanen K (2012) Formation of phenolic microbial metabolites and short-chain fatty acids from rye, wheat, and oat bran and their fractions in the metabolical *in vitro* colon model. J Agric Food Chem 60:8134–8145

Nordlund E, Katina K, Aura A-M, Poutanen K (2013) Changes in bran structure by bioprocessing with enzymes and yeast modifies the in vitro digestibility and fermentability of bran protein and dietary fibre complex. J Cereal Sci 58:200–208

Possemiers S, Bolca S, Eeckhaut E, Depypere H, Verstaete W (2007) Metabolism of isoflavones, lignans and prenylflavonoids by intestinal bacteria: producer phenotyping and relation with intestinal community. FEMS Microbiol Ecol 61:372–383

Rechner AR, Smith MA, Kuhnle G, Gibson GR, Debham ES, Srai SKS, Moore KP, Rice-Evans CA (2004) Colonic metabolism of dietary polyphenols: influence of structure on microbial fermentation products. Free Radic Biol Med 36:212–225

Venema K, van den Abbeele P (2013) Experimental models of the gut microbiome. Best Pract Res Clin Gastroenterol 27:115–126

Vetrani C, Rivellese AA, Annuzzi G, Mattila I, Meudec E, Hyötyläinen T, Oresic M, Aura A-M (2014) Phenolic metabolites as compliance biomarker for polyphenol intake in a randomized controlled human intervention. Food Res Int 63:233–238

Chapter 26
The TNO In Vitro Model of the Colon (TIM-2)

Koen Venema

Abstract This contribution describes the development and use of the TNO in vitro model of the colon (TIM-2). The unique features of this system are briefly discussed, as well as how these contribute to the predictability of this validated in vitro model for clinical trials. Several examples are provided of data where experiments in TIM-2 yielded similar data as clinical trials. Other examples highlight the use of the system to screen dietary components for certain functionality (e.g., increase in butyrate production), or determine likely mechanisms of action of dietary substrates (e.g., the bifidogenic nature of dietary carbohydrates). Moreover, examples are provided of metabolism of dietary compounds such as polyphenols of which the determination of in vivo metabolism is extremely difficult, due to the limited access we have to the colon. The system is inoculated with a microbiota originating from healthy volunteers (single donor or pooled microbiota) of different age, or from people with a disease (e.g., inflammatory bowel disease). Also, differences between the microbiota from lean and obese individuals have been studied. Recent developments in building hypotheses on the role of the microbiota in health and disease have been generated using substrates that were labeled with a stable isotope (i.e., ^{13}C), that allows tracing the label into microbial biomass and into metabolites produced by the gut microbiota. This has significantly advanced our knowledge on the role of the activity of the gut microbiota in health and disease and the members of the microbiota that are involved in this.

Keywords In vitro model • Colon • Microbiota • SCFA • Prebiotics

26.1 Description of TIM-2

This contribution focuses on food-application of TIM-2. Although numerous examples are available for pharmaceutical applications as well, these will not be exemplified here.

K. Venema (✉)
Beneficial Microbes Consultancy, Wageningen, The Netherlands
e-mail: koen.venema@outlook.com

© The Author(s) 2015   293
K. Verhoeckx et al. (eds.), *The Impact of Food Bio-Actives on Gut Health*,
DOI 10.1007/978-3-319-16104-4_26

26.1.1 History of the Model

The *T*NO computer-controlled, dynamic in vitro gastro-*I*ntestinal *M*odel of the colon (nick-named TIM-2; Fig. 26.1) was developed by TNO some 15 years ago (Minekus et al. 1999). It is based on the TIM-1 system (the TNO in vitro gastro-intestinal model of the stomach and small intestine) that is discussed elsewhere in this book.

Briefly, the model consists of four interconnected glass compartments, with a flexible membrane inside (Fig. 26.1a). In between the glass jacket and the membrane there is water of body temperature (37 °C for humans, 39 °C for pigs, 41 °C for birds, etc.). The temperature is controlled by a temperature sensor (Fig. 26.1j). By applying pressure on the water at regular intervals and in a certain sequence, the flexible membrane contracts and causes peristaltic waves which mixes the luminal contents and moves it through the system. This mixing is better than can be accomplished in a fermentor, where phase separation of fluids and solids occur. The system is kept at a pH of 5.8, the pH occurring in the proximal colon, or above by continuous measurement of the pH (Fig. 26.1b) and secretion of 1M NaOH (Fig. 26.1c) to neutralize the

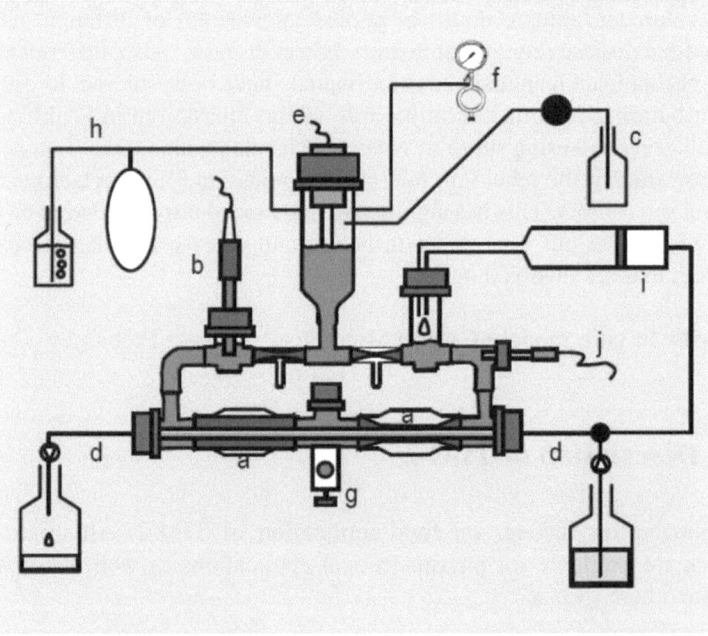

Fig. 26.1 Schematic figure of the TNO in vitro model of the colon (TIM-2). (a) peristaltic compartments containing faecal matter; (b) pH electrode; (c) alkali pump; (d) dialysis liquid circuit with hollow fibre membrane; (e) level sensor; (f) N₂ gas inlet; (g) sampling port; (h) gas outlet; (i) 'ileal efflux' container; (j) temperature sensor

acids that are produced by the microbiota. To prevent accumulation of microbial metabolites, which would lead to the inhibition or death of the microbes in the model when they would accumulate, the system is equipped with a dialysate system, consisting of a bottle with dialysate, a dialysis membrane running through the model and a bottle collecting the spent dialysate (Fig. 26.1d). This dialysis system is unique, and moreover, required to maintain an active microbiota for periods for up to 3 weeks. This was the longest the system has been tested. Normally experiments are performed over a period of 1 week (see Sect. 26.2).

The dialysis system has been tuned to keep physiological concentrations of microbial metabolites in the lumen of the model. For example, short-chain fatty acid (SCFA) concentrations are within the physiological range of 80–120 mM (Venema et al. 2000). The model is equipped with a level sensor to control the volume (Fig. 26.1e). When the volume in the system rises due to addition of the feed or through dialysis, this sensor activates the dial-out pump and maintains the volume at a constant level of ~120 mL. To keep the model anaerobic, it is flushed with gaseous nitrogen (Fig. 26.1f). This, plus the activity of the microbiota in the system, keeps the redox-potential at a level of ~ −300 mV, which is the value that has been reported to occur in the large intestine. Samples can be taken from the lumen through the sampling-port (Fig. 26.1g) or from the spent dialysate (Fig. 26.1d). This allows for a complete mass-balance to be calculated. Gases produced by the microbiota can be sampled too, although normally this is not done and the gas just escapes from the model through a bubbling-vial (Fig. 26.1h). Frequent sampling over time allows the study of the kinetics of production of certain microbial metabolites (e.g., SCFA) on specific substrates. It is obvious that such sampling is not possible in vivo, and hence TIM-2 allows studying the mode of action of certain compounds (foods or drugs) which cannot be studied in detail in man, nor in animals. The microbiota in the system is fed through a food syringe (Fig. 26.1i) which contains a simulated ileal efflux medium (SIEM), which in composition mimics the components that reach the colon from the terminal ileum through the ileal-cecal valve. It consists of some complex carbohydrates, some protein (not all protein is digested in the small intestine [SI]), some residual bile (not all bile is resorbed in the SI), and some minerals and vitamins [for details see Minekus et al. 1999; van Nuenen et al. 2003; Venema et al. 2000].

26.1.2   Special Features of the Model

Several features in TIM-2 are unique over other models. First of all, the peristaltic movements of the flexible membrane give a better mixing and movement of components through the entire model than would be accomplished by stirring (in a fermentor) or shaking (on a rocking-platform or otherwise). In TIM-2 there is no phase-separation of solids and liquids, which does occur in other systems. Viscous 'meals' or insoluble components can be used without problems in the system. Secondly, as eluded to above, the dialysis system is not only unique, but also required to maintain a highly active microbiota with a similar density as that found in the human large intestine. In batch incubations or less sophisticated models the microbiota is usually inoculated at a much lower density (100-fold or more) and is allowed to slowly grow to physiological densities. However, the metabolites produced by the microbiota start to inhibit further fermentation at these high densities in these systems. Since these metabolites are also taken up by the epithelial cells of the colon (colonocytes) in vivo, TIM-2 mimics much better the physiological situation in the large intestine. In fact, since all metabolites are collected and can be measured, a(n almost) complete mass-balance can be made, which is not possible in vivo, not even in animals, although in scientific studies usually animal are sacrificed to sample as much as possible, including blood samples. However, even in animals a mass-balance is not possible, as e.g., butyrate—one of the SCFA—is used by the colonocytes and does not reach the blood circulation. Therefore, TIM-2 allows studying (molecular) mechanisms. This is certainly the case if labeled substrates are used, as will be discussed in other sections of this chapter.

The model is incubated by using a fecal donation from volunteers. This fecal donation can be used in two ways: (a) a fecal donation from individual one can be introduced into one of the TIM-2 systems, the donation from individual 2 in a second unit, and so on. The composition and activity of the microbiota of the individuals can then be compared on say one and the same substrate. This has for instance been done for lactulose, with ten donors (Venema et al. 2003); or (b) the fecal donations of several donors are mixed to create a standardized microbiota that can subsequently be used for ~100 experiments. This allows comparison of multiple substrates or conditions starting with the same microbiota composition. This pool of fecal donations to a certain extent mimics a (small) population. It has been argued by many that mixing different fecal samples may disturb the microbial balance within a single fecal sample, but as far as we know, no direct comparison has been done to show this. Recently, we have set out to show that by mixing a number of different fecal donations, the functionality of the microbiota is not influenced (Aguirre et al. 2014). That is, the individual microbiotas showed the same functionality as the pool and produced very similar amounts and ratios of microbial metabolites (amongst others SCFA), despite being different in microbial composition. This underscores our hypothesis that there is an enormous functional overlap between microbes in the large intestine. This is not entirely surprising, as there are only a limited number of biochemical routes (let's say 5) from e.g., glucose to acetate. Since the microbiota is composed of approximately 200 species or more, there has to be enormous functional overlap between these microorganisms.

Upon introduction of the microbiota in the system, an adaptation period of ~16 h is applied, to allow the microbes to adapt to their new environment and the feed components. After that the experimental period starts, which normally takes 72 h (see Sect. 26.2). Fecal donations can be obtained from healthy volunteers of different age-classes (baby, adults, elderly), people with a disease or disorder [e.g., inflammatory bowel disease; (van Nuenen et al. 2004)], or from lean vs. obese individuals.

26.1.3   Stability and Reproducibility of the System

Since the model is computer-controlled it is highly reproducible. This has been shown e.g., by the clustering of the microbiota after the adaptation period referred to in the previous section (Fig. 26.2). In this example, the microbiotas originating from lean and obese individuals were compared. The figure shows that both micro-biotas clustered separately, but that each t = 0 sample of the lean microbiota clus-tered very closely together with the other lean samples, and similarly for the obese samples. It should be said that the microbiota in this adaptation period undergoes a change in composition, as it adapted to the model and feed (Rajilic-Stojanovic et al. 2010). This is not entirely surprising, as the conditions in the model are different to the colon, although we try to simulate them as good as possible. But, the system does not contain epithelial cells or immune cells that may influence the microbiota composition. This will be discussed in more detail later (see Sect. 26.5).

No matter whether the system is inoculated by a standardized microbiota or a microbiota from a single donor, it is highly reproducible. This has been shown in numerous studies (Kovatcheva-Datchary et al. 2009; Martinez et al. 2012; Rose et al. 2010). This is why experiments in TIM-2 are performed in duplicate.

In every project a control with the standard growth medium (SIEM) is performed. The other conditions applied in such projects are then compared to the effects obtained with the standard medium. This allows correcting for effects that occur simply through use of the in vitro model.

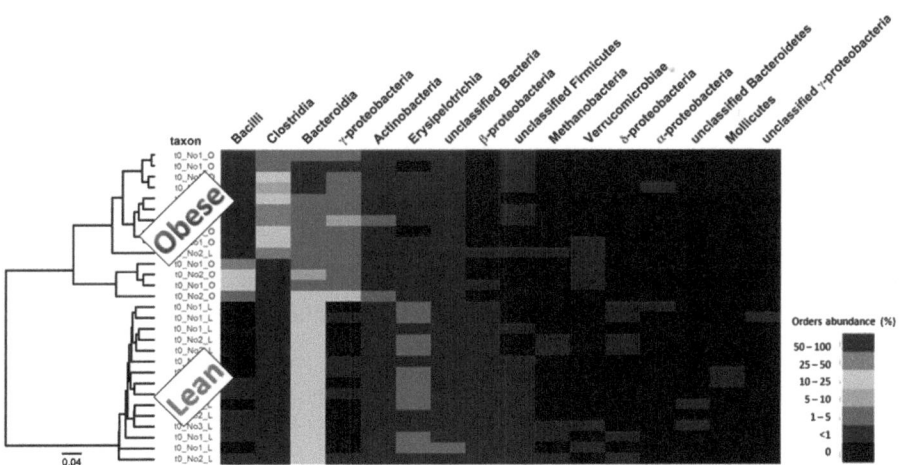

Fig. 26.2 Phylogentic tree (*left*) and heatmap (*right*) of samples from TIM-2 taken at t = 0 in an experiment where a lean and obese microbiota was compared. The lean and obese microbiotas cluster separately, although all t = 0 samples of each microbiota (lean or obese) cluster closely together

26.1.4   Relevance to Human In Vivo Situation

The model has been developed and optimized with the use of data from sudden-death individuals (Macfarlane et al. 1992). The data from these samples with respect to microbial metabolites and microbiota composition were used to develop TIM-2.

The effect of several food components that have been well-established in vivo have been confirmed in TIM-2. These include the bifidogenic nature of inulin (van Nuenen et al. 2003) and galacto-oligosaccharides (GOS) (Maathuis et al. 2012), or the high production of butyrate out of resistant starches (Rose et al. 2010). Due to this, the validated system is frequently used to test experimental substrates and is thought to be predictive for the in vivo situation.

Yet, in some situations the model is even capable of predicting results in the proximal colon that cannot be measured in a clinical trial. This has for instance been done in experiments using lactulose (Venema et al. 2003). Here, fecal donations were obtained from volunteers before they ingested lactulose for 10 days, and after. Both fecal donations were inoculated independently in TIM-2 and were fed lactulose in the system. The production of the major microbial metabolites, the SCFA, was analyzed both in fecal samples as well as in TIM-2 samples. At first surprisingly, the in vivo fecal samples before and after lactulose did not differ in SCFA content and ratio, while the samples in TIM-2 showed a marked reduction in butyrate production. However, since lactulose is a disaccharide, it is fermented very quickly by the gut microbiota (as also indicated by the accumulation of lactate), and this occurs in the proximal colon. We now hypothesize that the SCFA that were produced by the microbiota in vivo in the proximal colon were all taken up during transit to the distal colon, and that the SCFA present in the excreted fecal samples reflect those produced locally in the distal colon, rather than reflect the SCFA that were produced in the proximal colon. This is highly likely as it has been estimated that 95 % of the SCFA produced by the microbiota are absorbed. Therefore, TIM-2 gave insight in the processes at the site where the fermentation of lactulose happened—the proximal colon. This is data that could not be inferred from fecal samples taken from these volunteers.

26.1.5   Quality in Relation to Other Models
with the Same Applicability

Compared to other in vitro models mimicking the colon, TIM-2 has a number of features that are unique (see Sect. 26.1.2) and that allow to predict what would happen in a clinical trial. In the previous section the example of lactulose was given. To validate these results one would need to sample the proximal colon, which is only possible when using a long catheter that is inserted through the nose or throat and reaches the proximal colon (Venema 2011), or a tube that is stuck up the rectum to reach the proximal part of the large intestine. Although this has been done, it is very invasive and it is difficult to get ethical approval for testing functionality of numerous food components. Nevertheless, it is our belief that the results that were found in TIM-2 predict what happens in vivo, even though this could not be confirmed because the analytes in fecal matter in this case were not representative for those produced in the proximal colon, as they had all been absorbed by the colonic tissue during transit from proximal to distal colon, which may take anything from 24 to 72 h.

One of the reasons the system can predict what happens in real life so well is the presence of the dialysis system. As discussed above, this system prevents the accumulation of microbial metabolites, which normally are also taken up by the gut epithelium. In other systems metabolites tend to accumulate. Apart from slowing down or stopping further microbial breakdown of the substrates under study (Ramasamy et al. 2014), both the kinetics and true production of e.g., SCFA are inaccurate in these incubations. Since the metabolites are not removed, this allows for them to be converted into one another. For example, two molecules of acetate can be coupled to form butyrate, while lactate can be converted to propionate or butyrate. While this normally occurs to a certain extent in the gut microbiota, and is called cross-feeding (Kovatcheva-Datchary et al. 2009), in batch incubations this happens to a degree that is no longer physiological.

Also, the accumulation of the SCFA (in particular) leads to the inhibition of certain members of the microbiota, certainly if this is accompanied by a drop in pH. Because of this, the composition of the microbiota may change and no longer reflect that of the normal gut. It should be said that even in TIM-2 the microbiota will change after introduction of the inocula (Rajilic-Stojanovic et al. 2010).

Due to the fact that the system closely mimics physiological parameters, the experiments in TIM-2 usually take 1 week (see Sect. 26.2). In contrast, other models that mimic the large intestine usually take several weeks due to the need to reach steady state. Since multiple units of TIM-2 can be run at the same time (maximally 10), comparisons with a control can be made and steady state in TIM-2 is not strictly required. For a more extensive review of TIM-2 in comparison to other in vitro models mimicking the colon, please refer to Venema and van den Abbeele (2013).

26.2   General Protocol

After introduction of the microbiota into TIM-2 (at ~11 % w/w; 120 mL), the microbiota is allowed to adapt to the new situation over a period of 16 h and is fed with the standard SIEM medium, composed of complex carbohydrates, protein, some ox-bile, Tween 80 and vitamins and minerals (Maathuis et al. 2012). A total of 60 mL/day of this medium is fed to the microbiota. The medium is provided at a speed of 2.5 mL/h, reflecting the occasional opening and closing of the ileal-cecal valve in vivo.

After the 16 h adaptation period, there usually is a 2–4 h starvation period to allow the microbiota to ferment all the available carbohydrates in the system.

After this, the standard medium is replaced by the test carbohydrate. This test medium is fed for another 72 h. Just prior to the start of feeding the test medium, samples are taken from the lumen and dialysate at $t = 0$. Subsequently, every 24 h samples are taken from both the lumen and dialysate, and after analysis the cumulative production of microbial metabolites is calculated. Similarly, samples are taken every 24 h from the lumen to study changes in microbiota composition using microarray or, nowadays, next generation sequencing technology.

Although the above is considered to be a general protocol, every project may require the tailor-made adjustment of this protocol. Some experiments may not use a carbohydrate test substrate, but e.g. polyphenols (Gao et al. 2006; Bordonaro et al. 2014) or saponins (Kong et al. 2009), which requires addition of these compounds to the standard medium.

Over the past 5 years or so, we have also ran experiments in which we gave a single shot of 1 g of uniformly stable-isotope (^{13}C) labeled carbohydrates at $t = 0$ and then followed degradation of these substrates over time. Incorporation of label into microbial biomass using Stable Isotope Probing (SIP) (Maathuis et al. 2012) and into microbial metabolites using NMR and LC-MS (Binsl et al. 2010) were used to trace the fate of these labeled carbohydrates.

By choice no mucin is added to the system. The only commercial source of mucin is pig gastric mucin, which has a completely different composition than human colonic mucin, and therefore is not a good mimic for human colonic mucin. Therefore, the choice was made not to standardly add mucin to TIM-2 fermentations. In cases where addition of mucin would be essential, it could be added, although the above should be taken into consideration. Similarly, we do not add the insoluble and unfermentable cellulose. It is hardly fermentable to the human gut microbiota (Slavin et al. 1981), and microorganisms capable of using this substrate appear to only be present in fecal samples from methane-excretors while this community remained undetectable in non-methane-excretors (Robert and Bernalier-Donadille 2003). Sometimes however, it is used as a negative (non-fermentable) control.

Most of the times, the model simulates the proximal colon with a pH of 5.8, although sometimes the systems mimics the entire colon and the pH is programmed as a gradient over time from 5.8 in the proximal colon, to 6.4 in the transverse colon, to 7.0 in the distal colon. The lumen is then considered a plug which transits through the different parts of the large intestine.

26.3 Controls to Test Stability and Performance of the Model

TIM-2 is completely computer-controlled. All data with respect to the secretion of fluids (feed, NaOH, dialysate) as well as temperature and pH are monitored and stored. Moreover, these values are plotted on a computer-screen, such that the person operating the model is immediately aware of how the system is performing. Furthermore, the stability and performance of the system is monitored by studying these values after the experiment, in addition to monitoring the production of SCFA under standardized conditions. The production of SCFA should essentially be equal under identical conditions even if the experiments are performed over a stretch of several months. Unfortunately, SCFA cannot be measured online (yet), but the secretion of NaOH (to neutralize the drop in pH caused by the acids) is used as a proxy for SCFA production and this has worked very well to monitor the performance of the model. When a parameter is out of range, the run is terminated. Usually this happens because of a mechanical error in a pump or due to a leak in the system allowing oxygen to diffuse in, which would preclude a complete anaerobic environment.

The pH may not drop below 5.8 (the set-point value), but is allowed to rise above this value if for instance protein fermentation is expected or occurs. Protein fermentation leads to the production of putrefactive metabolites, amongst others ammonia, which raises the pH in the system. Under circumstances of excess protein it may in fact raise to a pH near neutrality.

The speed of dialysis is important to maintain physiological levels of microbial metabolites in the system. If dialysis does not occur (e.g., a pump broke down, or a tube got disconnected) this is immediately reflected in the production of acids (with NaOH as the proxy for that) and eventually in the microbiota composition, although the latter is only established days to weeks after the experiments, when the molecular analyses have been done.

Usually the experiments are very reproducible. The standard deviation (or range) normally falls within 10 %. This is why it was decided in the early 2000s to only perform duplicate experiments, and only repeat an experiment if the duplicates were way off.

26.4   Read-Out of the System and How This Information Can Be Used

Because of the health benefit of SCFA (den Besten et al. 2013), in particular butyrate (Havenaar 2011), most of the experiments carried out in the past 15 years focused on carbohydrate fermentation and the consequent production of the individual SCFA and their ratio. Experiments could screen numerous substrates for optimal SCFA production or would advance pre-established hypotheses that certain compounds would lead to the production of increased levels of certain SCFA, such as the case for butyrate when starch is fermented.

This requires the analysis and calculation for the cumulative production of these SCFA over time. Since all metabolites that are produced by the microbiota are collected in the system (from dialysate and lumen) this assessment of total SCFA production can be made.

Coupled to analyses for SCFA the (change in) composition of the microbiota is usually measured, to establish the link between SCFA produced and the microorganisms responsible for this. With the advent of the stable-isotope technology in gut microbiology this has become even easier, as there is a direct link between label incorporation into metabolites and into microbial biomass (Venema 2011).

However, the system has been used for numerous other applications as well. For instance, the effect of probiotics upon antibiotic treatment to quickly re-establish the microbial balance (Rehman et al. 2012), fermentation of polyphenols from tea, citrus or chocolate into more simple phenolic compounds, each with their own functionality (Bordonaro et al. 2014; Gao et al. 2006), or the metabolism of saponins from traditional Chinese medicines (Kong et al. 2009). No matter what the example, the aim has always been to increase the health of the host through the microbiota. Recently, the microbiota has also been shown to be involved in obesity (Venema 2010). This has led to the investigation in TIM-2 of the metabolism of microbiotas originating from lean and obese individuals (as yet unpublished). Numerous attempts are made to mine the microbiota for functional (healthy) activities (Roeselers et al. 2012). It is expected that TIM-2 may play a role in deciphering the beneficial microbes and metabolites that play a role in health and disease.

Sometimes samples from TIM-2 are incubated with other in vitro models described in this book, such as Caco-2 cell cultures (Lamers et al. 2003), immune cells (van Nuenen et al. 2005), or ex vivo pig intestinal tissue. In these cases, the effects of the microbial metabolites on parameters studied in these cells are of importance, such as barrier function, DNA damage, or induction of satiety hormones (such as GLP-1 or PYY). We have recently shown that samples from TIM-2 induce PYY, which is involved in appetite regulation, in pig intestinal tissue (Bussolo et al. 2014).

26.5 Advantages, Disadvantages and Limitations of the System

TNO has 10 units of TIM-2 available, allowing multiple parameters to be tested in parallel. In contrast to other multi-compartmental in vitro models, experiments are quick (usually three test days or less), yet still physiological and predictive. The advantages over other models are the presence of peristaltics and a dialysis system, the latter of which allows production of physiological concentrations of metabolites, and a highly active microbiota of normal density to be used. Another advantage is that a single parameter in the system can be changed, and the effect of that single parameter on microbiota activity can be studied. This has been done e.g., when studying the effect of different pH's on fermentation of carbohydrates (unpublished). Naturally, in vitro models have their limitations. As with every other in vitro model that mimics the colon, TIM-2 does not have epithelial or immune cells. However, as discussed in Sect. 26.4, samples can be incubated with these cells for even better predictability. Another limitation is that the model has been developed on the basis of literature data of mostly health individuals. Due to this, it is unclear exactly which parameters to simulate when simulating patient populations as discussed in van Nuenen et al. (2004). Another limitation is that (apart from volume and pH) there are no feed-back mechanisms in the system. Therefore, the experiments in such in vitro models will always be at most an indication of what may occur in real life, and the results need to be interpreted with care.

References

Aguirre A, Ramiro-Garcia J, Koenen ME, Venema K (2014) To pool or not to pool? Impact of the use of individual and pooled faecal samples for in vitro fermentation studies. J Microbiol Methods 107:1–7

Binsl TW, De Graaf AA, Venema K, Heringa J, Maathuis A, De Waard P, Van Beek JH (2010) Measuring non-steady-state metabolic fluxes in starch-converting faecal microbiota in vitro. Benefic Microbes 1:391–405

Bordonaro M, Venema K, Putri AK, Lazarova D (2014) Approaches that ascertain the role of dietary compounds in colonic cancer cells. World J Gastrointest Oncol 6:1–10

Bussolo CS, Roeselers G, Troost F, Jonkers D, Koenen ME, Venema K (2014) Prebiotic effects of cassava bagasse in TNO's in vitro model of the colon (TIM-2) in lean versus obese microbiota. J Funct Foods 11:210–220

den Besten G, van Eunen K, Groen AK, Venema K, Reijngoud DJ, Bakker BM (2013) The role of short-chain fatty acids in the interplay between diet, gut microbiota, and host energy metabolism. J Lipid Res 54:2325–2340

Gao K, Xu A, Krul C, Venema K, Liu Y, Niu Y, Lu J, Bensoussan L, Seeram NP, Heber D, Henning SM (2006) Of the major phenolic acids formed during human microbial fermentation of tea, citrus, and soy flavonoid supplements, only 3,4-dihydroxyphenylacetic acid has antiproliferative activity. J Nutr 136:52–57

Havenaar R (2011) Intestinal health functions of colonic microbial metabolites: a review. Benefic Microbes 2:103–114

Kong H, Wang M, Venema K, Maathuis A, van der Heijden R, van der Greef J, Xu G, Hankemeier T (2009) Bioconversion of red ginseng saponins in the gastro-intestinal tract in vitro model studied by high-performance liquid chromatography-high resolution Fourier transform ion cyclotron resonance mass spectrometry. J Chromatogr A 1216:2195–2203

Kovatcheva-Datchary P, Egert M, Maathuis A, Rajilic-Stojanovic M, de Graaf AA, Smidt H, de Vos WM, Venema K (2009) Linking phylogenetic identities of bacteria to starch fermentation in an in vitro model of the large intestine by RNA-based stable isotope probing. Environ Microbiol 11:914–926

Lamers RJ, Wessels EC, van de Sandt JJ, Venema K, Schaafsma G, van der Greef J, van Nesselrooij JH (2003) A pilot study to investigate effects of inulin on Caco-2 cells through in vitro metabolic fingerprinting. J Nutr 133:3080–3084

Maathuis AJ, van den Heuvel EG, Schoterman MH, Venema K (2012) Galacto-oligosaccharides have prebiotic activity in a dynamic in vitro colon model using a (13)C-labeling technique. J Nutr 142:1205–1212

Macfarlane GT, Gibson GR, Cummings JH (1992) Comparison of fermentation reactions in different regions of the human colon. J Appl Bacteriol 72:57–64

Martinez RC, Cardarelli HR, Borst W, Albrecht S, Schols H, Gutierrez OP, Maathuis AJ, de Melo Franco BD, De Martinis EC, Zoetendal EG, Venema K, Saad SM, Smidt H (2012) Effect of galactooligosaccharides and *Bifidobacterium animalis* Bb-12 on growth of *Lactobacillus amylovorus* DSM 16698, microbial community structure, and metabolite production in an in vitro colonic model set up with human or pig microbiota. FEMS Microbiol Ecol 84:110–123

Minekus M, Smeets-Peeters M, Bernalier A, Marol-Bonnin S, Havenaar R, Marteau P, Alric M, Fonty G, Huis in't Veld JH (1999) A computer-controlled system to simulate conditions of the large intestine with peristaltic mixing, water absorption and absorption of fermentation products. Appl Microbiol Biotechnol 53:108–114

Rajilic-Stojanovic M, Maathuis A, Heilig HG, Venema K, de Vos WM, Smidt H (2010) Evaluating the microbial diversity of an in vitro model of the human large intestine by phylogenetic microarray analysis. Microbiology 156:3270–3281

Ramasamy US, Venema K, Schols HA, Gruppen H (2014) The effect of soluble and insoluble fibers within the in vitro fermentation of chicory root pulp by human gut bacteria. J Agric Food Chem 62(28):6794–6802

Rehman A, Heinsen FA, Koenen ME, Venema K, Knecht H, Hellmig S, Schreiber S, Ott SJ (2012) Effects of probiotics and antibiotics on the intestinal homeostasis in a computer controlled model of the large intestine. BMC Microbiol 12:47

Robert C, Bernalier-Donadille A (2003) The cellulolytic microflora of the human colon: evidence of microcrystalline cellulose-degrading bacteria in methane-excreting subjects. FEMS Microbiol Ecol 46:81–89

Roeselers G, Bouwman J, Venema K, Montijn R (2012) The human gastrointestinal microbiota-an unexplored frontier for pharmaceutical discovery. Pharmacol Res 66:443–447

Rose DJ, Venema K, Keshavarzian A, Hamaker BR (2010) Starch-entrapped microspheres show a beneficial fermentation profile and decrease in potentially harmful bacteria during in vitro fermentation in faecal microbiota obtained from patients with inflammatory bowel disease. Br J Nutr 103:1514–1524

Slavin JL, Brauer PM, Marlett JA (1981) Neutral detergent fiber, hemicellulose and cellulose digestibility in human subjects. J Nutr 111:287–297

van Nuenen MHMC, Meyer PD, Venema K (2003) The effect of various inulins and *Clostridium difficile* on the metabolic activity of the human colonic microbiota in vitro. Microb Ecol Health Dis 15:137–144

van Nuenen MH, Venema K, van der Woude JC, Kuipers EJ (2004) The metabolic activity of fecal microbiota from healthy individuals and patients with inflammatory bowel disease. Dig Dis Sci 49:485–491

van Nuenen MH, de Ligt RA, Doornbos RP, van der Woude JC, Kuipers EJ, Venema K (2005) The influence of microbial metabolites on human intestinal epithelial cells and macrophages in vitro. FEMS Immunol Med Microbiol 45:183–189

Venema K (2010) Role of gut microbiota in the control of energy and carbohydrate metabolism. Curr Opin Clin Nutr Metab Care 13:432–438

Venema K (2011) Stable isotope probing and the human gut. In: Murrell JC, Whiteley AS (eds) Stable isotope probing and related technologies. ASM Press, Washington, DC, pp 233–257

Venema K, van den Abbeele P (2013) Experimental models of the gut microbiome. Best Pract Res Clin Gastroenterol 27:115–126

Venema K, van Nuenen HMC, Smeets-Peeters M, Minekus M, Havenaar R (2000) TNO's in vitro large intestinal model: an excellent screening tool for functional food and pharmaceutical research. Ernährung/Nutrition 24:558–564

Venema K, van Nuenen MHMC, van den Heuvel EG, Pool W, van der Vossen JMBM (2003) The effect of lactulose on the composition of the intestinal microbiota and short-chain fatty acid production in human volunteers and a computer-controlled model of the proximal large intestine. Microb Ecol Health Dis 15:94–105

Chapter 27
The Simulator of the Human Intestinal Microbial Ecosystem (SHIME®)

Tom Van de Wiele, Pieter Van den Abbeele, Wendy Ossieur, Sam Possemiers, and Massimo Marzorati

Abstract This chapter provides a general explanation of the Simulator of the Human Intestinal Microbial Ecosystem (SHIME®). The SHIME is one of the few gut models that mimics the entire gastrointestinal tract incorporating stomach, small intestine and different colon regions. After a general description of the model's development history and an overview of the specific features that distinguish SHIME from other gut models, we will give some insight in the general protocol of running SHIME experiments. However, with the SHIME being a highly flexible experimental setup, we also dedicate some part of this chapter discussing the modifications that can be performed, especially with respect to simulation of the mucosal (micro-)environment and interaction with host cells.

Keywords Multi-stage gut model • Colon microbiome • Mechanistic • Microbial adaptation

27.1 Description of SHIME®

'SHIME' is an acronym for the Simulator of the Human Intestinal Microbial Ecosystem and since 2010, the name has been jointly registered by ProDigest and Ghent University. This paragraph primarily describes the conventional experimental setup of the SHIME® system. Yet, because of its modular setup, the SHIME is highly flexible and it can be technically modified to target digestive conditions of interest: this will be briefly discussed further down this chapter.

T. Van de Wiele (✉) • M. Marzorati
Laboratory Microbial Ecology and Technology, Ghent University, Gent, Belgium
e-mail: tom.vandewiele@ugent.be

P.V.d. Abbeele • S. Possemiers
ProDigest, Gent, Belgium

W. Ossieur
Laboratory Microbial Ecology and Technology, Ghent University, Gent, Belgium

ProDigest, Gent, Belgium

© The Author(s) 2015   305
K. Verhoeckx et al. (eds.), *The Impact of Food Bio-Actives on Gut Health*,
DOI 10.1007/978-3-319-16104-4_27

27.1.1 History of the Model

The Simulator of the Human Intestinal Microbial Ecosystem (SHIME) is a multi-compartment dynamic simulator of the human gut developed in 1993 (Molly et al. 1993). The development of multi-compartment simulators of (parts of) the human gut originated from the awareness that fecal microbiota significantly differ from the in vivo colon microbiota in terms of community composition and metabolic activity. While inoculation of fecal microbiota into single-stage chemostats was a first attempt to mimic colon conditions, it was only useful for limited periods of time since environmental parameters such as pH, redox potential, available nutrients and microbial population dynamics constantly change. In order to maintain the inoculated intestinal microbiota over a longer timeframe, semi-continuous fermenters were developed where the intermittent supplementation of nutritional medium and the removal of microbial suspension could be simulated (Miller and Wolin 1981). While the latter systems typically make use of one single fermenter, the colon is a very heterogeneous region with clear differences in substrate availability, fermentation activity, microbial composition and several environmental conditions. This makes it impossible to simulate a representative culture of colon microbiota in one compartment. Several multi-compartment reactors were therefore developed to simulate the different conditions of the colon lumen (Macfarlane et al. 1989; Miller and Wolin 1981), from which the SHIME was one of the last in this generation of gut simulators.

Technically, the SHIME is an evolution of the simulator of the University of Reading introduced by Macfarlane et al. (1989) and mimics the conditions in the ascending, transverse and descending colon regions. The SHIME differentiates from the Reading model by incorporation of upper digestive tract conditions, leading to a succession of five compartments simulating the upper (stomach, small intestine) and the lower (ascending, transverse and descending colon) digestive tract.

The entire SHIME reactor operates at 37 °C. It contains double-jacketed glass vessels that are connected through peristaltic pumps (Fig. 27.1). The first two reactors

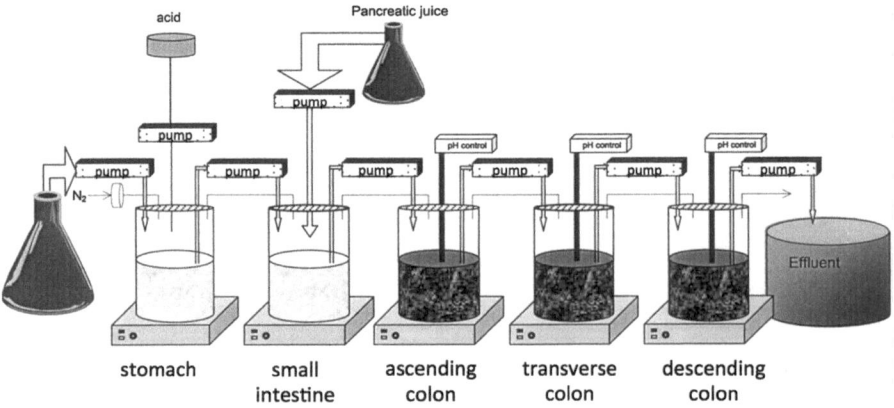

Fig. 27.1 Schematic representation of the SHIME®

follow a fill-and-draw principle adding three times a day a defined nutritional medium to a gastric compartment, and pancreatic and bile liquid to a small intestine compartment. The medium is composed of complex carbohydrate and protein sources with addition of mucins and a mineral and vitamin mix (Molly et al. 1993). Upon digestion in the gastric and intestine compartments, the slurry is pumped in the ascending colon vessel where colon digestion is initiated. The three colon compartments are continuously stirred with constant volume and pH control. Retention times in the upper digestive tract can be modulated by changing the flow rates from the gastric and intestine compartments, while retention times from the colon compartments are primarily modulated through a change in compartment volume. Depending on the human target group of interest the retention time may vary from 24 h to 72 h.

The pH of the gastric compartment used to work at a fixed pH of 2.0, yet with the advent of a completely computer-controlled SHIME system, specific pH profiles during gastric and intestine digestion can be established as well. While the small intestine compartment typically operates at slightly acidic to neutral conditions, the pH of the colon compartments is controlled between 5.6 and 5.9 in the ascending, 6.1–6.4 in the transverse and 6.6–6.9 in the descending colon. Mixing of the digestive slurry in the respective compartments is obtained with magnetic stir bars. The entire SHIME system is kept anaerobic by daily flushing the headspace of the respective compartments with N_2 gas or a 90/10 % N_2/CO_2 gas mixture.

27.1.2   Special Features of the Model

The emphasis of the SHIME® system is primarily put on the simulation of the colon microbial community. Because of the inaccessibility of the human colon region to take a representative microbial inoculum, the fecal microbiota is chosen as inoculum to the colon compartments of the SHIME reactor. The fecal microbiome is significantly different from the in vivo colon microbiome, both in terms of composition as metabolic activity. Yet, the colon being considered as a plug-flow system, the fecal microbiome is nothing else than a colon microbiome that has undergone community and metabolic shifts during transit from the proximal colon to the rectum. The idea of the SHIME system—and other multi-stage colon compartment reactors such as the Reading model—is to allow a suitable adaptation period for the fecal microbiome to adapt to the conditions that prevail in the respective colon compartments. From an engineering perspective a suitable adaptation time for starting up a reactor is around 5–10 times its residence time. Taking the example of a male individual with a gut residence time of 48 h and imposing that in the SHIME system, this would entail an adaptation time of 10–20 days upon inoculation with the fecal microbiota in the colon compartments. We will come back to this stabilization aspect in the next paragraph.

A second aspect is the choice of inoculum. The SHIME is typically inoculated with the fecal microbiome, derived from one individual. There has been or there still is a lot of debate on what is the most suitable inoculum for mimicking the human gut microbiome in the most representative way. Some research groups specifically opt for pooling fecal microbiota from for example ten different individuals (Minekus et al. 1999), thereby (partly) accounting for the huge interindividual variability that exists in microbiome composition and to incorporate properties from different microbiomes in order to create an 'average' 11th microbiome. Given the enormous functional redundancy of the gut microbiome such pooled microbiome will indeed take on a normal fermentation profile which generally not that different from the microbial fermentation profile of a single individual. While such approach may work when investigating hydrolysis and fermentation of carbohydrates, it fails to accurately mimic microbial processes that lack this functional redundancy. To exemplify, it has become clear that the microbial metabolic potency towards polyphenols such as daidzein, isoxanthohumol, catechins and others is highly dependent on an individual's microbiome (van Duynhoven et al. 2011). The existence of different 'metabotypes' has therefore even been proposed to distinguish a bioactive metabolite producing phenotype from a non-producing phenotype (Bolca et al. 2013). The above element is the primary reason why a SHIME reactor is inoculated with the fecal microbiome from one individual and succeeds in maintaining the microbial metabolic phenotype towards specific polyphenols during the in vivo/in vitro transition and several weeks after (Decroos et al. 2006; Possemiers et al. 2006).

The choice for individual inocula also appears to be a crucial aspect in recent studies where the SHIME was fundamentally optimized to enable colonization of the mucosal microbiome (Van den Abbeele et al. 2013). This concerns a third special feature of the SHIME model. The mucosal microbiome is this part of the gut microbial ecosystem that is able to colonize the mucus overlying the gut epithelium. Due to its close proximity to host epithelial cells, the mucosal microbiome is thought to have an intrinsically higher potency to modulate gut health, and by extension, human health (Van den Abbeele et al. 2011). The mucosal microbiome was already known to fundamentally differ from the luminal microbiome in composition and interestingly, presence of important mucosal colonizers such as *Faecalibacterium prausnitzii* seems to negatively correlate with occurrence (Willing et al. 2009) and postoperative recurrence of ileal Crohn's disease (Sokol et al. 2008). Given the difficult access to the mucosal environment, the development of gut simulators that accurately mimic mucosal microbial colonization is considered a strong asset to obtain a better understanding of the host-microbe interactome. It is in this philosophy that Van den Abbeele et al. (2013) decided to optimize the SHIME for mimicking mucosal microbial colonization by incorporation of mucin-covered microcosms. The major finding of this so-called M-SHIME, or mucosal SHIME, was that colonization of the mucosal environment was characterized by a higher abundance of butyrate producing Clostridium clusters IV and XIVa. This phylogenetic group is considered crucial for delivering butyrate as primary energy source to colonocytes and improves gut barrier function by strengthening the tight junctions. Coming back to the importance of individual inocula and avoiding pooled samples, it was also demonstrated that the M-SHIME was able to maintain the unique features of an individual's microbiome in terms of its mucosal composition (Van den Abbeele et al. 2013).

The fourth feature is the flexibility of the SHIME model and the ease with which reactor compartments can be added or left away. This modular setup is useful when a placebo-controlled study needs to be conducted, when different prebiotics need to be compared (Grootaert et al. 2009) or when a microbiome phenotype producing a bioactive metabolite is compared with a non-producing phenotype (Possemiers et al. 2006). Moreover, it is even possible to explore the interindividual variability in microbiome behavior upon specific treatments by having a common upper digestive tract simulation in the gastric and intestine compartments and subsequently split up the system in several parallel colon compartments, each of which are inoculated with the fecal microbiota from separate individuals.

A fifth feature of the SHIME model refers to the possibility of simulating the microbiome from different human target groups such as adult vs. infant, healthy vs. diseased (e.g. ulcerative colitis patients: Vermeiren et al. 2012) as well as the simulation of animal (pig, dog) microbiomes. In each of these specific cases, the microbial inocula, residence times of the different gastric compartments, composition of the gastric juices, region specific pH's, feed, feeding regimes and body temperatures are adapted in the SHIME set-up leading to an accurate and relevant simulation of the targeted human or animal host. Finally, other features of the SHIME include the gradual emptying of the gastric digest into the intestine compartment, the option of running a dynamic pH profiles in the gastric compartment and the possibility of putting a dialysis unit behind the intestine compartment to enable running experiments with real food matrices or food constituents that need to undergo predigestion and removal of sugar monomers or amino acids and peptides before the digest is transferred to the colon compartment.

27.1.3 Stability and Reproducibility of the System

As the SHIME® reactor is inoculated with a fecal microbiome, the latter needs an appropriate amount of time to adapt to the prevailing environmental conditions in the respective colon compartments. This adaptation process was studied more in detail by Possemiers et al. (2004) who monitored the increasing colonization of microbial groups of interest upon inoculation, as well as their metabolic activity. Highly-abundant groups such as *Bacteroides* obtained stable concentrations more easily, at 10 days, compared to less-abundant groups such as lactobacilli which needed 15 days of stabilization. To obtain a stable functionality in terms of short chain fatty acid production, an adaptation period of at least 15 days, even approaching 20 days, was needed. The length of stabilization obviously relates to the residence time that is imposed in the SHIME, but may also depend on the microbiome composition as such.

As the SHIME is a highly standardized system, and many digestive parameters are under control of the operator, it also leads to highly reproducible results. This is especially required when different products such as novel prebiotics, candidate drug components or new plant extracts, need to be compared with one another.

When it is not always possible to run different SHIMEs in parallel, for example because of the multitude of test compounds, SHIME experiments may need to be repeated. The scientist in charge needs to be cautious to conduct this experiment in exactly the same way and with the same inoculum to enable an adequate comparison between different compounds. Such reproducibility was previously tested for the conventional SHIME (without mucin-coated microcosms) and proven very effective in obtaining a reproducible microbial colonization process and accompanied metabolic activity (Van den Abbeele et al. 2010). Noteworthy, the authors found a small preferential colonization of *Bacteroidetes* and *Clostridium cluster* IX in the colon compartments in comparison with the stool sample. While the overall colonization profile was still representative of the in vivo colonization process, this slight bias has been a common observation for several dynamic models that work with luminal content only. Van den Abbeele et al. (2013) succeeded in removing this colonization selectivity by incorporating mucin-covered microcosms in the SHIME, thereby creating the M-SHIME or mucosal SHIME. Butyrate producing *Clostridium* clusters IV and XIVa were found to specifically colonize the mucosal environment thereby compensating for their lower abundance in the lumen.

27.1.4   Relevance to Human In Vivo Situation

The first validation paper of the conventional SHIME® setup concerned a study from Molly et al. (1994) where several microbe-associated characteristics were defined. The authors focused on fermentation profiles of pectin, xylan, arabinogalactan and starch and found these to be consistent with the results from incubations of the same products with fecal microbiota from human volunteers (Englyst et al. 1987). Similarly, an enzymatic profile focusing on glycosyl hydrolases (galactosidase, glucosidase, xylosidase) was recorded and no differences with the fecal incubations were found. Apart from the comparison with fermentation activity the metabolic potency towards sulphasalazin, a prodrug for ulcerative colitis treatment, and its conversion to the active compound 5-aminosalicylic acid (5-ASA) was also evaluated. Consistent with in vivo literature, no 5-ASA was detected in the gastric compartment, while small amounts of 5-ASA were detected in the intestine compartment and full sulphasalazin conversion was observed in the colon compartments (Peppercorn and Goldman 1972).

A second validation study discusses the preservation of metabolic phenotypes when transferring fecal microbiota from an individual to the in vitro SHIME system. Focusing on the microbial conversion of isoxanthohumol to 8-prenylnaringenin (8-PN), a metabolite with strong pseudo-estrogenic activity, Possemiers et al. (2006) demonstrated a one-on-one correlation between urinary 8-PN excretion by human individuals and their microbiome. Secondly, the microbiome from a 8-PN producing and non-producing individual could be stably transferred to the SHIME system and the metabolic phenotype was adequately preserved.

A third validation study concerns the application of the M-SHIME where both the luminal as mucosal microbiome from an individual were simulated in the SHIME colon compartment (Van den Abbeele et al. 2013). Analyzing the microbiome from the luminal and mucosal regions with the HIT-Chip micro-array revealed the largest variability in the microbial dataset to originate from the in vivo/in vitro transition and difference between the luminal and mucosal environment. Yet, those species that account for the unique profile of an individual's microbiome were preserved in the M-SHIME as well and accounted for 25 % of the variability in the microbial dataset.

27.1.5   Quality in Relation to Other Models with the Same Applicability

The SHIME® model is the sole in vitro model that integrates the entire gastrointestinal transit into one system. This is interesting to study for example digestibility of prebiotic substrates and its subsequent fermentability in the colon or the survival of pathogens or probiotics in the upper digestive tract before they reach the colon environment. Yet, one must consider that the conventional SHIME system operates without an absorption unit: this means that the nutritional medium for the SHIME must already be deprived of easily digestible carbohydrates or proteins that would normally be absorbed in the intestine.

The SHIME is the last of a generation of multi-compartment models that operate according to a semi-continuous stirred tank reactors setup. In comparison with the Reading model (Macfarlane et al. 1989) the SHIME connects the different compartments through peristaltic pumps. In terms of studying the gut microbiome over a long timeframe, both models would however be applicable.

In general, it must be stressed that the SHIME model has a strong emphasis on the ecological aspects of the colon microbiome. This entails that incubation experiments with the SHIME reactor are seldomly short and commonly take several weeks. This is put in place to look at the gradual adaptation of the microbiome to incoming substrates of interest (e.g. prebiotics or pharmaceuticals) or to evaluate the resilience of the microbial community against colonization by a (opportunistic) pathogen. Such research, which often necessitates an experimental period of more than a week to several weeks, strongly differs from the TIM-2 model, which monitors the microbiome on a shorter timeframe. One could conclude that short term experiments are highly suitable for evaluating the immediate metabolic potency of a carbohydrate or colonization ability of a microorganism of interest, while long-term experiments are primarily suitable to look at the adaptation of the microbial ecosystem to changing environmental conditions or inputs.

A last strong asset of the SHIME is its extension to M-SHIME—incorporating the mucosal microbiota (Van den Abbeele et al. 2013). While previous attempts primarily focused on the immediate and aspecific adhesion potency of gut microorganisms to mucin-covered glass slides, the integration of mucin-covered microcosms that can be replaced, to mimic desquamation, has been a breakthrough in the simulation of the mucosal microbiome and in the understanding of its dynamics.

27.2 General Protocol

Inoculation of the SHIME® colon compartments occurs with microbiota that has been isolated from fecal material of one individual. In contrast with other gut models, we deliberately choose not to work with the microbiome from pooled fecal samples from different human volunteers. The artificially high microbial diversity in pooled inocula creates disturbances in the cross-feeding processes between microorganisms that are adapted to one another in each of the separate microbiomes. We therefore advise to study interindividual variability through separate experiments. Upon inoculation of the colon compartments with fecal microbiota, the microbiome is given time to adapt itself to the prevailing conditions in the ascending, transverse and descending colon compartments.

A typical SHIME experiment consists of four stages: a stabilization period (2 weeks) to allow adaptation of the microbial community to the environmental conditions in the respective colon regions; a basal period (2 weeks) in which the reactor is operated under nominal conditions and baseline parameters are measured; a treatment period (2–4 weeks) where the effect of a specific treatment on the gastrointestinal microbial community is tested; and a washout period (2 weeks) to determine how long the changes induced by the tested substance can still be measured in the absence of the substance itself. This approach has mainly been used to investigate the activity and stability of probiotics and prebiotics during gastrointestinal transfer, the microbial conversion of bioactive food components (e.g. phytoestrogens), the metabolism of pharmaceutical components, the efficacy of colonic targeted delivery systems and the conversion and biological (in)activation of food and/or ingested environmental contaminants.

Shorter-term SHIME experiments are also a possibility. Specifically in the context of monitoring the initial stages of microbial colonization on the mucosal surface, 1-week experiments can be conducted. Distinguishing the mucosal from the luminal microbiome, Van den Abbeele et al. (2013) evaluated the colonization process of the microbiome derived from five human volunteers in M-SHIME systems over a timeframe of 5 days. The presence of mucins seems to play a specific role in the colonization process. In contrast with mucin surfaces, other contact surfaces, such as agar or plastic surfaces, did not result in a distinct colonization profile. Despite the fact that the applied mucins are derived from the porcine gut—and hence do not have the exact same composition as human mucins—the presence of this fairly similar glycoprotein surface already is an important driver in the colonization process.

Finally, by choosing the inoculum, tweaking digestive parameters, incorporating surface carriers or working with parallel reactor compartments, a SHIME operator is able to mimic the gastrointestinal conditions of a host target group or phenotype of interest, to incorporate internal control experiments or experimental replicates and to take into account the gastrointestinal colonization by mucosal microbiota.

27.3 Controls to Test Stability and Performance of the Model

As the SHIME® system can be operated over a longer timeframe, stability of the system is a crucial aspect. Investigating the modulation of the intestinal microbiome through certain treatments, also requires proper knowledge of a baseline situation and assumption of a stable microbiome. The microbial community composition and fermentation activity in the respective colon compartments is therefore closely monitored, especially during reactor startup, to evaluate the reactor's capacity to create a stable microbiome that is still representative for the human in vivo situation. Using moving window correlation, Possemiers et al. (2004) previously introduced a stability criterion based on measurements with PCR-DGGE for different microbial groups, short chain fatty acids and ammonium. Calculating the correlation between two consecutive days along reactor startup, the authors considered a community to be stable once 80 % correlation was measured. The remaining 15–20 % variability typically originates from normal biological fluctuations. Overall, community stability is reached after about 2 weeks, while functional stability is obtained after 3 weeks. This startup phase seems quite long for a conventional SHIME experiment, but it is desired when the microbiome from different colon regions is under study and when the research question concerns microbial adaptations to applied treatments. Yet, short-term SHIME experiments can be conducted as well. The mucosal colonization process in the M-SHIME occurs quite rapidly—within 5 days—when a clear distinction between luminal and mucosal microbiome composition is noted. The preferential colonization of butyrate producing clostridia in the mucosal environment is one of the characteristic features (Van den Abbeele et al. 2013).

27.4 Read-Out of the System and How This Information Can Be Used

Assessing fermentation activity is one of the most important SHIME® read-outs. During operation of the SHIME system, short chain fatty acid or ammonia production can be preliminary assessed by the amount of NaOH or HCl that is supplemented by the pH controllers to maintain proper pH values in the respective colon vessels. As the conventional setup of the SHIME system does not have an absorption unit, short chain fatty acids (SCFA) do accumulate throughout the distal colon compartments. However, the SCFA concentrations in the SHIME do not lead to values above 100 mM, which would eventually result in product inhibition of the fermentation process. As saccharolytic conditions are more abundant than proteolytic conditions in the proximal colon and vice versa, the operator can still deduce the drop in saccharolytic activity by subtracting the SCFA concentration in the proximal colon from the SCFA concentration in the distal colon and thereby obtain the net amount of SCFA that are produced in the distal colon. Such strategy has been used before to discern the colon regions where specific fibers or oligosaccharides are broken down (Grootaert et al. 2009).

While adaptation of microbiome fermentation profiles to changing nutritional conditions is often monitored over several weeks (Grootaert et al. 2009), the immediate metabolic potency of the colon microbiome towards specific compounds of interest can be evaluated within days, e.g. with the onset of isoxanthohumol conversion into 8-prenylnaringenin (Possemiers et al. 2006), or even hours, e.g. with the microbial conversion of tea catechins and wine polyphenols (Gross et al. 2010).

Apart from the general fermentation activity and metabolic potency, the SHIME is evaluated for the microbiome composition in the respective colon vessels. The common approach typically includes establishment of DGGE profiles, either at Eubacteria or at group-specific level, as a quick screening for evaluating what treatments or what sample times are interesting to study in more detail. This can then be complemented by a quantitative analysis with q-PCR or a high-throughput analysis at the phylogenetic level with next generation sequencing. One of the important read-outs with microbiome fingerprints and profiles is the determination of the so-called 'MRM-parameters' (microbial resource management) (Marzorati et al. 2009). These typically contain calculation of richness, evenness and diversity indices and of microbiome dynamics throughout the study period. Especially evenness seems to be a strong indicator of microbiome resilience against invasion by exogenous species (Wittebolle et al. 2009).

A final read-out consists of an evaluation of host-microbe interactions. Colon suspension or the derived intestinal water (supernatant after centrifugation) can be brought in indirect or direct contact with host epithelial cells. This allows assessing to what extent changes in microbiome composition, microbial metabolites, signaling molecules or antigens have differential effects at the level of the host. This can be performed both with epithelial cells (Grootaert et al. 2011) as with a combination of epithelial and immune cells (Possemiers et al. 2013). Interestingly, the SHIME has recently been coupled to the Host-Microbe Interaction Module (HMI), which is a bi-compartmental system containing mucosal microbiota one the one side and host cells on the other side of a semi-permeable membrane (Marzorati et al. 2014).

27.5   Advantages, Disadvantages and Limitations of the System

The advantages, disadvantages and limitations of the system are summarized in Table 27.1.

Table 27.1 Advantages and disadvantages to the SHIME® system

Advantages	• Integrates the entire gastrointestinal tract
	• Microbiome inoculation from different target groups: adult vs. infant, healthy vs. diseased (e.g. ulcerative colitis patients: Vermeiren et al. 2012) and animals (pig, dog)
	• Colon-region specific research (Possemiers et al. 2006)
	• Maintains microbiome stability over a long timeframe: possibility to monitor microbiome adaptation
	• Mechanistic research by multi-parametric control
	• Differentiation between mucosal and luminal microbiome in M-SHIME setup (Van den Abbeele et al. 2013)
	• Parallel control and treatment in TWIN-SHIME setups
	• Interindividual variability can be studied in a SHIME setup as unique features of an individual's microbiome are preserved. Limiting microbiome simulation to one colon region, eight different subjects can be simultaneously assessed
Disadvantages	• Conventional SHIME setup lacks dialysis. Incorporation of dialysis modules is possible after small intestine digestion (Ceuppens et al. 2012) and colon digestion
	• Lack of peristalsis, mixing is conducted by means of stirrers as normally performed in a standard dissolution apparatus
	• Absence of host cells in conventional SHIME. Solved by coupling to HMI module with epithelial or immune cells (Possemiers et al. 2013; Marzorati et al. 2014)

References

Bolca S, Van de Wiele T, Possemiers S (2013) Gut metabotypes govern health effects of dietary polyphenols. Curr Opin Biotechnol 24:220–225

Ceuppens S, Uyttendaele M, Drieskens K, Heyndrickx M, Rajkovic A, Boon N et al (2012) Survival and germination of *Bacillus cereus* spores without outgrowth or enterotoxin production during in vitro simulation of gastrointestinal transit. Appl Environ Microbiol 78:7698–7705

Decroos K, Eeckhaut E, Possemiers S, Verstraete W (2006) Administration of equol-producing bacteria alters the equol production status in the simulator of the gastrointestinal microbial ecosystem (SHIME). J Nutr 136:946–952

Englyst HN, Hay S, MacFarlane GT (1987) Polysaccharide breakdown by mixed populations of human fecal bacteria. FEMS Microbiol Ecol 95:163–171

Grootaert C, Van den Abbeele P, Marzorati M, Broekaert WF, Courtin CM, Delcour JA et al (2009) Comparison of prebiotic effects of arabinoxylan oligosaccharides and inulin in a simulator of the human intestinal microbial ecosystem. FEMS Microbiol Ecol 69:231–242

Grootaert C, Van de Wiele T, Van Roosbroeck I, Possemiers S, Vercoutter-Edouart A, Verstraete W, Bracke M, Vanhoecke B (2011) Bacterial monocultures, propionate, butyrate and H_2O_2 modulate the expression, secretion and structure of the fasting induced adipose factor in gut epithelial cells. Environ Microbiol 13:1778–1789

Gross G, Jacobs D, Peters S, Possemiers S, van Duynhoven J, Vaughan E et al (2010) In vitro bioconversion of polyphenols from black tea and red wine/grape juice by human intestinal microbiota displays strong inter-individual variability. J Agric Food Chem 58:10236–10246

Macfarlane GT, Cummings JH, Macfarlane S, Gibson GR (1989) Influence of retention time on degradation of pancreatic-enzymes by human colonic bacteria grown in a 3-stage continuous culture system. J Appl Bacteriol 67:521–527

Marzorati M, Wittebolle L, Boon N, Daffonchio D, Verstraete W (2009) How to get more out of molecular fingerprints: practical tools for microbial ecology. Environ Microbiol 10: 1571–1581

Marzorati M, Vanhoecke B, De Ryck T, Sadaghian Sabadad M, Pinheiro I, Possemiers S et al (2014) The HMI module: a new in vitro tool to study the host microbiome interactions from the human gastrointestinal tract. BMC Microbiol 14:133

Miller TL, Wolin MJ (1981) Fermentation by the human large-intestine microbial community in an in vitro semicontinuous culture system. Appl Environ Microbiol 42:400–407

Minekus M, Smeets-Peeters M, Bernalier A, Marol-Bonnin S, Havenaar R, Marteau P et al (1999) A computer-controlled system to simulate conditions of the large intestine with peristaltic mixing, water absorption and absorption of fermentation products. Appl Microbiol Biotechnol 53:108–114

Molly K, Vandewoestijne M, Verstraete W (1993) Development of a 5-step multichamber reactor as a simulation of the human intestinal microbial ecosystem. Appl Microbiol Biotechnol 39:254–258

Molly K, Vandewoestyne M, Desmet I, Verstraete W (1994) Validation of the simulator of the human intestinal microbial ecosystem (SHIME) reactor using microorganism-associated activities. Microb Ecol Health Dis 7:191–200

Peppercorn, Goldman (1972) The role of intestinal bacteria in the metabolism of salicyl azo-sulphapyridines. J Pharmacol Exp Ther 181:555–562

Possemiers S, Verthe K, Uyttendaele S, Verstraete W (2004) PCR-DGGE-based quantification of stability of the microbial community in a simulator of the human intestinal microbial ecosystem. FEMS Microbiol Ecol 49:495–507

Possemiers S, Bolca S, Grootaert C, Heyerick A, Decroos K, Dhooge W et al (2006) The prenyl-flavonoid isoxanthohumol from hops (*Humulus lupulus* L.) is activated into the potent phytoestrogen 8-prenylnaringenin in vitro and in the human intestine. J Nutr 136:1862–1867

Possemiers S, Pinheiro I, Verhelst A, Van den Abbeele P, Maignien L, Laukens D et al (2013) A dried yeast fermentate selectively modulates both the luminal and mucosal gut microbiota and protects against inflammation, as studied in an integrated in vitro approach. J Agric Food Chem 61:9380–9392

Sokol H, Pigneur B, Watterlot L et al (2008) *Faecalibacterium prausnitzii* is an anti-inflammatory commensal bacterium identified by gut microbiota analysis of Crohn disease patients. Proc Natl Acad Sci U S A 105:16731–16736

Van den Abbeele P, Grootaert C, Marzorati M, Possemiers S, Verstraete W, Gerard P et al (2010) Microbial community development in a dynamic gut model is reproducible, colon-region specific and selects for Bacteroidetes and Clostridium cluster IX. Appl Environ Microbiol 76:5237–5246

Van den Abbeele P, Van de Wiele T, Verstraete W, Possemiers S (2011) The host selects mucosal and luminal associations of co-evolved gut microbes: a novel concept. FEMS Microbiol Rev 35:681–704

Van den Abbeele P, Belzer C, Goossens M, Kleerebezem M, De Vos WM, Thas O et al (2013) Butyrate-producing Clostridium cluster XIVa species specifically colonize mucins in an in vitro gut model. ISME J 7:949–961

van Duynhoven J, Vaughan E, Jacobs D, Kemperman R, van Velzen E, Gross G et al (2011) The metabolic fate of polyphenols in the human superorganism. Proc Natl Acad Sci U S A 108:S4531–S4538

Vermeiren J, Van den Abbeele P, Laukens D, Visgnaes LK, De Vos M et al (2012) Decreased colonization of fecal *Clostridium coccoides/Eubacterium rectale* species from ulcerative colitis patients in an in vitro dynamic gut model with mucin environment. FEMS Microbiol Ecol 79:685–696

Willing B, Halfvarson J, Dicksved J, Rosenquist M, Järnerot G, Engstrand L et al (2009) Twin studies reveal specific imbalances in the mucosa-associated microbiota of patients with ileal Crohn's disease. Inflamm Bowel Dis 15:653–660

Wittebolle L, Marzorati M, Clement L, Balloi A, Daffonchio D, Heylen K et al (2009) Initial community evenness favours functionality under selective stress. Nature 458:623–626

Chapter 28
The Computer-Controlled Multicompartmental Dynamic Model of the Gastrointestinal System SIMGI

Elvira Barroso, Carolina Cueva, Carmen Peláez, M. Carmen Martínez-Cuesta, and Teresa Requena

Abstract The SIMGI (SIMulator Gastro-Intestinal) is an automated gastrointestinal in vitro model designed to dynamically simulate the physiological processes taking place during digestion in the stomach and small intestine, as well as to reproduce the colonic microbiota responsible for metabolic bioconversions in the large intestine. This computer-controlled system is a flexible modulating system that combines a gastric compartment that operates with peristaltic mixing movements, a reactor simulating the small intestine and three-stage continuous reactors that reproduce the colon region-specific microbiota. The compartments designed for digestion (stomach and small intestine) and fermentation (colon) can be connected to operate jointly. Alternatively, the digestion and fermentation processes can proceed independently. This section describes the conditions needed to inoculate, stabilize and differentiate the fecal microbiota in the SIMGI system, as well as the steps to follow in order to test the stabilized colonic microbiota with different food ingredients and/or by modifying the caloric intake in the nutrition media.

Keywords Automated in vitro dynamic model • Three-stage fermentation • SIMGI • Colon microbiota

E. Barroso • C. Cueva • C. Peláez • M.C. Martínez-Cuesta • T. Requena (✉)
Department of Food Biotechnology and Microbiology, Institute of Food Science Research, CIAL (CSIC-UAM), Madrid, Spain
e-mail: t.requena@csic.es

© The Author(s) 2015   319
K. Verhoeckx et al. (eds.), *The Impact of Food Bio-Actives on Gut Health*,
DOI 10.1007/978-3-319-16104-4_28

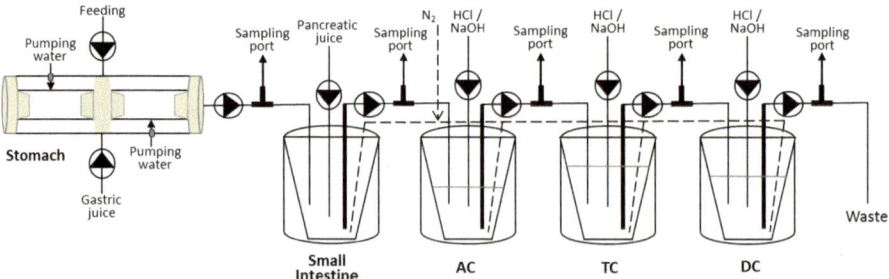

Fig. 28.1 Schematic diagram of the SIMulator Gastro-Intestinal SIMGI

28.1 Description of the Model

28.1.1 History and Special Features of the Model

The SIMGI model is a fully automated gastrointestinal multichamber simulator that has been recently developed at the Institute of Food Science Research CIAL (CSIC-UAM, Madrid, Spain). The SIMGI comprises five interconnected compartments (units) that simulate the stomach, small intestine and three stages of the large intestine (Fig. 28.1). The process of digestion is simulated in units Stomach and Small Intestine. Unit Stomach is comprised of two cylindrical transparent and rigid methacrylate plastic modules covering a reservoir of flexible silicone walls where the gastric content is mixed by peristaltic movements. The simulation of gastric peristalsis is achieved by changing the pressure of the water that flows in the jacket between the plastic modules and the reservoir. The pumped thermostated water keeps the temperature of the gastric content at 37 °C. The stomach compartment has different ports for input of experimental food components, gastric juice, and HCl. The pH decrease is computer-controlled to follow the curve resulting from a linear fit of data representing experimental in vivo conditions. The small intestine consists in a double jacket glass reactor vessel continuously stirred that receives the gastric content which is mixed with pancreatic juice and bile. The intestinal content is digested during 2 h at 37 °C and kept at pH 6.8. The stages of the large intestine are simulated in three double jacket glass reactors and the colon content is kept at 37 °C by pumping thermostated water into the space between the glass jackets. The pH in the colonic units (named ascending AC, transverse TC and descending DC colon) is controlled by addition of 0.5 M NaOH and 0.5 M HCl to keep values of 5.6±0.2 in the AC, 6.3±0.2 in the TC and 6.8±0.2 in the DC compartments. When the digested content of the small intestine is transferred to the proximal colon compartment (AC), the transit of colonic content between the AC, TC and DC compartments is simultaneously initiated at the same flow rate. The intestinal and colonic vessels contain ports for transit of intestinal content, sampling, continuous flushing of nitrogen and pH and temperature control.

The SIMGI design is aimed to dynamically operate with the five units simulating the whole gastrointestinal process. In addition, the SIMGI software allows the work of the stomach and the small intestine in a continuous way to study food digestion and at the same time running in parallel, direct feeding of the small intestine and the transit to the colonic vessels to study microbial community development and metabolism. In this way, the system is flexible and adaptable to each need of experimental approach.

28.1.2   Stability and Reproducibility of the System

The SIMGI model has been recently developed and, therefore, it has a short history in evaluating digestion of food components and/or the effects of diet on modulating the gut microbiota and its metabolic activity. The fermentative module of the system (AC, TC and DC compartments) follows the concept of multiple connected, anaerobic, pH-controlled vessels. Multi-compartment fermentation models are usually represented by three-stage culture reactors as initially designed by Gibson et al. (1988). This feature allows these models to reproduce differences from proximal (characterized by acidic pH and carbohydrate-excess conditions) to distal colonic regions (showing a carbohydrate-depleted and non-acidic environment). The validation of three-stage culture fermentation models to simulate the ascending, transverse and descending colon conditions have been earlier described by Macfarlane et al. (1998) and Molly et al. (1994).

The biological functioning of these fermentation models requires the development of a colon region-specific microbial community that needs to be stabilized before starting any experimental approach. This initial stage allows the microbial evolution in the three reactors from a fecal inoculum to a colon region-specific microbiota (Possemiers et al. 2004; Macfarlane et al. 1998). In addition, the stabilization period is required to provide a steady-state environment where the composition and metabolism of the microbial community can be evaluated during long-term experimental dietary interventions.

The evolution of the gut microbiota in the AC, TC and DC compartments of the SIMGI model was followed during a 2-week stabilization period (Barroso et al. 2015). Results indicated that from day 8 onwards the microbial DGGE fingerprints of samples from the same compartment clustered together and that at 14 days the microbial communities reached the steady state. The counts of some representative microbial groups analyzed by quantitative PCR (qPCR) in the AC, TC and DC compartments are shown in Fig. 28.2. Differences observed between the compartments indicated that *Bacteroides* were more representative in the AC and TC compartments than in the DC reactor, whereas the butyrate-producing groups *C. leptum* and *Ruminoccocus* were less represented in the proximal colon compartment (AC) than in the distal vessels (TC and DC). This specific microbiota composition during the stabilization period has also been observed in the three-stage culture model SHIME (Barroso et al. 2014; Van den Abbeele et al. 2010).

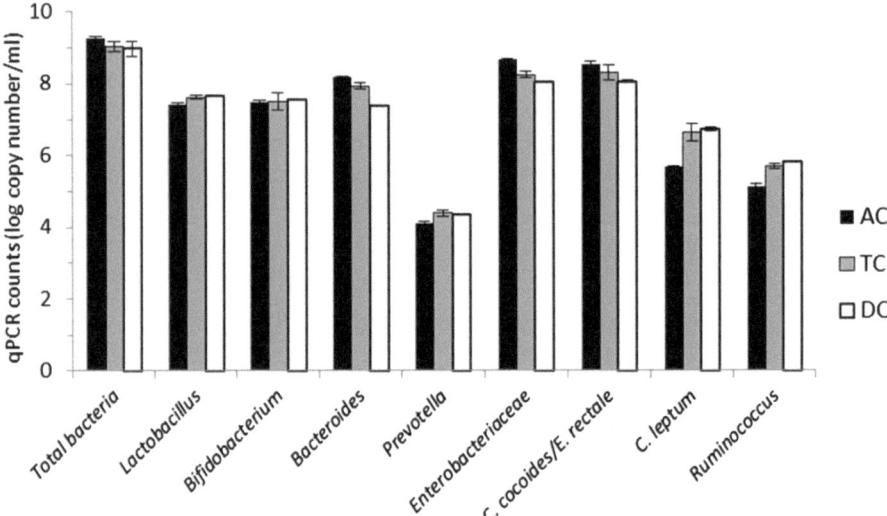

Fig. 28.2 Mean ± SD of qPCR counts (log copy number/mL) for the microbial groups analyzed in the ascending (AC), transverse (TC) and descending colon (DC) compartments of the SIMGI after 2 weeks of stabilization

28.1.3   Relevance to Human In Vivo Situation

The suitability of the SIMGI fermentation model to reproduce human conditions associated to changes in dietary lifestyles has been evaluated by the simulation of an obesity-associated microbiota in this in vitro system (unpublished results). For this purpose, the three colonic reactors were inoculated with fecal microbiota from an overweight individual after which the system was feeding daily with a high energy nutritive medium for 2 weeks (microbial stabilization period). Increase of energy in the nutritive medium was achieved by increasing the content of simple sugars (fructose), simulating high consumption of sugar-sweetened beverages, and of carbohydrates from simple starches (maize and potato). After the stabilization period a dietary intervention during 7 days was performed by lowering energy of the nutritive medium in order to observe possible changes in microbial composition and metabolism induced by a sharp shift in the diet. This low energy diet was obtained by suppression of simple carbohydrates and reduction of the content of readily fermentable starches. The selection of components of the nutritive media was based on the rational design of media with different energy content described by Payne et al. (2012) to compare the impact of dietary energy on gut microbiota in a three-stage in vitro continuous fermentation model that uses fecal microbiota immobilized within a porous, non-biodegradable polysaccharide matrix (Cinquin et al. 2006).

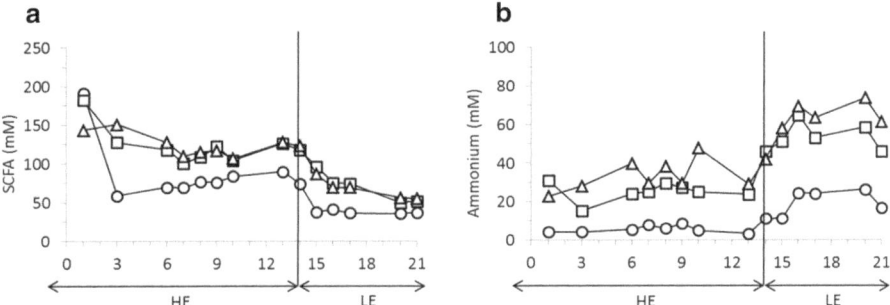

Fig. 28.3 Changes in concentration (mM) of SCFAs (**a**) and ammonium (**b**) in the ascending (AC; *circles*), transverse (TC; *squares*) and descending colon (DC; *triangles*) compartments of the SIMGI at different times after inoculation and feeding with high energy (HE) and low energy (LE) diets

The comparison of short chain fatty acids (SCFA) and ammonium formation under high energy (microbiota stabilization period) and low energy (dietary intervention) diets is shown in Fig. 28.3. Shift to a low energy diet resulted in a twofold decrease in the average content of total SCFA of the three colon compartments compared to the high energy intake period. Additionally, the shift from high to low energy medium caused a twofold increase in the ammonium content of the distal colon compartments (TC and DC) and a remarkable sixfold increase in the proximal colon compartment (AC). The SFCA and ammonium results could be compared with in vivo data from obese subjects, where a significant decrease of SCFA and increase of proteolytic products were observed when the individuals consumed diets high in protein and reduced in total carbohydrates (Russell et al. 2011).

28.1.4   Quality in Relation to Other Models with the Same Applicability

The SIMGI is designed to simulate not only the fermentation process as described above, but also the process of digestion. This is a competitive advantage compared to other systems that simulate digestion and fermentation operating in separate modules (TIM-1 and TIM-2). The whole system is computer controlled through an operator panel and programmable logic controller. It can be set up to sequentially proceed (continuously or feeding the system from 1 to 6 times daily) from the operation of food intake into the stomach throughout the delivery of distal colon content to waste. The system differs from the SHIME model in that the stomach uses peristaltic movements for mixing the ingested food with gastric fluids. Additionally, the computer software allows the definition of pH acidification curves that can be set up according to results obtained from in vivo studies (Marteau et al. 1990). The stomach emptying is programmable to follow the equation described by Elashoff et al. (1982) that allows the modification of the shape of the emptying curve depending on liquid, semisolid or solid foods. This process implies that the gastric content will be delivered to the small intestine at different pH values (up to pH 2). This is an important feature to take into account when evaluating probiotic survival growth under gastric and duodenal conditions (Fernández de Palencia et al. 2008) that would enable their eventual functionality when reaching the large intestine. Another distinctive characteristic of the SIMGI is that the small intestine and the three colon reactors are continuously flushed with nitrogen inducing anaerobic conditions for the oxygen-sensitive intestinal microbial communities, thus allowing a permanent anaerobic atmosphere.

28.2   General Protocol

The operation of the SIMGI as a fermentation model mainly follows the protocols established for multi-compartment culture reactors. For most of dietary interventions, the choice of the nutritive medium can be based on the medium developed by Macfarlane et al. (1998). Therefore, the three colonic vessels which will be inoculated with fecal samples are filled and pre-conditioned with the nutritive medium that will be fed to the system during the microbial stabilization period in a volume of 250, 400 and 300 mL for the AC, TC and DC compartments, respectively. The volumes of the colonic reactors and the transit of colonic content between compartments, three times daily at a flow rate of 5 mL/min, are intended to give an overall residence time of 76 h, adapting the conditions already standardized in the SHIME model (De Boever et al. 2000; Molly et al. 1993; Van den Abbeele et al. 2010). Further on, the AC, TC and DC vessels are inoculated with 20 mL of a fresh 20 % (w/v) human fecal slurry prepared in anaerobic conditions with sodium phosphate buffer (0.1 M, pH 7.0), containing 1 g/L sodium thioglycolate as reducing agent, as described by De Boever et al. (2000). The inoculated AC, TC and DC vessels are allowed to equilibrate overnight in batch conditions at 37 °C. The pH in the colonic compartments is controlled by addition of 0.5 M NaOH and 0.5 M HCl to keep values of 5.6 ± 0.2 in the AC, 6.3 ± 0.2 in the TC and 6.8 ± 0.2 in the DC. The temperature is kept at 37 °C by pumping water into the space between the double glass jackets of the reactor vessels. The development and stabilization of the microbial community until steady-state conditions in the three colon vessels is approached by feeding the small intestine compartment with nutritive medium (75 mL, pH 2) mixed with pancreatic juice (40 mL of a solution of 12 g/L $NaHCO_3$, 6 g/L oxgall dehydrate fresh bile and 0.9 g/L porcine pancreatine) three times a day during 14 days (Van den Abbeele et al. 2010). The small intestine digestion is performed during 2 h at 37 °C and the whole content of the vessel is automatically transferred at a flow rate of 5 mL/min to the proximal colon compartment (AC). The transfer of intestinal content between the reactors is initiated to operate simultaneously in order to keep constant the AC, TC and DC reactor volumes. The process of stabilization of the colon microbiota in the three-stage reactors of the SIMGI model is usually reached after 14 days of running the system under the described conditions. This 2-week stabilization period allows the development of the fecal microbiota to colon region-specific microbial communities. After this period, the system provides a steady-state microbial environment ready to be evaluated under different experimental dietary interventions.

28.3   Controls to Test Stability and Performance of the Model

The SIMGI model is entirely under computer control that is conducted through an operator panel and a programmable logic controller Unitronics' Vision120™. The software offers several operating modes: (1) the whole gastrointestinal tract; (2) independent experiments in parallel with stomach and small intestine in one module and the three-stage colon reactors in the other; and (3) the same but with the small intestine working with the colon reactors. If needed, each compartment can operate individually one by one. Instructions for setting up the working pH values, temperature, peristaltic or rotation frequency, volume and flow rates of fluids, etc., are all set before starting the experiments as well as the number of daily repetitions of the process. All the parameters are registered and errors (if any) are documented. For the fermentation process, the three reactors are equipped with pumps that supply NaOH and HCl to control the consigned pH values. The representation on line of pH curves and volumes of acid and base added to the compartments are indicative of microbial development and metabolism. This information is crucial as control of the process since the analyses of metabolites and fermentation products are delayed and not measured online. Occasionally, general plate counts can be performed as an additional control of microbial viability. The system, however, can be upgraded to analyze on line the formation of gas (ammonium, H_2, CO_2 and CH_3).

28.4   Read-Out of the System

The system displays sampling ports between compartments to remove aliquots of colon content without disrupting the anaerobic conditions. The primary information searched during the set-up of the system and the diet interventions is the general evolution of the gut microbial community and the quantification of the targeted microbial groups and their metabolites. Analyses of colon microbiota is based on molecular biology tools (PCR-DGGE, qPCR, metagenomics, etc.), whereas general plate counts are only occasionally performed as a control of microbial viability. The choice of non-culturable microbial methods is in accordance with the large number of gut bacterial groups that are non-culturable by conventional techniques due to their generally fastidious growth requirements (Allen-Vercoe 2013).

The stabilization of the microbial community until steady-state conditions in the three colonic reactors and its further development during dietary interventions are evaluated by sampling and measuring the production of SCFA, branched chain fatty acids (BCFA) and ammonium over time. These metabolites are considered representative of fermentative and proteolytic microbial metabolism. Depending on the conditions assayed (diet interventions, development of microbial communities representative of dysbiosis associated to intestinal pathologies, etc.), specialized analytical methodologies targeting the formation of specific intermediate and end products are currently being developed to be incorporated to the system (unpublished results).

28.5 Advantages, Disadvantages and Limitations of the System

The advantage of the SIMGI model is associated to its flexible-modulating characteristics and the automated control of the working parameters that can be adjusted to simulate physiological conditions. The combination of peristaltic movements and controlled emptying of the gastric and small intestine compartments are advantages of the system in comparison to other multicompartmental models. The SIMGI model has not yet incorporated devices simulating the gut microbiota-host interactions. Therefore, assays for evaluating this type of crucial interaction is currently approached by co-culturing colon-region specific microbiota suspensions from the AC, TC and/or DC vessels with epithelial or immune cells. The incubation of this SIMGI complex colonic microbiota with Caco-2 cells has shown not to disturb the epithelial barrier integrity (unpublished results). Moreover, the SIMGI microbiota has demonstrated to induce the phenotypical maturation of human monocyte-derived dendritic cells (unpublished results). However, a limitation of the SIMGI model is the lack of devices to evaluate the formation of microbial biofilms adhering to the colonic epithelium. The simulation of intestinal absorption to remove end products of microbial metabolism is also a limitation of the system to prevent inhibition of the colon microbiota. Both drawbacks of the system are on the way to be overcome by setting up microbial/mucosa interfaces in the SIMGI lumen and by including dialysis devices between compartments.

In summary, the fully automation of the SIMGI model allows precise control of the environmental parameters that simulate the gastrointestinal tract. This multistage dynamic model has demonstrated to reproduce complex and stable microbial communities and it can be used as a tool for studying the effects of diet or food components on modulating the gut microbiota and its metabolic activity.

References

Allen-Vercoe E (2013) Bringing the gut microbiota into focus through microbial culture: recent progress and future perspective. Curr Opin Microbiol 16:625–629

Barroso E, Cueva C, Peláez C, Martínez-Cuesta MC, Requena T (2015) Development of human colonic microbiota in the computer-controlled dynamic SIMulator of the gastrointestinal tract SIMGI. LWT Food Sci Technol 61:283–289

Barroso E, Van de Wiele T, Jiménez-Girón A, Muñoz-González I, Martín-Alvárez PJ, Moreno-Arribas MV, Bartolomé B, Peláez C, Martínez-Cuesta MC, Requena T (2014) *Lactobacillus plantarum* IFPL935 impacts colonic metabolism in a simulator of the human gut microbiota during feeding with red wine polyphenols. Appl Microbiol Biotechnol 98:6805–6815

Cinquin C, Le Blay G, Fliss I, Lacroix C (2006) Comparative effects of exopolysaccharides from lactic acid bacteria and fructooligosaccharides on infant gut microbiota tested in an in vitro colonic model with immobilized cells. FEMS Microbiol Ecol 57:226–238

De Boever P, Deplancke B, Verstraete W (2000) Fermentation by gut microbiota cultured in a simulator of the human intestinal microbial ecosystem is improved by supplementing a soygerm powder. J Nutr 130:2599–2606

Elashoff JD, Reedy TJ, Meyer JH (1982) Analysis of gastric-emptying data. Gastroenterology 83:1306–1312

Fernández de Palencia P, López P, Corbí AL, Peláez C, Requena T (2008) Probiotic strains: survival under simulated gastrointestinal conditions, in vitro adhesion to Caco-2 cells and effect on cytokine secretion. Eur Food Res Technol 227:1475–1484

Gibson GR, Cummings JH, Macfarlane GT (1988) Use of a three-stage continuous culture system to study the effect of mucin on dissimilatory sulfate reduction and methanogenesis by mixed populations of human gut bacteria. Appl Environ Microbiol 54:2750–2755

Macfarlane GT, Macfarlane S, Gibson GR (1998) Validation of a three-stage compound continuous culture system for investigating the effect of retention time on the ecology and metabolism of bacteria in the human colon. Microb Ecol 35:180–187

Marteau P, Flourié B, Pochart P, Chastang C, Desjeux JF, Rambaud JC (1990) Effect of the microbial lactase (EC 3.2.1.23) activity in yoghurt on the intestinal absorption of lactose: an in vivo study in lactase-deficient humans. Br J Nutr 64:71–79

Molly K, Van de Woestyne M, Verstraete W (1993) Development of a 5-step multichamber reactor as a simulation of the human intestinal microbial ecosystem. Appl Microbiol Biotechnol 39:254–258

Molly K, Van de Woestyne M, De Smet I, Verstraete W (1994) Validation of the simulator of the human intestinal microbial ecosystem (SHIME) reactor using microorganism-associated activities. Microb Ecol Health Dis 7:191–200

Payne AN, Chassard C, Banz Y, Lacroix C (2012) The composition and metabolic activity of child gut microbiota demonstrate differential adaptation to varied nutrient loads in an in vitro model of colonic fermentation. FEMS Microbiol Ecol 80:608–623

Possemiers S, Verthé K, Uyttendaele S, Verstraete W (2004) PCR-DGGE-based quantification of stability of the microbial community in a simulator of the human intestinal microbial ecosystem. FEMS Microbiol Ecol 49:495–507

Russell WR, Gratz SW, Duncan SH, Holtrop G, Ince J, Scobbie L, Duncan G, Johnstone AM, Lobley GE, Wallace RJ, Duthie GG, Flint HJ (2011) High-protein, reduced-carbohydrate weight-loss diets promote metabolite profiles likely to be detrimental to colonic health. Am J Clin Nutr 93:1062–1072

Van den Abbeele P, Grootaert C, Marzorati M, Possemiers S, Verstraete W, Gérard P, Rabot S, Bruneau A, El Aidy S, Derrien M, Zoetendal E, Kleerebezem M, Smidt H, Van de Wiele T (2010) Microbial community development in a dynamic gut model is reproducible, colon region specific, and selective for *Bacteroidetes* and *Clostridium* cluster IX. Appl Environ Microbiol 76:5237–5246

Index

© The Author(s) 2015   329
K. Verhoeckx et al. (eds.), *The Impact of Food Bio-Actives on Gut Health*,
DOI 10.1007/978-3-319-16104-4